CRYSTALS POLARISING MICROSCOPE

A HANDBOOK FOR CHEMISTS AND OTHERS

BY

N. H. HARTSHORNE, Ph.D., M.Sc.
READER IN CHEMICAL MICROSCOPY, UNIVERSITY OF LEEDS

AND

A. STUART, M.Sc., F.G.S.
HEAD OF THE DEPARTMENT OF GEOLOGY, UNIVERSITY COLLEGE OF THE SOUTH WEST, EXETER

CONTENTS

CHAP.		PAGE
	INTRODUCTION	ix
I	THE CRYSTALLINE STATE	1

Typical Crystal Structures. Crystalline Form and Chemical Constitution. Structure Defects in Real Crystals. Isotropism and Anisotropism. Crystals and Glasses. Polymorphism. Isomorphism. Mixed Crystals.

II	THE STEREOGRAPHIC PROJECTION	26

Fundamental Principles. Great Circles and Small Circles. Properties of the Stereographic Projection. Some Necessary Constructions. Generalised Stereographic Nets.

III	THE MORPHOLOGY OF CRYSTALS	38

The Space Lattice. Point Group and Space Group. Crystal Growth and Crystal Habit. Measurement of Crystal Angles. Crystal Axes, Indices, and the Law of Rational Indices. The Symmetry of Crystals. Point Group and Space Group Nomenclature. Relationship between Point Group and Space Group Symmetry. Forms. Holohedrism, Hemihedrism, and Tetartohedrism. Hemimorphism. Enantiomorphism. Crystal Systems and the 32 Crystal Classes. Cleavage. Twinned Crystals. Multiple and Mimetic Twinning. Summary.

IV	THE OPTICAL PROPERTIES OF CRYSTALS	89

Double Refraction. Wave Surfaces. Wave Front. Wave Velocity and Ray Velocity. Application of Huygens' Construction to Birefringent Crystals. Refractive Index. Polarisation of Light by Crystals. The Uniaxial Indicatrix. Circularly and Elliptically Polarised Light. Biaxial Crystals. The Biaxial Wave Surfaces. External and Internal Conical Refraction. The Biaxial Indicatrix. Orientation and Dispersion of the Indicatrix. Optical Sections under the Microscope. Absorbing Crystals. Optical Activity. Relationship between Optical Properties and Crystal Structure. Summary.

V	THE POLARISING MICROSCOPE	160

General Construction. The Nicol Prism. The Objective. The Eyepiece. The Stand. Diaphragms. Additions to Stage Equipment. Typical Polarising Microscopes. Illumination and Illuminating Apparatus. Sources of Monochromatic Light. Use and Care of the Polarising Microscope.

VI	PREPARATION AND MOUNTING OF MATERIAL	192

Slides and Cover Slips. Apparatus for Preparation and Mounting of " Chemical " Material. Preparation and Mounting without Recrystallisation. Methods which include Recrystallisation. Methods for Hard Compact Materials. Permanent Mounts of Crystals. Methods of Changing the Orientation of Crystals under Examination. Section Cutting (Single Crystals).

CONTENTS

CHAP.		PAGE
VII	THE MICROSCOPIC EXAMINATION OF CRYSTALS. (i) PARALLEL LIGHT	219

Colour. Crystal Form. Cleavage. Measurement of Edge (Profile) Angles. Measurement of Size and Thickness. Determination of Refractive Index (Isotropic Crystals). Immersion Media and their Standardisation. Index Variation Methods. Pleochroism. Variation in Relief. Isotropism and Anisotropism. Relative Retardation. Normal and Anomalous Interference Colours. Extinction and Extinction Angles. Graphical Determination of Extinction Angles. Measurement of Extinction Angles. Determination of "Fast" and "Slow" Directions in Crystal Sections. Determination of Birefringence. Crystal Aggregates and Twinned Crystals. Optical Anomalies.

VIII THE MICROSCOPIC EXAMINATION OF CRYSTALS. (ii) CONVERGENT LIGHT 270

The Microscope as a Conoscope. Methods of Viewing the Interference Figure. Methods of Isolating the Interference Figures of Small Crystals. Interference Figures of Uniaxial Crystals. Interference Figures of Biaxial Crystals. Dispersion. Determination of the Optical Sign from Observations upon Interference Figures. Measurement of the Optic Axial Angle. Determination of the Refractive Indices of Birefringent Crystals.

IX LIQUID CRYSTALS. 322

The Mesomorphic State. Smectic Mesophases. Nematic Mesophases. Cholesteric Mesophases. Gradation of Mesomorphic Properties in Homologous Series.

X METHODS OF ATTACK AND EXPERIMENTS 353

The Systematic Examination of a Single Compound. Presentation of Optical Crystallographic Data. Points of Importance and Methods of Attack in Typical Applications of the Polarising Microscope. Experiments.

XI SPECIAL METHODS 386

Index Variation Methods. Single Variation Method by Temperature Control. The Double Variation Method. The Universal Stage. Examination of (i) Uniaxial Crystals, (ii) Biaxial Crystals on the Universal Stage. Determination of Refractive Indices with the Universal Stage. Use of the Universal Stage to Determine the Orientation of the Indicatrix.

XII EXAMPLES OF THE USE OF THE POLARISING MICROSCOPE . . 422

Systematic Identification by Means of Optical Properties. Characterisation and Identification of Single Compounds by Optical Properties. Investigation of Heterogeneous Equilibria. Slags and Refractories. Textile Fibres. Cell Walls of Plants.

APPENDIX 458

INDEX 462

FOLDING PLATE

NOMOGRAM FOR DETERMINATION OF REFRACTIVE INDICES . . . . . . AT END OF BOOK

INTRODUCTION

Every transparent crystalline compound is characterised by a unique set of optical properties. The polarising microscope is an instrument by means of which these optical properties may be determined with a degree of precision which depends on a number of factors, such as the shape and size of the particles to be examined, the refinement of the methods, and the quality of the instrument used.

Theoretically it should be possible to identify any crystalline substance selected at random solely by means of its optical properties, if these have been previously recorded. In practice this cannot be done because of limitations imposed by the factors mentioned above. It is usually found, however, that any two crystal species can be differentiated by means of the polarising microscope with the greatest readiness, and that any crystalline compound, the optical properties of which are known, can be identified, once the field of enquiry has been narrowed by the knowledge that it belongs to a limited group.

It might be thought by many that the use of the polarising microscope for purposes of identification would have been largely superseded by the X-ray powder method. This is by no means the case. The X-ray method of identification is indeed a very valuable one, and, where the necessary apparatus is in regular use, convenient. It is also particularly useful where the state of division of the material is below the limit of microscopic resolution, or in the case of opaque substances which cannot be studied by ordinary optical crystallographic methods. But for the normal run of transparent crystalline material, it will commonly be found that the polarising microscope will give the required answer more simply and more quickly. Moreover, and this is important, the microscope gives direct visual information about the outward form, texture, and state of division of the specimen, and, if it is a mixture, the manner of aggregation of its ingredients, none of which properties (which may be of prime importance in the problem) is to be ascertained, at least directly, by the essentially " blind " X-ray method of attack. It should be said at once, however, lest the above should have given the contrary impression, that it is no part of our thesis to suggest that the two methods are in competition. They are in fact complementary, each having its appropriate application, and in many problems the one supplements information gained by means of the other. In this connection it should be said that the determination of optical properties is not only useful for the identification of substances, but makes an important contribution to the study of crystal structures by means of X-rays.

The applications of the polarising microscope in chemical and related problems may be stated in somewhat more detail as follows :

(*i*) *It is a general aid to the examination of all kinds of solid material.* The examination of solid material under the polarising microscope forms, in many cases, a useful preliminary to its analysis, since it may indicate whether it is homogeneous or heterogeneous, whether impurities are present, etc., and may even suggest the best lines upon which to carry out the analysis.

(*ii*) *It can frequently be used to identify and characterise solid compounds in cases where ordinary analytical and physical methods fail or are tedious, or where confirmation of the results obtained by these methods seems desirable.* Thus, in many cases, it can be used to identify the constituents of salt mixtures, the chemical analysis of which gives only the ions present ; to identify the solid phases in phase equilibrium studies, thereby saving much of the analytical work, and settling questions which the analytical work has left open ; to differentiate between isomers having no (or indefinite) melting-points, or lacking other distinctive properties ; and to control the progress of fractional crystallisations and trial syntheses more simply and effectively than could be done by chemical methods.

(*iii*) *It affords a ready means of obtaining definite characteristic data with which to describe the crystals of new compounds for record purposes.* Descriptions limited to " white needles," " yellow leaflets " and the like, which are all too common in chemical literature, are really of little value, partly because they are meagre and vague, and partly because the same substance may exhibit many different habits when crystallised under a variety of conditions.

(iv) *It is essential for the determination of optical data required in certain X-ray structural studies.* Since the optical properties of a crystal are determined by its structure, they may be used to throw light on the way in which the crystal is built up. This may result in much saving of time in X-ray structure analysis, since certain arrangements of the building units may be at once ruled out as being incompatible with the optical evidence.

Concrete examples of these applications are collected and discussed in Chapter XII rather than here, as they will be more fully appreciated at that stage.

The polarising microscope may be immediately applied to crystals almost irrespective of their size, outward form, or of whether they are mixed with other matter. Crystals as small as 0·01 mm. in average diameter may be examined satisfactorily, and those that are too big to be mounted on a microscope slide may be crushed and the fragments examined. This is not to say that better results cannot often be obtained by studying a substance after it has been recrystallised, but it may be truthfully asserted that by means of the polarising microscope, some optical crystallographic data can be obtained on any specimen of a crystalline substance, without special preparation, unless its state of division is considerably less than 0·01 mm. The method is thus eminently suitable for examining crystals just as they happen to occur. It should also be noted that the quantity of material needed for an examination is usually extremely small.

As a guide to the beginner it may be as well to give an outline of the arrangement of this book. The subject matter of Chapter I which deals with the general properties of the crystalline state will be mostly familiar to the trained chemist, but it was thought desirable to include it for the benefit of others and in order to bring the reader into a " crystallographic frame of mind " at the outset. Chapter II deals with the methods of representing morphological and optical vectors by means of the stereographic projection, and while this is important in advanced optical studies, it may be omitted at a first reading. So too might Chapter IX on the somewhat special topic of liquid crystals, and Chapter XI which deals with special methods.

It is important that the subject matter of Chapters III and IV should be fully grasped because these contain the foundations upon which the optical method depends. To assist the beginner the basic facts are collected in summaries at the ends of these chapters.

It has been assumed that the reader is familiar with the elements of the wave theory of light, the phenomenon of interference, and the properties of simple lenses.

CRYSTALS AND THE POLARISING MICROSCOPE

CHAPTER I

THE CRYSTALLINE STATE

A well-formed crystal attracts attention because it is bounded by plane surfaces (*faces*) and usually shows some degree of symmetry. The external form of such a crystal, as expressed by certain of its interfacial angles, is entirely characteristic of the substance of which it is composed, unless it belongs to the cubic system of crystal classification. All crystals belonging to this system, which comprises the highest orders of crystal symmetry, have the same angles between corresponding pairs of faces.

These external properties of crystals result from the orderly arrangement in space of their atoms, the pattern of the arrangement being in every case characteristic of the substance when the interatomic distances are taken into account. The atoms may exist in the structure as individual units, neutral or ionised, or they may be combined as discrete molecules or polyatomic ions.

The internal structures of a very large number of crystalline substances have now been determined by the method of X-ray analysis, originally developed by W. H. and W. L. Bragg, and based on the suggestion by Laue in 1912 (confirmed by Friedrich and Knipping) that the regular array of atoms in a crystal could act as a diffraction grating for X-rays. These investigations may be said to have opened up an entirely new field in chemistry—the chemistry of the solid state. Some of the more important results of the application of this method are as follows. Previously it had been generally assumed that the ultimate particle in all chemical compounds was the molecule, although this could not be proved for the majority of solid inorganic compounds since the methods of determining molecular weights could not be unambiguously applied to them. The X-ray method has shown that in many types of inorganic compounds the molecule cannot in fact be distinguished as a structural unit in the crystal, which consists instead of a *continuous* array of atoms or ions. Examples of such structures are given below. Obviously for the term " molecule " to have any meaning at all in such cases, it must be applied to the whole crystal, which forms as it were one giant molecule. In crystals in which discrete molecules do occur, or in which polyatomic ions are present, the X-ray method has given valuable information regarding the structures of these units. The method has also yielded important data on interatomic distances, which have thrown much light on the nature of the bonds between atoms.

It is beyond the scope of this book to deal with the internal structures of crystals in detail, and for fuller information the reader is referred to the list of works at the end of this chapter. Some typical examples of the simpler structures are, however, shown in Figs. 1 to 8, and their essential features given below.

It should first be realised that the structure of a crystal is the result of its building units having packed themselves together so that the potential energy of the system has a minimum value at the prevailing temperature and pressure. The factors which determine the details of the structure are : (*a*) the nature of the bonds between the atoms, and in particular whether these bonds are directional (covalent) or non-directional (van der Waals, metallic, ionic), and (*b*) geometrical considerations depending on the effective sizes, and in the case of molecules and polyatomic ions, the shapes of the building units.

In a crystal, adjacent atoms are in close proximity to one another, and some at least are to be thought of as being in " contact ", though an atom has no fixed radius irrespective of its state of combination. In Figs. 1 to 8, the atoms are shown widely separated from one another, but this is purely with the object of bringing out more clearly their relative positions.

Xenon (Fig. 1*). Xenon is one of the inert gases, and the atoms of this group of elements are characterised by possessing extremely

* The thin continuous and broken lines in Figs. 1–8 merely serve to emphasise the relative positions of the atoms ; they have no chemical significance. Directed chemical bonds, where these occur, are indicated by *thick* continuous lines.

stable electronic structures. In consequence they have no tendency to enter into chemical combination, for this involves either a transfer or a sharing of electrons between the atoms concerned. The only attractive forces between the atoms of an inert gas are of the undirected van der Waals type, and these forces are extremely weak. These elements therefore only exist in the solid state at very low temperatures (the melting-point of xenon, for example, is $-140°$ C). The non-directional character of the forces results in a structure like that shown by a set of equal spheres packed so as to occupy the minimum space. Actually there are two ways of packing spheres to meet this requirement, and the one adopted in the xenon structure is that known as *cubic close packing*. This is based on an arrangement of eight atoms situated at the corners of a cube, with six others at the centres of the cube faces, as may

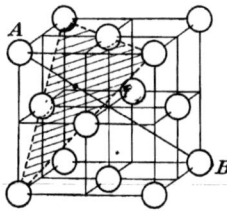 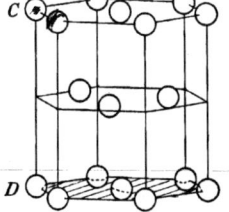

FIG. 1.—Xenon [Atomic—van der Waals].   FIG. 2.—Magnesium [Metallic].

be seen from the elementary cell * of the structure shown in the figure. An example of the other way, known as *hexagonal close packing*, is given below.

Magnesium (Fig. 2). Magnesium has a typical metallic structure. It is generally considered that the atoms in a solid metal have lost some or all of their valency electrons to form an electron "atmosphere" which permeates the whole structure. The structure therefore consists essentially of an assemblage of positive ions and electrons, the former determining the structure and the latter wandering more or less freely through it and cementing the structure together. Under the influence of an electrical

* There are different ways of dividing up the space occupied by a crystal structure into cells which are representative of its architecture. The one shown in the figure, though not the true *unit* cell (see p. 38), is that which shows the essential symmetry of the structure most clearly.

potential difference, the electrons drift towards the positive pole, and this constitutes the familiar property of electrical conductivity which is typical of metals. The forces holding the structure together are either non-directional, or nearly so, and they are very much stronger than van der Waals forces. As a result most metals adopt structures based on the close-packing of equal spheres (or slightly distorted versions of these structures) and, owing to the stronger forces involved, they melt at much higher temperatures than do the inert gases. Some metals adopt a somewhat less compact arrangement, known as the *body-centred cubic* structure, the unit cell of which consists of eight atoms at the corners of a cube with another atom at its centre.

In magnesium the atoms are arranged according to the hexagonal method of close-packing. The arrangement will be clear from the figure. The relationship between this and the cubic method of close-packing may be seen by considering the shaded layers of atoms in Figs. 1 and 2. The packing *in these layers* is the same in both structures, which also resemble one another in that they are built up by piling a series of such layers parallel to one another so that the atoms in any layer nestle in the depressions between the atoms of adjoining layers. The difference lies in the fact that in the direction normal to the layers, the hexagonal type repeats itself every *third* layer, whilst the cubic type repeats itself every *fourth* layer. Thus along this direction, in Fig. 2 atom C projects on to atom D, but in Fig. 1 atom A projects on to atom B and here there are two layers of atoms in between. The reason why these alternative methods of close-packing are geometrically possible is that the atoms in any layer occupy only one half of the depressions between the atoms in an adjoining layer. Therefore when two layers have been packed together, the third layer has two alternative positions relative to the first. It can either occupy a position such that its atoms, viewed along the direction normal to the layers, are superposed on those in the first layer, in which case the hexagonal structure results, or its atoms can occupy the other set of depressions in the second layer, in which case the cubic structure is obtained.

Sodium Chloride, NaCl (Fig. 3). This is a typical ionic compound of formula type AX, that is with equal numbers of positive (Na) and negative (Cl) ions. The attractive forces between these ions are electrostatic and non-directional. Consequently the crystal structure is based on the close-packing of spheres as in the cases described above, only here it is a question of packing together equal numbers of spheres of two different sizes. The sodium ions are the smaller with an effective radius of about 0·95 Å, and the effective radius of the chloride ions is about 1·8 Å.

In a system of this type, the maximum stability is achieved by an arrangement in which each positive ion is surrounded by a certain number n of negative ions situated at equal distances from it, and each negative ion is similarly surrounded by n positive ions. The value of n depends very largely on the ratio between the radii of the two kinds of ions, the principle being that the smaller ions do not have as nearest neighbours so many of the larger species that they have room to " rattle about " in the interstices, since such an arrangement would correspond to a separation of oppositely charged ions beyond their equilibrium distance of approach.

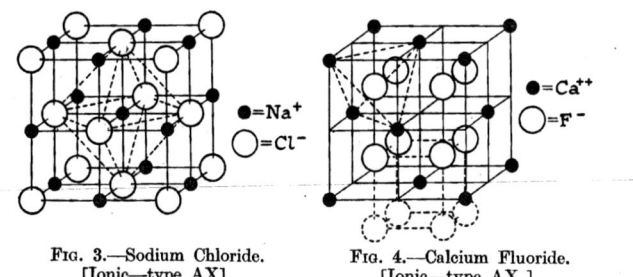

Fig. 3.—Sodium Chloride. [Ionic—type AX].

Fig. 4.—Calcium Fluoride. [Ionic—type AX_2].

In the case of sodium chloride, n is 6. Each ion is surrounded by six of the opposite species arranged at the corners of a regular octahedron. This is clear from Fig. 3, in which the octahedral disposition of the chloride ions surrounding the central sodium ion of the unit cell is indicated by broken lines.

Calcium Fluoride, CaF_2 (Fig. 4). This is an ionic compound like that just considered, but of the formula type AX_2. The same principles determine the structure but it is now necessary to accommodate twice as many negative ions as positive ions, to preserve electrical neutrality. In calcium fluoride this is achieved by an arrangement in which each calcium ion is surrounded by eight fluoride ions situated at the corners of a cube, and each fluoride ion is surrounded by four calcium ions situated at the corners of a regular tetrahedron.

Mercuric Iodide, HgI_2, red modification (Fig. 5). In

common with a number of other salts possessing a marked degree of covalent character (notably chlorides, bromides, and iodides of di- and tri-valent metals), this substance crystallises in what is known as a *layer* type of structure, which may be thought of as being intermediate between the typically ionic structures (*e.g.* sodium chloride and calcium fluoride above) in which the ions pack together as independent units without any association to form molecules, and the molecular structures to be mentioned below, in which discrete molecules appear as units in the pattern.

The structure of mercuric iodide consists of parallel crumpled sheets composed of mercury and iodine atoms, running right

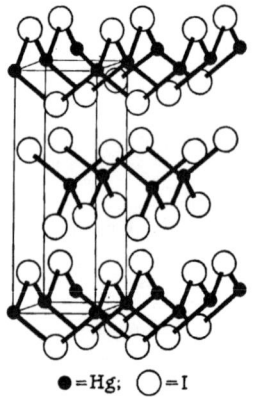 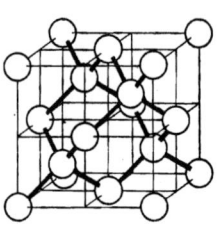

● = Hg;   ○ = I

Fig. 5.—Mercuric Iodide (red modification), [Layer—type AX_2].

Fig. 6.—Diamond [Covalent—endless].

through the crystal, the sheets being held together by van der Waals forces. Each mercury atom is attached to four iodine atoms arranged at the corners of a regular tetrahedron, and each iodine is attached to two mercury atoms. The bonds between the atoms are partly ionic and partly covalent, and so the crystal may be regarded as something between a pile of two-dimensional ionic crystals and a pile of molecules of formula $(HgI_2)_n$, where n is very large and is determined by the cross-sectional area of the crystals parallel to the layers.

On account of the weak forces holding the layers together, crystals with this type of structure always cleave readily in the direction parallel to the layers.

Diamond (Fig. 6). This structure consists of a continuous array of carbon atoms, each one of which is surrounded by four others situated at the corners of a regular tetrahedron. The bonds joining the atoms are covalent, and so the whole crystal forms one " giant " molecule. This accounts for the resistance of diamond to high temperatures, and its extreme hardness. In the figure the bonds are indicated by heavy lines.

Iodine, I_2 (Fig. 7). In this structure the atoms occur in pairs, having an interatomic distance of 2·70 Å, whilst the smallest distance between atoms belonging to adjacent pairs is 3·54 Å. The pairs are iodine molecules, in which the two atoms are joined by a single covalent bond. The forces holding the molecules together in the crystal are of the weak van der Waals type ; hence the much greater distance, 3·54 Å, between atoms belonging to adjacent molecules than between the two atoms in the same molecule. These weak forces are responsible for the fact that iodine and molecular crystals in general melt at comparatively low temperatures and are readily volatile. The melting-point of iodine is 113° and it sublimes freely at temperatures well below this.

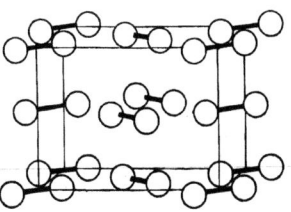

FIG. 7.—Iodine [Molecular].

Benzene, C_6H_6 (Fig. 8). This is given as an example of the crystal structure of an organic compound. In all such compounds the identity of the molecule is preserved in the crystal. The nature of the intermolecular forces, however, depends on the presence or absence of polar groups in the molecule, and this influences the structure.

In the case of non-polar molecules such as those of the hydrocarbons, including the present example, the intermolecular forces are of the undirected van der Waals type, and the structure is determined mainly by the shape of the molecule. Thus in Fig. 8 it will be seen that the benzene molecules are packed in a manner recalling the cubic close-packed arrangement of Fig. 1 (a consequence of the non-directional character of the attractive forces), but that they are tilted in different directions in adjacent layers. This latter feature of the structure is imposed by the shape of the molecules, and represents the only way in which the force fields of the six-membered rings can be mutually accommodated to achieve minimum potential energy. (Similar considerations apply to the packing of the dumbell-shaped iodine molecules in Fig. 7.)

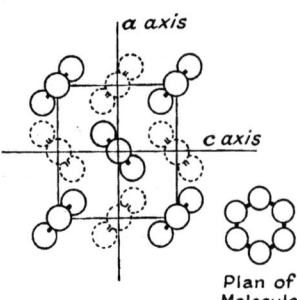

FIG. 8.—Benzene [Molecular].

The structure is shown as an elevation on the plane normal to the b axis. The molecules present an edgewise view. The layers of molecules represented respectively by full and broken lines succeed one another alternately along the direction of this axis. Hydrogen atoms are not shown.

When strongly polar groups such as $>C=O$, $-OH$, or $-NO_2$ are present, their mutual interaction establishes directed forces between the molecules, and the shape of these is no longer the sole factor determining the structure. The effect is particularly marked with certain compounds containing the $-OH$ group, since here the possibility of linkage through hydrogen bonds arises. Thus resorcinol, $C_6H_4(OH)_2$, below 70° forms a very open structure in which the aromatic rings are prevented from packing closely together by the fact that they are linked by directed hydrogen bonds. The density of this structure is actually less than that of the substance in the liquid form, since when it melts these bonds are broken, and the molecules can then approach one another more closely.

Crystalline Form and Chemical Constitution. Attempts to relate crystalline form with chemical constitution engaged the attention of chemically-minded crystallographers for many years before the discovery of the X-ray method of crystal analysis. Progress during this period was, however, mainly limited to the tracing of gradations in crystal properties in certain groups of closely related chemical compounds. As examples may be mentioned the classical researches of A. E. H. Tutton on the isomorphous sulphates and selenates of the alkali metals, and the double sulphates and selenates of these and the bivalent metals. Tutton established by measurements of the most refined nature that the crystallographic axial ratios (see Chapter III) and physical properties of a series such as $(K, Rb, Cs)_2SO_4$ alter in a regular manner with the weight of the cation. (See, for example, the book by this author, *Crystalline Form and Chemical Constitution*, Macmillan, 1926.)

In the absence of any knowledge of the internal structures and of the true nature of the different types of chemical bonds, important advances along these lines of investigation were not to be expected, more especially since it was tacitly assumed by most crystallographers that the fundamental chemical unit in all crystals was a discrete molecule. In the light of modern knowledge it is clear that there is in general no very direct relationship between chemical constitution and external crystalline form. There is, as we have seen, a marked relationship between the chemical constitution of a crystalline substance and its internal structure, and the latter in turn determines the external symmetry, but, as will appear later (Chapter III), a given set of external symmetry elements can arise from a number of different types of internal structure. This means that the external form of a crystal is, to say the least, a most ambiguous guide to the chemical character of the substance composing it.

The relationship between internal structure and external form will be referred to again in Chapter III, but for the moment it is possible to draw some simple conclusions from the cases given in Figs. 1 to 8, and it will be instructive to do so. Thus in Figs. 1, 3, 4, and 6, the succession of atoms is the same in the vertical direction as in the back-to-front and side-to-side directions, and the result is that the external form shows cubic symmetry. Fig. 2 shows a somewhat different case, but since it is a structure based on the close-packing of equal spheres, it also results in an external form having a high order of symmetry. In the layer structure of mercuric iodide (Fig. 5), the sequences of atoms in the back-to-front and side-to-side directions are the same, but different from that in the vertical direction. The symmetry is therefore lower than cubic, but it is still of the " right-angle type ". The substance belongs in fact to the tetragonal system (see Chapter III).

In molecular crystals, as already explained, the shape of the molecules exerts a considerable influence in determining the crystal pattern (in the absence of strong directed forces), and so unless the shape is of a high order of symmetry, the internal symmetry and therefore the external symmetry of the crystal is usually low.

Structure Defects in Real Crystals. The picture presented above has shown the structure of a crystal as consisting of a rigid network of atoms, each species of which has its assigned positions in the pattern, these positions being fixed (apart from the small displacements due to thermal vibrations). It has also been implied that the pattern extends without interruption throughout the whole crystal.

Investigation has shown, however, that this picture, though essentially correct, needs some qualification, since most real crystals show minor departures of one kind or another from it. These departures are classified below. It may be said at once that they are not such as to affect the constancy of the interfacial angles or of the optical properties of a given substance, at least as measured by standard methods,* and that their existence has only been revealed by X-ray analysis or has been inferred from certain physical properties of crystals.

(a) *Mosaic Structure.* With most crystals, measurements of the angular widths of X-ray beams reflected from both external and cleavage surfaces give values that are appreciably greater than those calculated for a perfect lattice structure. From this and from other evidence, it can be shown that the crystal is built up of " mosaic " blocks, about 10^{-5} cm. in diameter, which may deviate from perfect alignment with one another by as much as several minutes of arc. When a beam of X-rays strikes the surface, each block in the area illuminated reflects in a slightly different direction and the angular width of the resultant beam is greater than would be the case if the structure were continuous.

The existence of a mosaic structure in most real crystals makes it very difficult to predict the values of certain physical properties, such as the mechanical strength, from calculations based on a perfect structure, for although the discontinuities between the mosaic blocks are for the most part only of atomic dimensions, they are nevertheless flaws. Thus it is possible to calculate the strength of a sodium chloride crystal from the interionic forces assuming a perfect structure. Real sodium chloride crystals, however, prove to have a much lower strength.

(b) *Defects within the Mosaic Block.* In some crystals the mosaic blocks themselves fail to conform in all respects with the characteristics of a perfect structure. These so-called " defect structures " may be of the following types.

(i) *Frenkel and Schottky defects.* These have been postulated to

---

* An exception to this statement must be made in the case of certain intermetallic " superlattices " and the corresponding disordered structures (see b (iv) below). Thus the superlattice CuAu is tetragonal whilst the disordered structure having this composition belongs to the cubic system.

account for certain properties of ionic crystals, notably their electrical conductivity. They are " holes " in the structure, that is, vacant lattice positions, which migrate through the crystal under the influence of an applied electric field. In defects of the type postulated by Frenkel, the ion which originally occupied the hole is supposed to have been displaced to an interstitial position in the structure some distance away. In the Schottky defect, the ion from the hole is assumed to have migrated to the surface of the crystal.

The existence of such defects cannot be detected by X-ray methods, since the time taken for an ion to migrate from one lattice position to another (which determines the rate of movement of the holes) is believed to be of the order of 10^{-11} seconds, and this is only a minute fraction of the time required to record X-ray reflections.

(ii) *Lattices containing wandering atoms.* In this type of defect structure, atoms of one or more of the species present move about freely in the interstices of a rigid structure formed by the remaining atoms, or at least occupy the interstitial positions in this structure in a random manner, which implies some degree of mobility. For example, above 146° C., the silver ions in silver iodide can wander freely through the rigid structure formed by the iodide ions. A somewhat similar defect is shown by silver mercuric iodide, Ag_2HgI_4, above 51°. Here again the iodide ions form a rigid lattice, and the silver and mercury ions occupy certain interstitial positions in this structure, but in a wholly random manner.

Salts with defect structures of this kind usually possess an abnormally high electrical conductivity, since the mobile ions drift in one direction under the influence of an applied electrical field. In the two cases quoted above, both compounds show a greatly enhanced conductivity as compared with solid salts in general. Moreover the conductivity of silver iodide is much greater than that of the silver mercuric iodide, indicating that the mobility in the latter case is more restricted, as is shown by the fact that the silver and mercury ions do not occupy all the interstitial positions available.

(iii) *Structures containing rotating molecules or polyatomic ions.* In crystals of many compounds which contain molecules or polyatomic ions as structural units, the molecules or ions rotate freely about an axis, or in all directions about their centres, when a temperature characteristic of the substance is exceeded. An increase in the rotational motion of the structural units with rise of temperature must occur in all crystals as part of the increased thermal energy of the units, but in most substances this has not passed the stage of an oscillation about the axis or centre of rotation when the melting point of the crystal is reached. In the cases now considered, free rotation is established without collapse of the whole structure, *i.e.* the crystal forces are strong enough to maintain the freely rotating units in their mean positions in the structure.

The phenomenon occurs in many long-chain organic compounds, *e.g.* the higher paraffins (above $C_{20}H_{42}$), the molecules of which rotate about the axis of the chain in the temperature range just below the melting point. It also happens with the triangular CO_3^- and NO_3^- ions, which in some cases rotate about the centre in all directions, as for example in ammonium nitrate above 125° C. In this case the tetrahedral ammonium ions also rotate about their centres.

The establishment of free rotation results statistically in an increase in the symmetry of the molecule or ion, and so is frequently accompanied by a change in the crystal structure of the substance (see *polymorphism* below). Thus in the last example cited, the rotation of the NH_4^+ and NO_3^- ions makes them in effect spherical, and the crystal structure adopted is one of the simple cubic types which result from the packing together of equal numbers of spheres of two different sizes.

(iv) *Defect structures in mixed crystals.* When two substances form a series of solid solutions, the resulting crystals must necessarily show departures from a perfect structure (except in some cases at certain definite compositions) in that with continuously changing composition one or more sets of equivalent positions in the structure must be occupied by more than one kind of atom. For example, in mixed crystals of NaCl and KCl, sodium and potassium ions are distributed at random among the positions which are occupied solely by sodium ions in the NaCl structure (Fig. 3), and by potassium ions in the KCl structure.

Another type of defect is shown by mixed crystals of two compounds of different formula type, for example $MgCl_2$ and LiCl. When a small amount of lithium chloride dissolves in the magnesium chloride structure, the arrangement of the chloride ions remains the same (except for a change in the lattice dimensions) and determines the structure, but *two* lithium ions must enter the structure for every magnesium ion displaced, in order to preserve electrical neutrality. The structure then contains, as it were, an excess of positive ions, and these " excess " ions are accommodated in the interstices of the original pattern. At the other end of the series, magnesium chloride dissolves in the lithium chloride structure, which is determined by the arrangement of the chloride ions. Every magnesium ion which enters the structure replaces two lithium ions, so that a number of the lithium positions must become vacant.

Solid solutions formed by metals may be based either on the structures of the pure metals, or in many systems on other structures which make their appearance at intermediate compositions. In either case the atoms of the different species may be distributed at random among the lattice positions, and generally are so distributed if the alloy is rapidly cooled from the molten condition. In many cases, however, solid solutions of certain definite compositions adopt an ordered arrangement of their atoms if they are annealed, or if they are made by slow cooling from the liquid state. For example a copper-gold alloy of composition CuAu obtained by quenching the corresponding liquid mixture has a cubic structure like that of solid xenon (Fig. 1) with copper and gold atoms distributed at random among the atomic positions. On being annealed, the alloy develops an ordered arrangement in which the horizontal layers of atoms in the structure (as drawn in the figure) are alternately gold and copper. Owing to the somewhat larger size of the gold atom as compared with the copper atom, this causes a slight sideways expansion of the structure which thus becomes tetragonal. Similarly at the composition Cu_3Au, annealing results in the development of an ordered arrangement in which gold atoms occupy the corners of the cubic unit cell and copper atoms the centres of the faces. In this case there is no change in crystal system in passing from the random or disordered structure to the ordered one.

These ordered structures are termed " superlattices ". Annealed alloys having compositions near that of a superlattice show in general the structure of the latter with some of the atomic positions occupied by the " wrong " atom. They can therefore be regarded as solid solutions of one or other of the parent metals in the superlattice, and are defect structures like all phases in which there is a random arrangement of one or more of the atomic constituents.

Intermediate phases in alloy systems, *i.e.* those having different structures from those of the parent metals are known formally as intermetallic compounds on the ground that they behave as compounds from the standpoint of the freezing point diagram of the system. When they extend over a range of composition, however, there is nothing which sharply distinguishes them from the solid solutions based on the structures of the parent metals. In both cases there is random arrangement, with or without the formation of superlattices on appropriate heat treatment.

(v) *Interstitial and non-stoichiometric compounds.* Crystals of certain of the heavy metals possess the property of taking up variable amounts of the lighter elements (H, C, N, B) without any other change in the structure than a slight expansion. The small atoms of the light element occupy the interstices between the metal atoms. To this class of compound belong, for example, the carbides of tantalum, zirconium, titanium, and vanadium. A stoichiometric relation between the numbers of metal and light element atoms only occurs if all structurally equivalent interstices, or a simple fraction of them, which are of a suitable size to accommodate the light atoms, are occupied.

Yet another example of a non-stoichiometric compound, but of a different type, is furnished by ferrous sulphide. It is found that this substance hardly ever has a composition corresponding exactly to the formula FeS, there being a deficiency of iron. Analysis of the structure leads to the conclusion that the Fe positions are not completely occupied.

Isotropism and Anistropism. In a crystal belonging to the cubic system, the sequence of atoms along the three crystallographic axes (that is, the directions perpendicular to the cube faces) is the same. Any physical property therefore has the same value along these three mutually perpendicular directions. By the electromagnetic theory of light it can be shown that in such a system the refractive index will be the same not only along these three special directions, but along *all* directions of propagation (see also p. 138). This is true of certain other physical properties of cubic crystals, for example magnetic susceptibility, and with respect to these properties such crystals are said to be *isotropic*; their behaviour in this respect resembles that of a piece of unstrained glass.

In crystals of other systems the sequence of atoms varies with direction in such a way that it is impossible to find three mutually perpendicular directions along which it is the same. As a result all physical properties, including the refractive index, vary with direction, and such crystals are termed *anisotropic*.

Crystals and Glasses. A pure crystalline element or compound, which melts without undergoing chemical change, does so at one temperature if the prevailing pressure is constant. Mixed crystals melt over a range of temperature, but this range has perfectly definite limits for a given composition and pressure, again provided that no chemical change takes place. In both cases the melting process is characterised by the fact that two distinct phases are present, namely the liquid and the unmelted solid.

There are, however, a number of solid materials which on heating pass into the liquid state by way of a gradual softening, there being no single temperature at which it may be said that the properties of a solid cease and those of a liquid begin, and at no stage in the process are two phases present side by side. Some of these materials are single compounds, *e.g.* vitreous silica and boric oxide, whilst others are mixtures such as the familiar mixtures of silicates known generically as " glass ". When in the solid state, such substances have no characteristic outward form, and if free from strain are isotropic. They are said to be in the *amorphous* or *glassy* state. Some of them change to the crystalline state in the course of time, or under the action of suitable stimuli (shock, scratching, rise of temperature, contact with a crystal of the substance), and some substances which normally are crystalline can be brought into the glassy condition by rapidly cooling a melt, particularly if they are impure.

Glasses are formed typically by substances which in their crystalline state contain endless chains or networks of linked atoms (for example, many of the silicates), or which possess a somewhat complex molecular structure (for example, the alkaloid piperine). They have a high viscosity in the liquid state, particularly in the neighbourhood of the freezing-point. It is therefore understandable that when the liquid is cooled and enters the region in which the crystalline state is the more stable, the movements of the molecules or ions are too sluggish, or their shapes are too awkward for them readily to take up their correct positions in the crystal structure, particularly if the rate of cooling is rapid. The random arrangement characteristic of the liquid state therefore persists down to temperatures at which the viscosity increases so much that it merges into the rigidity of a solid. Bonds characteristic of the crystalline state of the substance are probably formed wherever appropriate atoms find themselves in juxtaposition, but there is no regular pattern. This random arrangement accounts for the isotropism of glasses since *on the average* the sequence of atoms is the same in one direction as in another, as in a liquid.

Many crystalline substances lose their shape to some extent at temperatures just below the melting-point, because they become soft enough to yield to the surface tension, which strives to make the surface area as small as possible. Thus copper crystals above 800° C. show a decided rounding of their faces and corners. Such substances, however, like other crystals, pass into the liquid state quite sharply, and their behaviour need not therefore be confused with the softening of amorphous solids.

Polymorphism. It frequently happens that a single substance has more than one crystalline form, *i.e.* exhibits *polymorphism*, such substances being termed di-, tri-, etc. -morphic according as they possess two, three, or more forms. When the substance is an element the same phenomenon is often denoted by the term *allotropy*, but this also includes all cases in which an element exists in different modifications whether they be solid or not. The polymorphic modifications of a substance differ from one another in all physical properties, but they yield identical liquids and vapours, because they are merely different crystal structures built up from the same atoms, molecules or ions. They are thus to be distinguished from *isomeric* molecular compounds, so frequently met with in organic chemistry, for in these the molecules themselves are different arrangements of the same atoms. Differences of this kind persist when the substances are melted or vaporised. There are, however, many cases of isomeric substances which are more or less rapidly convertible the one into the other, either spontaneously or by the aid of a catalyst (*dynamic isomerism*), and the final result of melting or vaporising either is the same equilibrium mixture of the two, *i.e.* the behaviour is to all appearances the same as that which would be shown by two modifications of a polymorphic substance.* Where this possibility is present it is not always easy in practice to distinguish between it and poly-

* Except that in polymorphic substances the unstable form always has the lower melting-point, and this is not necessarily the case with dynamic isomers.

POLYMORPHISM

morphism, in the absence of a knowledge of the crystal structures derived from X-ray analysis.

Polymorphism may be either *enantiotropic* or *monotropic*. To explain these terms a substance having two forms only will be considered. If each form is stable within some region of temperature and pressure, the other being unstable under these conditions, *i.e.* if òne form can be converted into the other and *vice versa* by altering the temperature or pressure, then the polymorphism is said to be enantiotropic. But if one form is unstable under all conditions, or at any rate under all conditions at which the substance has been studied, the polymorphism is said to be monotropic. This important difference can be made clear by means of pressure-

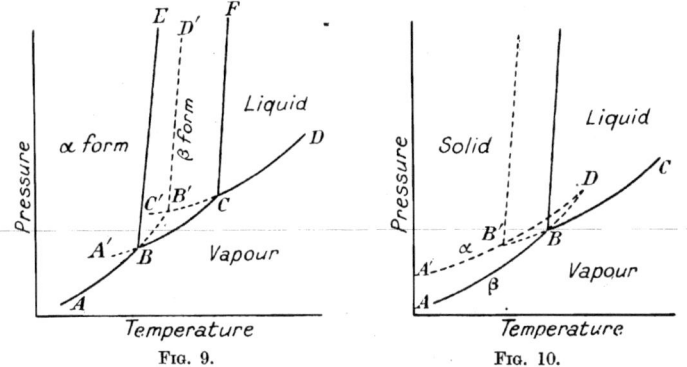

Fig. 9. Fig. 10.

temperature diagrams. Fig. 9 shows such a diagram for an enantiotropic substance. The area ABE gives the conditions under which the α modification is stable, EBCF is the corresponding area for the β form, FCD that for the liquid state, and below ABCD the substance exists entirely as vapour. The curves AB, BC, and CD are thus the vapour pressure curves for the α form, β form, and liquid respectively. The point B is called the *transition point* between the α and β forms, since it marks the temperature below which the α form and above which the β form are stable, when the substance is under the pressure of its own vapour. At higher pressures the transition temperature is given by points on the line BE. The slope of this line is usually very small,* and the

* The slope of this and all other curves in the diagram is given by the Clausius-Clapeyron equation $dp/dt = Q/T\Delta V$, where p = pressure, t = tem-

difference between the transition temperature at, say, atmospheric pressure and that corresponding to the point B some small fraction of a degree. Similarly, the point C is the melting-point of the substance when under its own vapour pressure. The melting-point at higher pressures is given by points on the line CF, which like BE is nearly vertical.

It will be seen from the foregoing that the liquid can only exist in stable equilibrium with the β form. But owing to the well-known fact that a liquid may be supercooled, *i.e.* cooled below its freezing-point without solid separating, provided that none of the solid is present, it is possible to obtain an extension of the curve DC, namely CC'. A similar reluctance to change is observed to a greater or less degree in the transformation of enantiotropic forms if all traces of the form towards which the change is directed are excluded, so that the curves AB and CB may be followed to B' and A' respectively. The extent to which these so-called *metastable* equilibria can be realised in practice depends mainly on the structural relationship of the two forms (see below), and on the care taken to exclude the stable modification. Should the point B' reach the curve CC' (it cannot go beyond it for the superheating of solids without melting is unknown), the B' will be the melting-point of the α form, and the line B'D' will give the effect of pressure upon it.

Fig. 10 gives a typical pressure-temperature diagram for a substance having two monotropic modifications. AB is the vapour pressure curve for the stable (β) form and A'B' that for the unstable (α) form, B and B' being the respective melting-points when the substance is under the pressure of its own vapour. The two curves may be imagined to intersect at the point D, which is therefore the transition point, but since this lies at a higher temperature than either of the melting-points it is not realised in practice. Below the melting-points the α form is always unstable with respect to the β form. It may happen that, instead of cutting at D, the curves do so at some temperature which is lower than A and A', which we have tacitly assumed to be the lowest temperatures at which the substance has been studied. If this is so, the case is that of

perature, Q = the latent heat of the transition, T = the temperature in absolute units at which the transition occurs, and ΔV = the change in volume accompanying the transition. Since ΔV for solid-solid and solid-liquid transitions is small, dp/dt is large and so BE and also CF show very little deviation from the vertical.

POLYMORPHISM

a system which is monotropic within the range of temperature studied, but is actually enantiotropic.

When a substance possesses more than two polymorphic forms, they may be all enantiotropic, all monotropic, or some may be enantiotropic and the rest monotropic towards them. The pressure-temperature diagram in such a case is merely an elaboration of one or other, or both of those already given. Thus if a substance has three enantiotropic modifications, there will be two transition points, one between the α and β forms, another between the β and γ forms, and there will possibly be a metastable transition point between the α and γ forms.

From the standpoint of the structural changes involved, polymorphic transformations are of many different kinds. In some cases the change is of quite a minor character; at the other extreme are transformations involving a radical alteration in the arrangement of the building units of the crystal. Silica furnishes examples of both types. This substance has at least six different crystalline forms, all of which consist of a three-dimensional network of tetrahedral SiO_4 groups, in which each O atom is shared with a neighbouring group. The three principal forms are quartz, tridymite, and cristobalite, which are enantiotropically related to one another as follows:

$$\text{quartz} \underset{}{\overset{870°}{\rightleftharpoons}} \text{tridymite} \underset{}{\overset{1470°}{\rightleftharpoons}} \text{cristobalite}.$$

These differ fundamentally in the manner of arrangement of the SiO_4 groups, and the change from one to another involves the breaking of —Si—O—Si— bonds. The structure has, so to speak, to be taken to pieces and built up again on a different plan. Each of these forms also exists in a low temperature (β), and a high temperature (α), modification, which differ merely by a small shift in the relative positions of neighbouring SiO_4 groups, and the transformation from one to the other takes place without any bonds being broken. It is therefore readily understandable that the quartz-tridymite-cristobalite transformations occur sluggishly, and consequently all these forms can exist in the metastable condition for long periods; the α-β transformations on the other hand take place readily and reversibly when their respective transition temperatures are reached.

Another type of polymorphic transformation is that between the hexagonal and cubic close-packed structures. Certain metals exhibit both of these structures (cobalt and nickel, for example), and some compounds also exist in two forms based respectively on interpenetrating hexagonal and cubic close-packed arrangements of the constituent atomic species (zinc sulphide is one). The transformation from one structure to the other involves the shifting of the close-packed layers parallel to one another, as will be seen from the explanation of the difference between these two types of structure given on p. 4. In metals at least, this type of change takes place readily when the transition point is reached, since it does not involve any major reorganisation of the structure.

Lastly may be mentioned those polymorphic transformations that are due to the development of a defect structure above a certain temperature. Some cases have already been touched on above. Thus it has been shown that silver iodide above 146° and silver mercuric iodide above 51° depend for their rigidity and form on the iodine constituent alone, whilst the metal atoms are distributed at random and are mobile. At all temperatures below 146°, silver iodide possesses a normal type of crystal structure in which both silver and iodide ions are arranged in a regular manner in fixed and definite positions. This structure is different from that adopted by the iodide ions above 146° C., and so at this temperature there is a sharp heterogeneous transformation from one crystal form to the other. The case of silver mercuric iodide is rather different, for although there is a heterogeneous transition from one form to the other at 51° C., the random arrangement of the silver and mercury ions to which the high temperature form principally owes its properties, has already begun to appear at about 40° C., though it is not complete until the transition point is reached.

It has also been mentioned above that polymorphism may arise through the free rotation of molecules or polyatomic ions at higher temperatures, and the example of ammonium nitrate has been cited.

Isomorphism. Before the development of the X-ray method of structure analysis, the only geometrical basis for the comparison of crystals was their external form. Crystals of different substances were said to be *isomorphous* if their corresponding interfacial angles were precisely the same (this can only happen for all the angles when the crystals belong to the cubic system), or differed by very small amounts, say 1 or 2 degrees.

The first cases of isomorphism were discovered by Mitscherlich in 1819 in the course of a crystallographic study of the phosphates and arsenates of sodium, potassium, and ammonium. He found that crystals of the corresponding phosphates and arsenates, *e.g.* KH_2PO_4 and KH_2AsO_4, resembled each other so closely that he was unable to detect any difference in their interfacial angles. (Actually small differences do exist, but these were probably within Mitscherlich's errors of measurement.) Many other cases of a similar kind were subsequently found, and this led to the view that if two compounds were isomorphous, there was a strong probability that their chemical formulæ differed only by the substitution of one atom by another of similar chemical nature. Chemists are familiar with the fact that the application of this principle played a not unimportant part in the early determinations of atomic weights. Thus the isomorphism of potassium perchlorate and potassium permanganate indicated the correct atomic weight for manganese. The formula of the perchlorate was known from other evidence, and so if that of the permanganate could be written merely by substituting manganese for chlorine, the atomic weight of the former followed at once from the relative weights of manganese and chlorine in the formula weights of the salts. Similarly the isomorphism of potassium selenate and potassium chromate with potassium sulphate led to the correct atomic weights being assigned to selenium and chromium, the formula of potassium sulphate being already known.

There are, however, very many exceptions to this principle. Thus calcium sulphate, $CaSO_4$, is not isomorphous with strontium sulphate, $SrSO_4$, but is isomorphous with sodium perchlorate, $NaClO_4$, whilst strontium sulphate is isomorphous with potassium perchlorate, $KClO_4$. Sodium nitrate, $NaNO_3$, is not isomorphous with potassium nitrate, KNO_3 (at least at ordinary temperatures), but is isomorphous with calcium carbonate, $CaCO_3$. Again the complex salts $K_2[NbOF_5],H_2O$ and $K_2[WO_2F_4],H_2O$ are isomorphous, and many other such cases could be quoted.

The reasons for these facts become clear when we consider the matter from the standpoint of crystal structure. It has already been indicated in the early part of this chapter that geometrical considerations connected with the shapes and sizes of the building units play an important part in determining the structure of a crystal. For example, suppose that two salts are composed of ions of types A^{n+} and $(XY_m)^{n-}$. If both anions have the same shape and the ratios of the size of the cation to that of the anion agree within certain limits, it will frequently be found that the salts adopt similar structures, irrespective of whether they have the same value of n, the ionic charge, and therefore largely irrespective of the natures of the atoms A, X, and Y. Geometrical factors obviously favour the same way of packing the ions together for both the salts. Further, if similar structures are adopted in such a case, the shapes of the respective unit cells are likely to be so similar that the external forms of the crystals will resemble one another very closely, or exactly, if the structures are of the cubic type.

It is obvious that such conditions will often arise when the two salts are of the same chemical type, but that they will also arise when the salts are chemically unrelated. Moreover, in two compounds of similar chemical formula type, the relative sizes of cation and anion may differ to such a degree that the structure adopted by one is unstable for the other, particularly if the cation : anion size ratio for the first is near the permissible limit for that particular structure. Thus let us take the case of $CaSO_4$, $SrSO_4$, $NaClO_4$, and $KClO_4$ mentioned above. The Sr^{++} ion is larger than the Ca^{++} ion and by an amount that oversteps the limit of stability for the $CaSO_4$ structure. $SrSO_4$ therefore crystallises in another structure. On the other hand the ClO_4^- ion is of the same tetrahedral shape and of about the same size as the $SO_4^=$ ion, and the Na^+ and Ca^{++} ions are very nearly of the same size. $CaSO_4$ and $NaClO_4$ are therefore isomorphous. Similarly the K^+ ion is larger than the Na^+ ion and cannot accommodate itself to the $NaClO_4$ structure : $KClO_4$ therefore adopts the $SrSO_4$ structure.

The isomorphism between $NaNO_3$ and $CaCO_3$ and between $K_2(NbOF_5),H_2O$ and $K_2(WO_2F_4),H_2O$ may be explained in a similar way. The NO_3^- and $CO_3^=$ ions both consist of a triangle of oxygen atoms with the nitrogen or carbon at the centre, and they are nearly the same size. As stated above, the Na^+ and Ca^{++} are also of nearly equal size, and the two salts have almost identical structures. The $(NbOF_5)^=$ and $(WO_2F_4)^=$ ions are both octahedral in shape with the niobium or tungsten atom at the centre and the oxygen and fluorine atoms at the corners, and their sizes are nearly the same. Hence conditions are favourable for their combinations with K^+ ions to adopt similar structures.

It should be emphasised that in the above examples we have

MIXED CRYSTALS AND ISOMORPHISM 23

been comparing compounds of the same chemical bond type, and that only in such cases do geometrical factors predominate. With compounds of different bond type, it is the nature of the bonds which exerts the main influence in deciding the crystal structure. To illustrate this, let us consider the compounds MgO and ZnO. On geometrical considerations alone it would be expected that they would have the same structure, since the effective radii of the Mg^{++} and Zn^{++} ions as deduced from a study of other structures do not differ much. The Mg—O bond is, however, mainly ionic, whilst the Zn—O bond is much more covalent in character, and the crystal structures are quite different. MgO has the sodium chloride structure (Fig. 3) in which each atom is surrounded by six of the other kind. ZnO crystallises in a structure in which each zinc atom is surrounded by *four* oxygen atoms, and each oxygen by four zinc atoms.

In Chapter III it will be made clear that for each type of external form there are usually several types of internal structure which could have given rise to it. For example, the structures shown in Figs. 1, 3, 4, and 6 all produce crystals with the same external symmetry. There is no chemical relationship between the four substances (beyond the fact that sodium chloride and calcium fluoride are typical salts), and their structures differ significantly. There is in fact no property other than that of external crystalline form (and the isotropism common to all cubic crystals) which justifies their being classed together. For such reasons the original definition of the term isomorphism has tended nowadays to become replaced by a more useful one which states that *isomorphous substances are those in which geometrically similar structural units are arranged in the crystal in the same way*. According to this definition, xenon, sodium chloride, calcium fluoride, and diamond are not isomorphous, but all cases in which there is a structural similarity, with or without chemical similarity, are covered by it.

Mixed Crystals and Isomorphism. The formation of mixed crystals was once regarded as being closely connected with isomorphism. However, many pairs of isomorphous substances do not form mixed crystals, and there are many cases of mixed crystals being formed by substances that are not isomorphous in the sense of the definition just given.

In general, mixed crystal formation requires that the structures of the two ingredients be compatible both as regards type of structure and also dimensions of the structural units. If the agreement between these properties is close, the two substances may be miscible in all proportions; if not so close, each may dissolve the other to a limited extent only; whilst if the structures or the size of the structural units, or both, are very different, mixed crystals will not usually be formed.

For example, molybdenum and tungsten both crystallise in the same type of cubic structure and their effective atomic radii differ by less than 1%. They form an unbroken series of mixed crystals in which the atoms of molybdenum and tungsten occupy the lattice positions at random. Similarly potassium chloride, KCl, and potassium bromide, KBr, are miscible in all proportions. Both these salts crystallise with the sodium chloride structure (Fig. 3) and the radii of the anions differ by about 8% only of that of the smaller chloride ion. This difference in size is insufficient to make the structure unstable at any stage as one anion replaces the other.

Potassium iodide, KI, also crystallises with the sodium chloride structure, but only enters into limited miscibility with potassium chloride since the radius of the iodide ion is about 19% greater than that of the chloride ion. The same is true of NaCl and KCl where there is an even bigger percentage difference between the radii of the two cations. On the other hand, NaCl and silver chloride, AgCl, form a wide range of mixed crystals, despite the fact that the chemical relationship between sodium and silver is not so close as that between sodium and potassium; the size ratio $Na^+ : Ag^+$ is more favourable than is that for sodium and potassium ions.

In some cases where the geometrical factors would seem to favour the formation of mixed crystals, this does not happen because one of the salts is insoluble. For this reason $NaNO_3$ and $CaCO_3$, already mentioned, which geometrically are highly compatible, cannot form mixed crystals because $CaCO_3$ is insoluble. Crystals of $NaNO_3$ deposited from solution on crystals of $CaCO_3$, however, orient themselves in conformity with the $CaCO_3$ structure, and the same thing occurs in other similar cases. This phenomenon is usually referred to as the formation of *oriented overgrowths*.

As examples of mixed crystal formation between non-isomorphous substances, mention may be made of the case of silver bromide, AgBr, and silver iodide, AgI. The former crystallises with the sodium chloride structure, whilst the latter has quite a different structure under ordinary conditions, though it is interesting to note that at very high temperatures it changes to the sodium chloride structure. AgBr will dissolve up to 70% of AgI, or to put it another way, the AgBr lattice will "stand" 70% of its bromide ions being replaced by iodide ions before it becomes unstable. On the other hand, AgI will dissolve very little AgBr.

Another example which has already been mentioned under "defect structures" is that of LiCl and $MgCl_2$. The former crystallises with the sodium chloride structure, whilst the latter has a layer structure of a somewhat similar type to that shown by mercuric iodide (Fig. 5). CaF_2 and yttrium fluoride, YF_3, present a similar example. The structure of the former is given in Fig. 4, and YF_3 has a structure like this but with additional fluoride ions at the centre of the cell and at the middle of each of its twelve edges. The radius of the Ca^{++} ion is only 6–7% greater than that of the Y^{+++} ion. Thus conditions for mixed crystal formation are more favourable than would appear from the formulæ, and in fact CaF_2 can dissolve up to about 40 formula weight per cent of YF_3.

RECOMMENDED FOR FURTHER READING

The Crystalline State, W. L. Bragg (Bell, 1933).
An Introduction to Crystal Chemistry, R. C. Evans (Cambridge, 1939).
Structural Inorganic Chemistry, A. F. Wells (Oxford, 1945).
Chemical Crystallography, C. W. Bunn (Oxford, 1945).

CHAPTER II

THE STEREOGRAPHIC PROJECTION

Drawings of various kinds, either in perspective or in plan, are useful for giving a picture of the external appearance or habit of crystals, but it is convenient to be able to show accurately the angular relations of the crystal faces (and hence the crystal symmetry), by means of the more highly abstract graphical representation afforded by geometrical projections. Various types of these have been used, but as an effective means of displaying crystal symmetry and the geometrical relationships of the optical properties to external morphology, as well as being an indispensable tool in many optical investigations, the stereographic projection occupies a unique place. For these reasons, and as it is easily constructed with a few simple drawing instruments, it is the only one that will be used in this book. Examples of its use will be encountered in the following pages, so only the fundamental principles upon which it is based, and the necessary geometrical constructions in which the student should become proficient, are dealt with in this chapter.

Fundamental Principles of the Stereographic Projection. Consider a sphere (Fig. 11), of which NS is the vertical diameter, and ABCD the plane normal to this diameter through the centre of the sphere. This plane cuts the sphere in a circle XYX'Y' known as the *primitive circle* or briefly, the *primitive*. If any point p on the surface of the sphere is joined to S by a straight line, the intersection P of this line with the equatorial plane is the stereographic projection of p. Points such as p situated on the surface of the upper hemisphere are always projected in this way, but points on the lower hemisphere such as p' (which for the purpose of this demonstration is regarded as being vertically below p), may be projected by joining to N, the pole of the upper hemisphere (in which case it plots also at P), or by being projected from S on to the plane ABCD as at P'. The first method is adopted in showing the relationships of crystal faces, and the second is necessary for certain geometrical constructions, as will be shown later. Fig. 12 is a vertical section through Npp'S of Fig. 11 and will make the

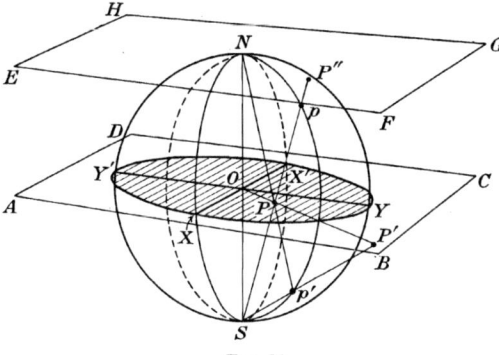

Fig. 11.

two methods of projection plain. It should be noted that projection on any plane parallel to ABCD (Figs. 11 and 12), by the methods detailed above will give a stereographic projection, but on a different scale.

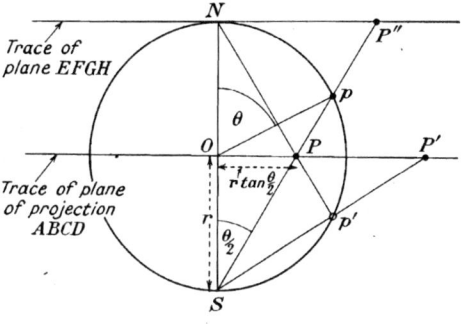

Fig. 12.

For the cartographer who wishes to map the surface of the earth by using this method of projection, the features to be shown are already disposed on the surface of a sphere, but for the crystallographer it is necessary first of all to make a *spherical projection* of the relevant crystal properties. For this purpose the crystal is

THE STEREOGRAPHIC PROJECTION

imagined to be oriented within a sphere so that for example an important set of edges is parallel to the N–S axis as in Fig. 13. From the centre of the sphere lines are imagined to radiate in directions normal to the planes represented by each crystal face, and where these *face normals* pierce the sphere, *face poles* are located. These face poles are then projected on to the plane of projection within the primitive circle in the manner already described and are referred to as face poles in the projection. Faces occurring on the upper half of the crystal are represented conventionally in the projection by points, and those on the lower half by small open circles.

The face normals will not necessarily pass through the crystal face which they represent, but often through the face produced, as shown in Fig. 14. As a matter of fact, it is better to regard each

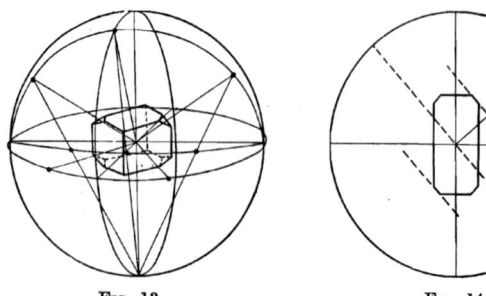

Fig. 13.   Fig. 14.

crystal face as being translated parallel to itself until its plane intersects the centre of the sphere of projection for the following reason. It is often necessary to represent a crystal face or other plane not only by its pole, but also by projecting the trace of the intersection of the plane with the sphere. This is shown in Fig. 15. The plane LXMX' is represented by its face pole p, which is projected to P, and also by the arcs XlX' and XmX' which are the stereographic projections of the upper and lower parts of the circle XLX'M respectively. Further, P is not only the projection of the face pole p, but may also be taken to represent the *direction* Op, which might be, for example, an optic axis of the crystal (see p. 113) or the pole of a zone of crystal faces (see below). The representation of a plane by projecting the circle in which it intersects the sphere is often called the *cyclographic projection* of the plane.

Great Circles and Small Circles. Great circles are the

PROPERTIES OF THE STEREOGRAPHIC PROJECTION   29

intersections with the sphere of planes which pass through its centre. In the discussion above it was pointed out that crystal faces and other planes are to be regarded as moving parallel to themselves until they pass through the centre of the sphere of projection. It will be clear therefore that all planes such as crystal faces, cleavage planes, symmetry planes, and the like, when shown in cyclographic projection will be the stereographic projections of great circles.

Small circles are formed by the intersection of the sphere by planes which do not pass through its centre (Fig. 15). They form the locus of points at an equal angular distance from a point on the surface of the sphere, or alternatively, the intersection with the

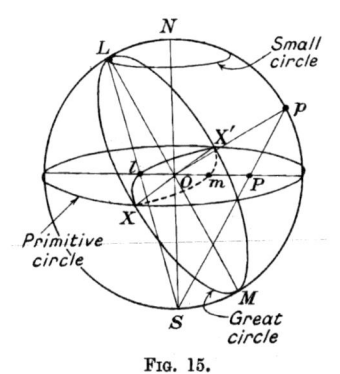

FIG. 15.

sphere of a cone of directions making equal angles with the axis of the cone.

Properties of the Stereographic Projection. 1. *The projection of a circle on the sphere is a circle.* This is a property which makes the drawing of the projection easy. Vertical great circles will project as straight lines (circles of infinite radius), through the centre of the projection. Other great circles will project within the primitive circle as arcs of circles, and the line joining diametrically opposite points where these arcs cut the primitive will pass through the centre, being a projection of the diameter of the great circle. Small circles entirely within either the north or south hemisphere will project as complete circles within the primitive; those which lie in both hemispheres will project as arcs of circles.

2. *Angular truth is preserved in the projection.* This means that

C.P.M.   B* the angle between the planes of two great circles which intersect on the sphere is the same after projection on to the equatorial plane. It is this property which makes the stereographic projection so valuable a means of checking calculations involving the solution of spherical triangles, and in such problems as the estimation of the extinction angle on any face of a crystal (see page 253), as well as being a more rapid substitute for calculations where the utmost accuracy is not required. It will be seen from Fig. 16 that the *linear* value of a degree in the projection is greater near the primitive than near the centre of the projection, but this causes no difficulty in measurement.

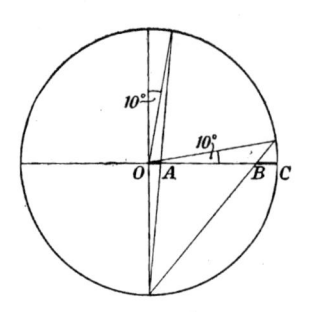

FIG. 16.—Distances OA and BC represent 10° at centre and margin of projection, respectively.

The proofs of the above statements are not presented here, but the reader who is interested will find them in the *Manual of Petrographic Methods* by A. Johannsen, 2nd Edition, 1918, pp. 8–9.

Some Necessary Constructions. The following graphical constructions are sufficient to enable the student to make for himself stereographic projections from angular values of the face poles, to solve problems such as the graphical determination of extinction angles, and to understand the construction of stereographic nets. The only *necessary* instruments are a pair of compasses with an extension bar, a protractor of not less than 5" diameter, a scale in millimetres and a pair of set-squares. Special protractors and scales have been devised to help in the drawing of projections. They are not necessary, but effect a saving of time. However, after a time, linear scales engraved on celluloid become inaccurate because of contraction. In any case, it is good that the student should first undergo the discipline of going through all the constructions without these special aids. Reference may be made to the bibliography at the end of this chapter for descriptions of these instruments.

All the constructions may be done on the plane of the primitive circle. This is possible because any great circle can be imagined to be revolved into the plane of the primitive. Consider Fig. 17.

SOME NECESSARY CONSTRUCTIONS

Suppose it is desired to plot a pole p which lies 45° above the primitive on the East–West vertical great circle. This plots at P. The same result is obtained by doing the construction on the primitive circle, which for this purpose may be regarded as the vertical circle revolved around the horizontal diameter BOB' which it has in common with the vertical circle. X becomes the "south pole" for the purpose of the construction, and p', 45° from B on the primitive, plots at P also.

1. *The Primitive Circle.* This should be drawn with a radius of 5 cm. Any larger size becomes inconvenient when constructing circles with a small inclination from the centre of the projection, and on so small a circle with careful work an accuracy of half a

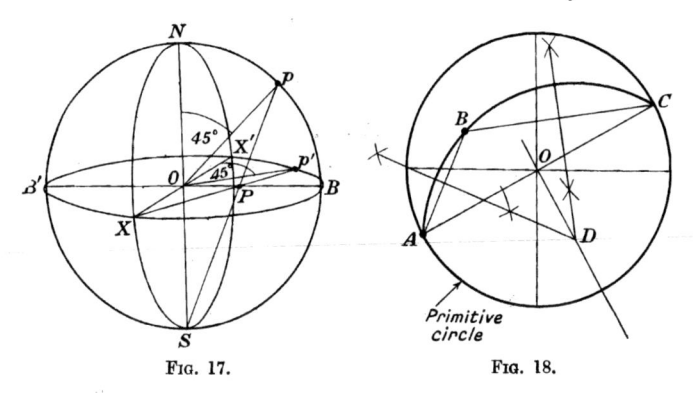

FIG. 17.   FIG. 18.

degree may be attained. Generalised nets to be described later, may be used with a much greater radius.

2. *To draw a great circle through two poles, one lying on the primitive.* In Fig. 18, which is a projection on the equatorial plane, A and B are the two poles. Join A to O at the centre and produce to C on the primitive. AOC is the projected diameter of the required circle. Bisect the chords AB and BC. The intersection D of the perpendicular bisectors then marks the centre of the projected circle. This centre, around which the circle ABC may be drawn, lies on the normal to AC from O.

3. *To draw a great circle through two poles, both lying within the primitive.* The principle underlying this construction is to find a point diametrically opposite to either of the given poles. In Fig. 19, which, like Fig. 18, shows the projection on the equatorial plane,

A and B are the given poles. Through A and O draw a diameter and produce, then construct the normal to AO from O, meeting the primitive at S, which becomes the "south pole" for the purpose of the construction. Project A to the primitive at D from S. Draw

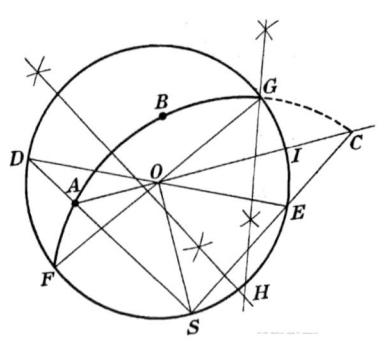

FIG. 19.

the diameter DOE to find E which is thus 180° from D. The stereographic projection of E is located at C. Thus three points on the required great circle have been found and its centre is located at H by the usual construction and the circle drawn in.

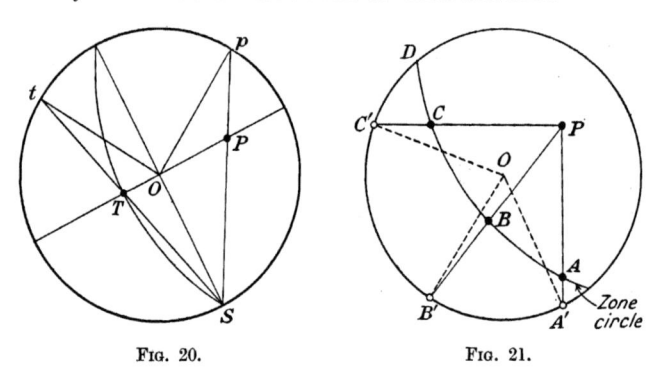

FIG. 20.   FIG. 21.

4. *Given a pole, to draw the corresponding great circle*, i.e. *the great circle to which it is normal*. In Fig. 20, P is the given pole. Draw the diameter through P and O and the normal to it, OS. Project P to the primitive at p. Locate t, 90° from p and project back to S thus finding T, 90° from P in the projection. The great circle

SOME NECESSARY CONSTRUCTIONS

through T and S can now be found by the construction shown in Fig. 18. The reverse of this problem, to locate P given a great circle, is obvious from the figure.

The faces of crystals occur in sets which intersect (or their planes produced intersect) in parallel edges. Such a group of faces is called a *zone*. It will be appreciated that all the face poles of these faces will lie on a great circle. It is appropriate to point out here an important property of the pole of a zone, which is the projection of the normal to the circle on which the face poles lie. In Fig. 21, P is the pole of the zone circle ABCD, on which A, B, and C are poles of the faces of a zone. If now the poles are projected from P to the primitive, the angles between the faces may be measured

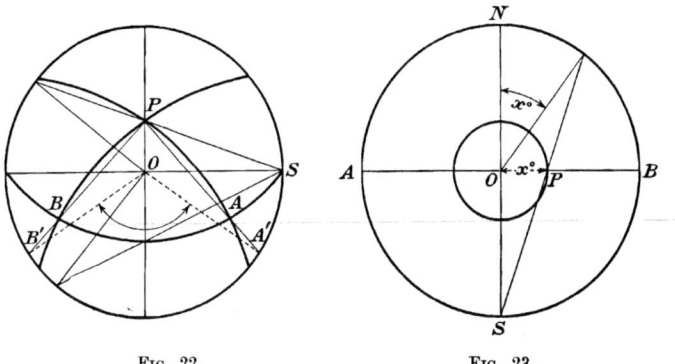

Fig. 22.   Fig. 23.

around the primitive circle as shown by the dotted lines from O. The converse of this construction is a very useful way of plotting the faces of a zone if the polar angles and the inclination of the zone circle are known.

5. *To measure the angle between two poles.* Pass a great circle through the two poles as described in sections 2 and 3. Locate the pole of this circle as shown in 4 above, and measure the angle by projecting to the primitive from the pole as in Fig. 21.

6. *To measure the angle between two great circles.* In Fig. 22, two great circles are shown passing through P. The method is to pass a circle 90° from P over the two circles between which the angle is to be measured. This cuts off two equal segments PA and PB each of 90° arc. Now P is the pole of the circle just drawn so that

THE STEREOGRAPHIC PROJECTION

the angle will be given by projecting A and B from P to the primitive, the angle being measured in the manner described above, as shown by the dotted lines from O in the figure.

Small Circles.

7. *To construct a small circle $x°$ from the centre of the projection.* Measure the angle $x°$ along any diameter, such as AOB in Fig. 23, by laying off from N, normal to the diameter as shown, and project on to the diameter by joining to S. The required radius is OP.

8. *To draw a small circle with a radius of $x°$ from a point on the primitive* (Fig. 24). Draw the diameter from the point (P), and the normal to this diameter from O to S. Lay off $x°$ in opposite

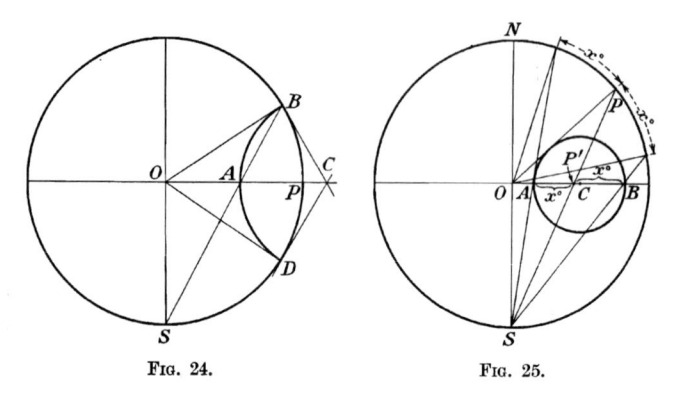

FIG. 24.   FIG. 25.

directions from P (*i.e.* in the Fig. $\lfloor BOP = \lfloor DPO = x°$), and project one of these points, say B, on to the diameter. Three points, B, A, and D, on the required circle have now been found, and the centre C is found by the method of bisection of the chords BA and AD. Another method is to draw tangents at B and D to the diameter produced. This method is in practice not as accurate as the first described.

9. *To draw a small circle of $x°$ around a point within the primitive* (Fig. 25). P' is the given point. Draw the diameter through it and the normal OS. Project P' to p. Lay off in opposite directions from p around the primitive, angles of $x°$, and project the angle $2x°$ back on to the diameter. This marks the diameter AB of the small circle, and it is only necessary to bisect this line to get the

centre C of the required circle. Notice that the geometrical centre is not coincident with the angular centre P'.

10. *To locate a pole given its angular distance from two other points.* The two given points A and B may be on the primitive, as shown in Fig. 26. Draw diameters from each point, and as already described (sections 8 and 9) construct small circles of the required angular radius around each point. In the case shown in the figure, the required pole P is located 70° from B, and 50° from A.

When the two given points are within the primitive, draw the two small circles corresponding to the required angles $x°$ and $y°$ around the two points, as demonstrated in section 9. These small circles intersect at two points. This means that geometrically

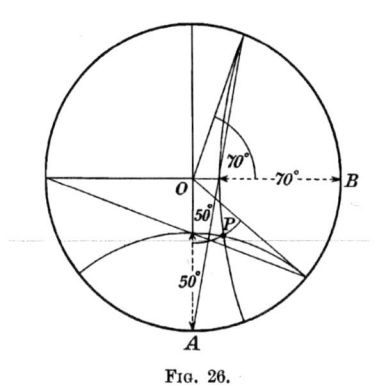

FIG. 26.

there are two solutions to the construction, one or both of which may represent real poles in the particular case under consideration. There is usually no difficulty in deciding this point from a knowledge of the other measurements.

Generalised Stereographic Nets. The labour of construction and measurement is greatly reduced by the use of printed stereographic nets, two types of which are in general use. The first, the so called Wulff net (Fig. 27), consists of a series of great circles, usually at 2° intervals, drawn between two diametrically opposite points on the primitive, and around these two points is a series of small circles at the same angular interval. The second, the Federov net (Fig. 28), is like two Wulff nets at right angles, superposed upon one another, and has in addition a series of small circles described about the centre of the projection as well as a series of vertical great circles. The angular distance between adjacent circles on the Federov net is 5°; closer spacing would result in confusion. The nets may be printed on card or paper and the plotting done upon them directly, or tracing paper may be pinned to the centre of the net, the drawing being revolved above it to bring the appropriate

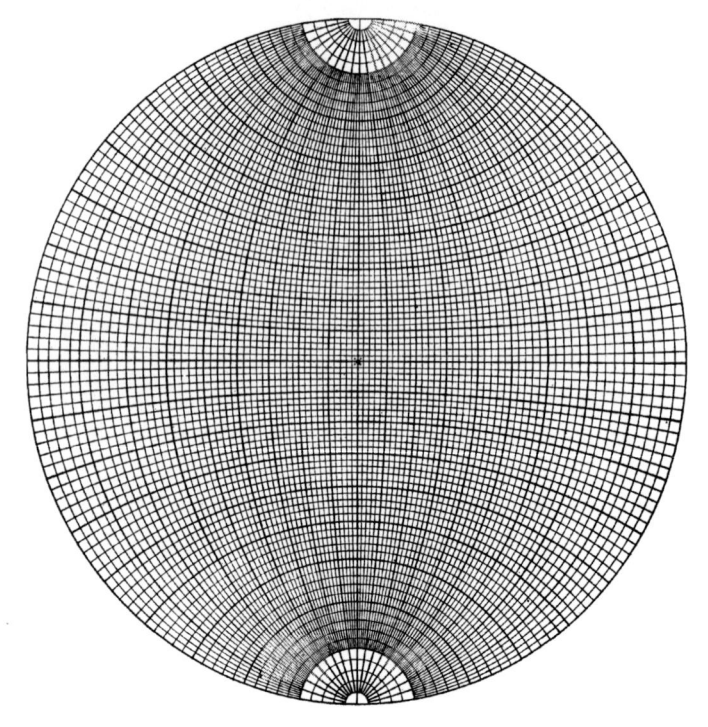

Fig. 27.

great or small circle into position, when it may be traced. Alternatively, the nets themselves are printed on tracing paper, and pinned by the centre to a sheet of stout paper below, when the appropriate circles or poles may be pricked through and subsequently inked in.

It is sometimes necessary to rotate a projection through a given angle around a horizontal axis. The generalised net is very useful

FIG. 28.

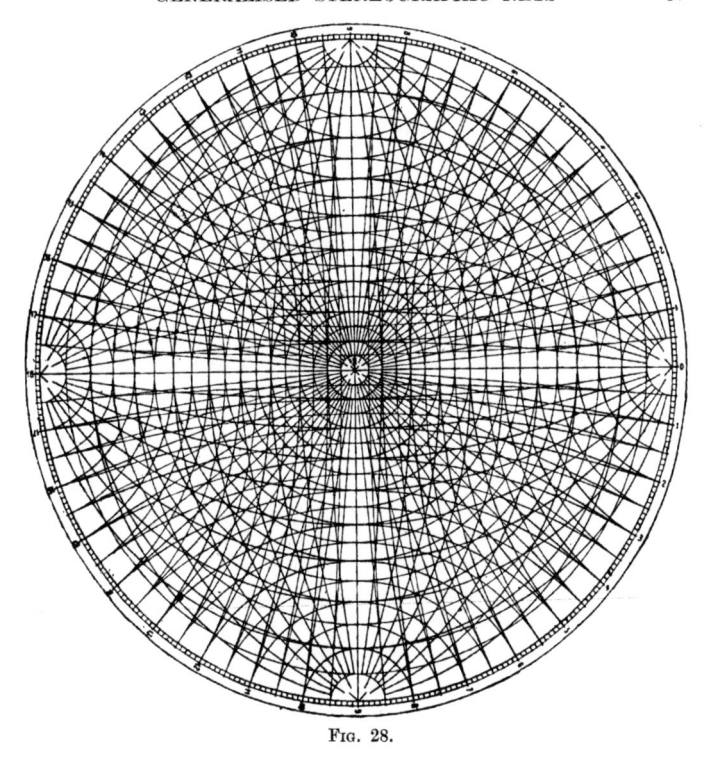

for this purpose. Each pole in the projection is turned through the appropriate angle along a small circle lying normal to the axis of rotation.

RECOMMENDED FOR FURTHER READING

Graphical and Tabular Methods in Crystallography, T. V. Barker (Murby & Co., London, 1922).

" On a Protractor for use in constructing stereographic and gnomonic projections of the sphere ", A. Hutchinson, *Mineralogical Magazine*, 1908, **15**, 93–112.

" The stereographic projection and its possibilities from a graphical standpoint ", S. L. Penfield, *Amer. J. Sci.*, 1901, d, 1–24, 115–44.

" The Stereographic Projection ", C. S. Barrett, *Amer. Inst. of Mining and Metallurgical Engineers*, 1937 ; *Metals Technology Technical Publication* No. 819.

CHAPTER III

THE MORPHOLOGY OF CRYSTALS

The early crystallographers, beginning with Hooke and Haüy,* arguing from the fact that many crystals can be split along cleavage planes into smaller and smaller units of which the larger crystal may be built, came to the conclusion that crystals were composed of minute parallelepipeda, which were in fact the primitive molecules of the substance. The idea was subsequently modified, and the concept of representing each unit of the crystal structure by a point in a space lattice took its place.

A *space lattice* is a regular arrangement of points in three dimensions, such that (*i*) the line joining any two of the points, when produced in either direction, will pass through a series of points spaced as were the original pair, and (*ii*) all parallel lines of the structure passing through points of the lattice present a similar spacing of points. In such a lattice *each point has a like environment*, and the structure is homogeneous.

Following the work of Frankenheim, Bravais in 1848 proved that there can exist only 14 space lattices fulfilling the condition stated above. These are shown in Fig. 29.

It is evident that a space lattice can be divided into elementary cells in many different ways, as shown in Fig. 30. The cells shown in Fig. 29 are those selected by Bravais himself on the basis of the following principles. The cell with the shortest sides and if possible of rectangular shape was chosen. In most cases this results in a cell which contains the equivalent of one lattice point, *i.e.* a true *unit cell*—Nos. 1, 4, 6, 7, 8, 10, 12, 13 and 14. (Each point is situated at the junction of 8 similar cells, and therefore contributes $\frac{1}{8}$ to each. Since there are 8 corners to a cell, it contains the equivalent of $8 \times \frac{1}{8} = 1$ point.) In the cells selected for the

* Haüy, *Essai d'une Theorie sur la Structure des Cristaux*, 1784.

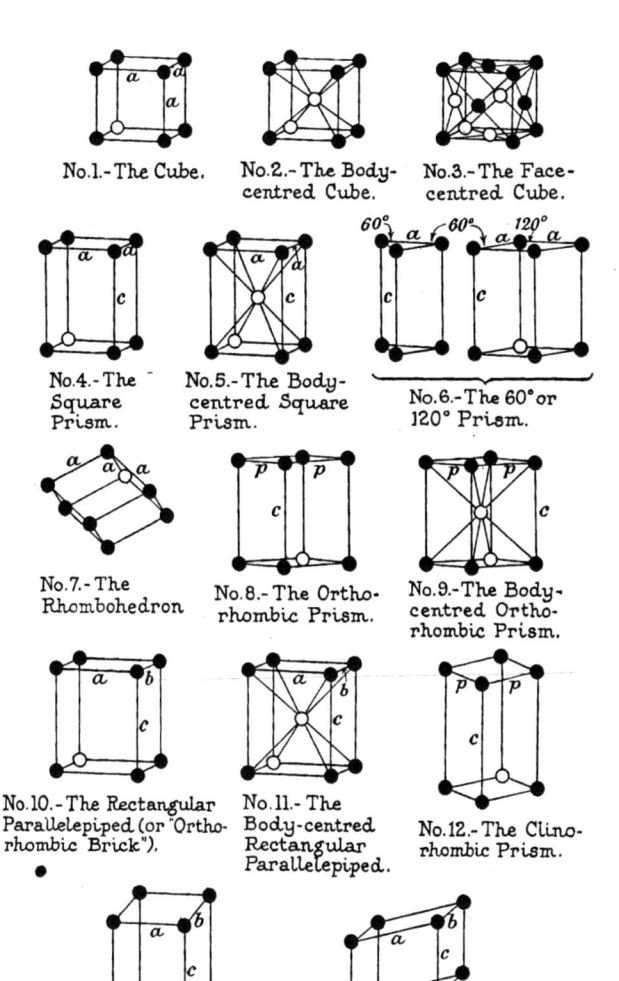

No. 1.—The Cube.
No. 2.—The Body-centred Cube.
No. 3.—The Face-centred Cube.

No. 4.—The Square Prism.
No. 5.—The Body-centred Square Prism.
No. 6.—The 60° or 120° Prism.

No. 7.—The Rhombohedron.
No. 8.—The Orthorhombic Prism.
No. 9.—The Body-centred Orthorhombic Prism.

No. 10.—The Rectangular Parallelepiped (or "Orthorhombic Brick").
No. 11.—The Body-centred Rectangular Parallelepiped.
No. 12.—The Clinorhombic Prism.

No. 13.—The Monoclinic Parallelepiped.
No. 14.—The Triclinic Parallelepiped.

NOTES. With the object of making clear the arrangement of the points, those which would be invisible if the cells were solid are represented by open circles.
- No. 7. Angles $a:a:a$ equal but not 90°.
- No. 8. Diagonals of top and bottom faces unequal.
- No. 12. Diagonals of top and bottom faces unequal, and these faces not at right angles to the vertical faces.
- No. 13. Angle $a:c = 90°$; angle $b:c$ not 90°.
- No. 14. Angles $a:b:c$ unequal and not 90°.

FIG. 29.—The Fourteen Space Lattices.

remaining lattices, there is an additional point at the centre or, in No. 3, in the middle of each face. These are therefore not true unit cells since they contain more than the equivalent of one point.

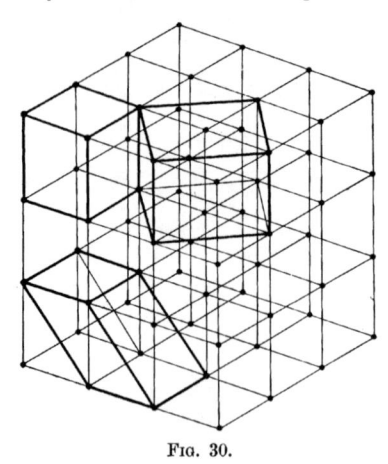

Fig. 30.

They are termed *compound cells*, and were chosen by Bravais in order to bring out more clearly the relationship between the lattice and the external symmetry.

The cells shown in Fig. 29 are not in all cases the ones now used

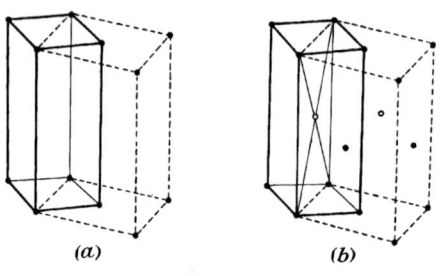

(a)   (b)

Fig. 31.

by X-ray crystallographers, who for greater convenience in structure analysis describe Nos. 8 and 12 in terms of a side-centred type of cell, and No. 9 as a face-centred cell (Fig. 31).

The Bravais cells can be classified according to the symmetry of their shape into 7 groups which correspond closely to the 7 groups,

THE MORPHOLOGY OF CRYSTALS 41

or *systems*, into which crystals have been divided on the same grounds, thus :

Bravais Lattice Number	Corresponding Crystal System
1, 2, 3	Cubic
4, 5	Tetragonal
6	{Hexagonal / Trigonal *
7	Trigonal
8, 9, 10, 11	Orthorhombic
12, 13	Monoclinic
14	Triclinic

The shapes of the Bravais cells show the highest symmetry possible (*holosymmetry* or *holohedrism*) in the system to which they belong. In each crystal system, however, there are classes of crystals which display lower symmetry than that of the normal or holohedral class, and the simple Bravais arrangement of the structural units cannot account for these. There are 32 of these classes in all.

The work of Sohncke in 1879 went far in providing a solution to the problem. He investigated the homogeneous point systems which result when two or more identical space lattices are made to interpenetrate one another in a regular manner. The resulting point systems do not fulfil the requirements of a space lattice as given above, but are homogeneous in the sense that every point belonging to a selected lattice has a similar environment.

The regular interpenetration of the two or more lattices is accomplished by a series of so-called " coincidence movements " upon the points of a selected lattice, these movements being of the following kinds :

(*i*) translation through a definite distance in a fixed direction, or
(*ii*) rotation through a definite angle around a stated axis, or
(*iii*) a combination of (*i*) and (*ii*).

Sixty-five arrangements are found to be possible, and their symmetry accounts for most of the symmetry classes of crystals. Figs. 32 to 35 show some of the simpler Sohncke point systems.

* It should be noted that crystals of the hexagonal system are based on the hexagonal lattice, the unit cell of which is a 60°-angle rhombus-based right prism. Crystals of the trigonal system are based on *either* the hexagonal lattice, *e.g.* quartz, *or* on the rhombohedral lattice the unit cell of which is a rhombohedron, *e.g.* calcite. X-ray analysis is necessary to decide whether a trigonal crystal has a hexagonal or a rhombohedral lattice.

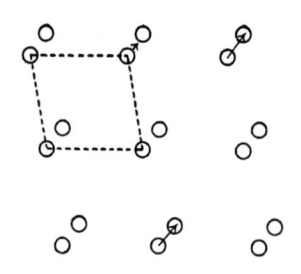

FIG. 32.—Two-fold Prism System. (Simple translation in direction of arrow.)

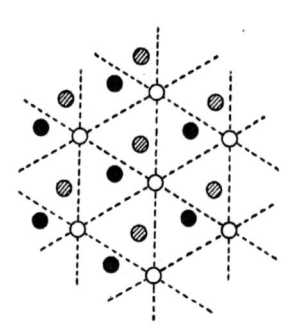

FIG. 33.—Sohncke System, No. 17, in Plan. Three trigonal lattices in one plane.

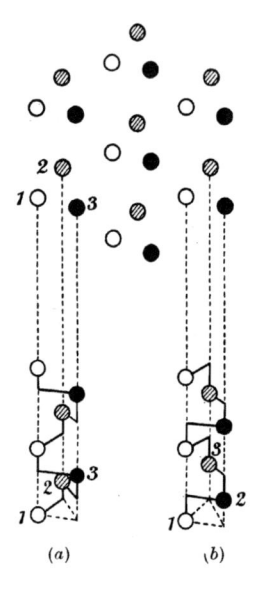

(a)   (b)

FIG. 34.—Sohncke Systems, Nos. 15 and 16, in Plan (top) and Perspective. Left-hand Screw (a), Right-hand Screw (b).

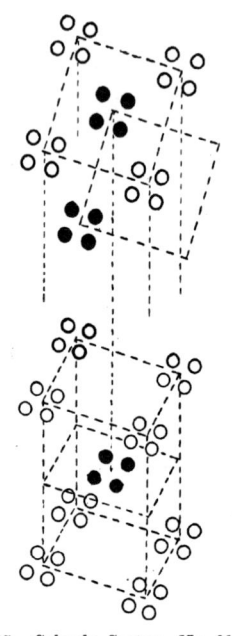

FIG. 35.—Sohncke System, No. 31, in Plan (above) and Perspective. Two parallel interpenetrating square prism systems.

THE MORPHOLOGY OF CRYSTALS

It will be seen that each lattice of the complex system is identical, and that the centres of gravity of the clusters of points form an exactly similar type of lattice. These clusters may be taken to represent the major structural units of a crystal, *i.e.* the units by the repetition of which in space the whole crystal structure is built up. The arrangement of the separate points in each cluster expresses the symmetry according to which the atoms are arranged in the structural unit.

The 65 point systems of Sohncke do not account for (*i*) hemimorphous crystals, *i.e.* crystals terminated differently at opposite ends of an axis; nor (*ii*) certain classes of enantiomorphous crystals, *i.e.* crystals which can occur in two different forms related as an object and its mirror image. Schönflies, Fedorov, and Barlow, working independently between 1890 and 1894, showed that these cases could be covered by introducing additional coincidence movements consisting of the *reflection* of a set of points, *regarded as having no symmetry in themselves*, across a plane, followed either by a translation, or by a rotation about an axis normal to the reflection plane, or by repetition about a centre of symmetry. It was found that there exist 230 possible ways of combining all these symmetry operations. Each of these combinations, termed a *space group*, results in a structure which has the symmetry of one or other of 32 crystal classes or *point groups* as they are called. This latter term is used because the symmetry of each crystal class may be defined by a group of symmetry elements operating about a centre so as to bring any point or unit of the structure back to its original position. The operations of a space group, on the other hand, result in the translation of a point to a corresponding point in the structure and therefore by their repeated operations building up the whole structure.

This work on crystal structure was entirely mathematical, and at the time no means of determining the space group of the crystal existed. The development of X-ray methods of investigating crystal-structure has resulted in the complete confirmation of the early mathematical theory.

When the conditions in which crystallisation takes place are favourable, the regular arrangement of the structural units results in the formation of a regular polyhedron bounded by perfectly plane surfaces. When, however, circumstances do not favour quiet and unhindered crystallisation, a true crystal may be produced with curved faces, or extremely complicated forms without any plane faces may result, as for example crystals of quartz in granite, or metallic crystals produced from a melt. In these cases the mutual interference of the growing crystals has prevented any expression of the regular internal structure, which still exists, as can be inferred, for example, from the occurrence of regular cleavage planes, or by the optical properties, and may be revealed by X-ray examination.

Crystals which are bounded by plane faces are described as *idiomorphic* or *euhedral*, and irregular ones as *allotriomorphic* or *anhedral* (see Fig. 36).

The fact that each chemical substance has its own architectural structure means that every substance is unique in the sum of its crystallographic properties. In regard to one property only, that

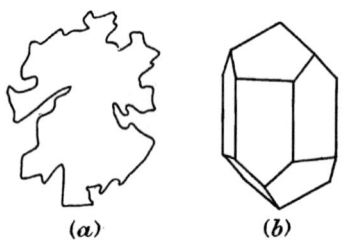

(a) (b)

FIG. 36.—Anhedral and Euhedral Quartz Crystals.
(a) Section of anhedral crystal from granite.
(b) Euhedral crystal.

of outward form, substances crystallising in the cubic crystal system may exhibit the same form and identical interfacial angles. But even in cases like this, difficulty is rarely encountered in differentiating between substances, when other properties such as density, cleavage, hardness, colour, refractive index, etc., are taken into account. The fact of the crystallographic uniqueness of each chemical substance has been used by Fedorov [*] and some of his followers, notably Barker,[†] to identify substances by morphological means alone, and both have suggested methods by which all known crystalline solids could be classified with this end in view.

The above statement must not be taken to imply that every idiomorphic crystal of one substance will present the same appear-

[*] See A. E. H. Tutton, *Crystalline Form and Chemical Constitution* (1926), p. 227.
[†] T. V. Barker, *Systematic Crystallography* (Murby, 1930).

THE MORPHOLOGY OF CRYSTALS 45

ance to the eye. It will be seen from Fig. 37, which represents a network (*i.e.* one plane of an imaginary space lattice) that although the number of possible lines passing through the points of the network is limited (and so therefore is the number of possible planes

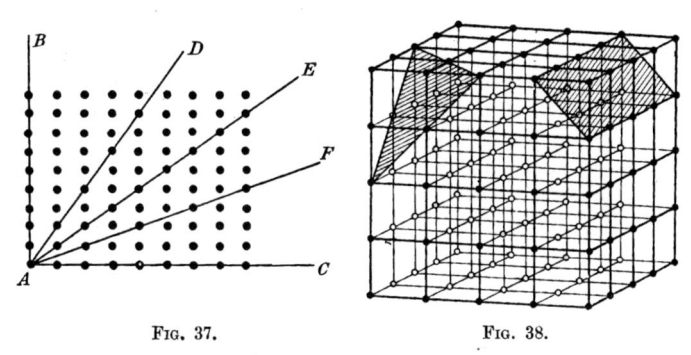

Fig. 37.

Fig. 38.

in the space lattice), yet a great number of combinations of planes enclosing solid figures is possible. Each different combination of such planes is called a *habit*. Fig. 38 shows a simple cubic lattice, and the manner in which such forms as the cube, octahedron, and dodecahedron arise. The octahedral planes cut symmetrically

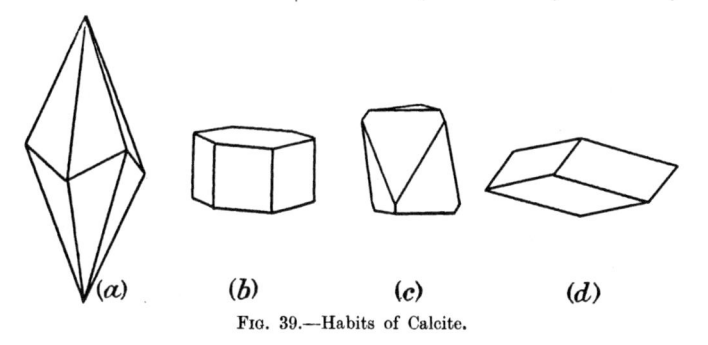

(a)   (b)   (c)   (d)

Fig. 39.—Habits of Calcite.

(*truncate*) the corners, and the dodecahedral ones the edges, of the cube.

Calcite, $CaCO_3$, which crystallises in the trigonal system, displays a remarkable variety of habits. The four drawings in Fig. 39 illustrate the diversity of appearance which is possible with crystals of this substance. Each crystal represented belongs not only to the same crystal system, but also to the same class (3̄m, see p. 57) of the system, that is each habit is developed on the same internal structure. It is convenient to notice here a number of general terms descriptive of habit.

The moderate development of a pair of parallel faces at the expense of others produces a *tabular* crystal (Fig. 40, a) and excessive development of this kind produces a *platy* crystal (Fig. 40, b), as in mica, lead iodide, which occurs in thin hexagonal plates, and potassium chlorate, which occurs in thin rhombic plates. Crystals of columnar form are known as *prismatic* (Fig. 40, c), unless the prism is so elongated as to be needle-like, when the term *acicular*

Tabular (a)

Platy (b)

Prismatic (c)

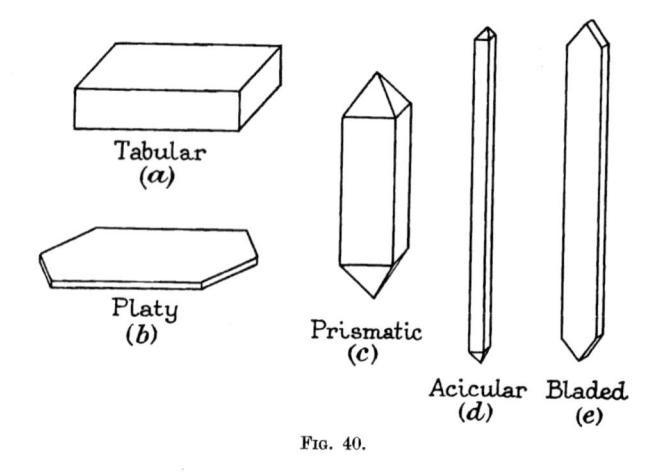

Acicular (d)

Bladed (e)

Fig. 40.

(Fig. 40, d) may be used. Typical prismatic forms are developed by potassium dithionate, magnesium sulphate heptahydrate, and ammonium dihydrogen phosphate, whilst acicular forms are given by the unstable monoclinic modification of sulphur, by potassium permanganate, potassium nitrate, and by most oxalates and tartrates. Flattened needles may be termed *bladed* crystals (Fig. 40, e).

The uneven development of like faces may mask a similarity of habit. Fig. 41 shows the variety of appearance due to this cause presented by octahedra of magnetite (Fe_3O_4). Such distortion is very common in crystals, for during growth they are usually resting on at least one face, which is thereby prevented to a large extent from receiving molecules from the solution, except at its edges.

THE MORPHOLOGY OF CRYSTALS 47

It is worth noting that this impedance results in the face or faces that are masked becoming abnormally *large*.

The various habits of calcite, mentioned above (Fig. 39), are the result of the development of an entirely different set of faces in each case, but crystals like those shown in Fig. 40, c and d, only differ in the relative development of the same set of faces. In this

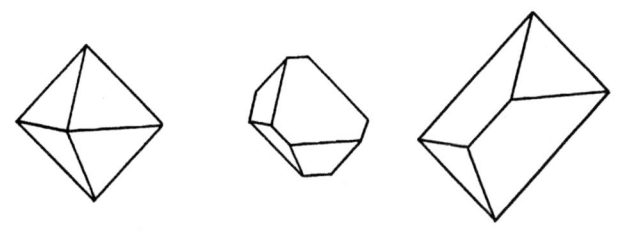

FIG. 41.—Unevenly Developed Octahedra of Magnetite.

case the only difference in the crystallographic description would be in the use of one of the descriptive terms already mentioned.

The fact that from one crystal structure such a great variety of habits may be developed has given rise to much research in order to elucidate the relationships between crystal structure, the environment of the growing crystal, and the resulting habit. It has been known for a long time that crystals which grow rapidly tend to develop extreme shapes such as thin plates or needles. If, however, a substance is grown very slowly in a number of different pure solvents, a different habit may be characteristic for a given solvent.

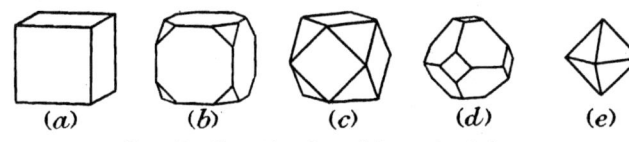

(a)　(b)　(c)　(d)　(e)

FIG. 42. Transition from Cube to Octahedron.

Again, the presence of impurities in a solution may influence the habit. Thus, for example, the addition of caustic soda to the pure solution from which alum crystallises in octahedra (Fig. 42, *e*), will induce the formation of faces of the cube, which truncate the corners as shown in Fig. 42, *d*. Sodium chloride crystallises in cubes from a pure aqueous solution, but the addition of a little urea induces the development of small octahedral faces on the corners of the cube (Fig. 42, b; see also Expt. 2, Chap. X), and the addition of larger amounts may so alter the relative importance of the cube and octahedral faces that the salt can be induced to crystallise in unmodified octahedra. It seems evident that this effect is due to the preferential adsorption of urea on the (111) planes of the crystal, thus hindering the deposition of the sodium and chloride ions on these planes.

A. F. Wells, in a series of papers * deals with the problem of crystal habit and internal structure, and the student is referred to these for further examples similar to those mentioned above, and for a very complete and critical discussion in the light of the most recent research. Important references to the literature on this subject are to be found in Wells's papers.

It only remains to point out here, that in describing the crystal morphology of any substance, all the relevant details as to conditions of crystallisation should be stated.

From the foregoing discussion it will be seen that the only criterion of crystallographic form is the data derived from the measurement of interfacial angles, which, measured between like faces, are constant for crystals of the same substance. From these angles may be calculated the relative lengths of the imaginary axes which are taken to intersect in the centre of the crystal, and to which the positions of the faces are referred, as will shortly be described. It follows that the relative lengths of the axes, together with the angles between them, must be constant for any one substance, and also characteristic of it, unless it belongs to the cubic system, in which the interfacial angles are determined by the symmetry, and the axes are equal in length.

The Measurement of Crystal Angles. The interfacial angles of crystals are measured by instruments known as *goniometers*. The simplest form of these is the *contact goniometer*, which is a semicircular protractor with a movable arm pivoted at the centre of the circle. When the two faces enclosing the angle to be measured are fitted closely to the base and the movable arm of the instrument, the value of the angle may read off directly on the graduated half circle.

Contact goniometers are now only used on large crystals with dull faces, which do not lend themselves to examination upon a *reflecting goniometer*. This extremely accurate instrument, of which

* *Phil. Mag.*, 1946, Series 7, 37, 184, 217, 605.

there are now various types, was invented by Wollaston in 1809. In the ordinary "single circle" type, the crystal to be measured (which need not be larger than about 1 mm. in average diameter, but must have smooth, well-developed faces) is mounted on a stout needle attached to the spindle of a graduated rotatable drum, which can be read to one minute of arc by means of a vernier. The needle is provided with lateral and tilting adjustments, by means of which any *zone* of the crystal may be brought parallel to the axis of rotation of the drum. (A zone is a group of faces, produced if necessary, the intersections of which are parallel lines, see p. 33). This having been done for one zone, parallel light from a collimator, fitted with a slit, is directed at the crystal, and, by turning the drum, is reflected from each face of the zone in turn into a telescope provided with cross hairs, the reading on the drum being taken every time the image of the slit coincides with the intersection of the cross hairs. The differences between these readings give the angles between the normals to the faces. These polar angles are the ones used in crystallographic data, as they are more useful than the included angles for the purpose of representing the crystal form in the stereographic projection.

With the above type of reflecting goniometer, a new setting of the crystal is required for each zone. (The angular relationships between the different zones can be calculated, because some faces are common to more than one zone.) There are more complicated "two circle" and "three circle" types by means of which all the interfacial angles can be measured with one setting of the crystal.

Crystal Axes, Indices and the Law of Rational Indices. The directions of the axes (already mentioned), to which the positions of crystal faces are referred, are determined by the intersections of three prominent faces at right angles, or as nearly at right angles as possible. The general case where none of the angles between any pair of axes is a right angle, as in the triclinic system, will be taken. In Fig. 43, XOX', YOY', and ZOZ' represent the three axes intersecting at O. They are parallel to the three pairs of pinacoid faces (Fig. 61, p. 66, and p. 62 for definition) which make up the simple crystal. The axes are arranged in the conventional way, the XOX' or a axis pointing to the left front of the observer, the YOY' or b axis pointing to the right of the observer, and the ZOZ' or c axis being arranged vertically. The angles between the axes are designated α, β, and γ, α being between the axes b and c,

β lying between a and c, and γ between a and b. The three planes containing the axes divide the space occupied by the crystal into eight parts known as *octants*.

For each substance a face is selected, which is inclined to all three axes, and makes intercepts on them which are as nearly equal as possible. Such a face is ABC in the figure, the intercepts on a, b, and c being OA, OB, and OC respectively. The face chosen is called the *parametral plane*, and its intercepts, expressed in terms of the intercept on b as unity, are called the *parameters*, or *relative lengths of the axes* of the crystal. These intercepts are usually given the same symbols, a, b, and c, as the axes to which they respectively refer. The ratios $a:b$ and $c:b$ are called the *axial ratios*, and these are equal to a and c respectively since $b = 1$. The accuracy of goniometrical measurements on good crystals is such that these quantities can usually be expressed to four decimal places. For example, for $CuSO_4, 5H_2O$ (triclinic system), $a:b:c = 0.5715 : 1 : 0.5575$. The angles between the axes for this substance are α 82° 16′, β 107° 26′, γ 102° 40′. The axial ratios and interaxial angles of a crystal are collectively termed its *crystal elements*.

Four axes are convenient in defining the positions of the faces in crystals of the hexagonal system (and may be used for crystals of the trigonal system too); this case will be dealt with later.

The parameters or relative lengths of the axes having been defined, the position of any other face on the crystal may be fixed by means of the intercepts, expressed in terms of the parameters, which it makes on the axes. For example, in Fig. 43 the face HKL makes intercepts on a, b, and c respectively of $\frac{1a}{3}$, $\frac{1b}{2}$, and $\frac{1c}{2}$. These may themselves be taken as the symbol of the face, as was proposed by Weiss in 1818. A much more convenient way of expressing the face symbol is that of Miller (1839), who obtained it by taking the reciprocals of the intercepts and clearing away fractions by multiplying throughout by a small integer. The symbol of the face HKL in the figure would be $\frac{3}{1}, \frac{2}{1}, \frac{2}{1}$ or 3,2,2. These numbers or *indices* are conventionally placed in brackets without commas, thus—(322). To state the case generally, if hkl are the Millerian indices, the intercepts along the axes are $\frac{a}{h}, \frac{b}{k}, \frac{c}{l}$.

CRYSTAL AXES AND INDICES

It will be seen that the symbol for the parametral plane is (111), and that all parallel faces have the same indices, because the ratios of the intercepts in each case will be the same. The intercept on an axis to which a face is parallel is infinity, and the corresponding Millerian symbol is 0. The faces which in Fig. 61 were used to determine the directions of the crystal axes will therefore have symbols as follows : front face, (100) ; right side face, (010) ; top face, (001).* Faces which cut the horizontal axes and are parallel to the vertical axis will have the general symbol ($hk0$).

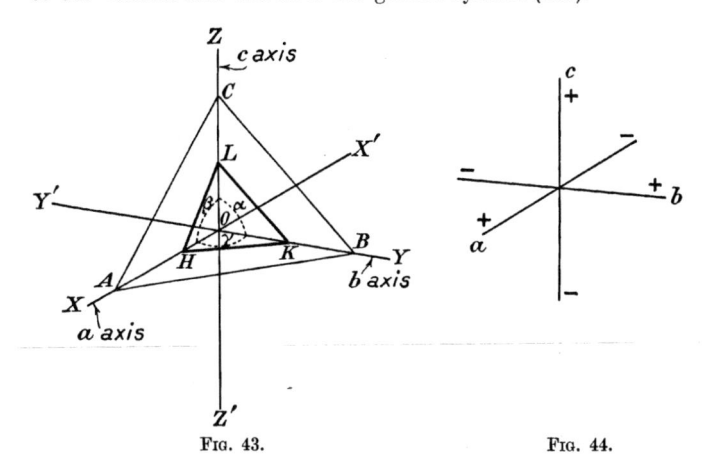

FIG. 43.   FIG. 44.

In order fully to describe the faces of a crystal, it is necessary to attribute different signs to opposite ends of the same axis. This is shown in Fig. 44. Where a face makes a negative intercept, the fact is indicated by placing a negative sign above the appropriate index figure as shown in the diagram. The symbol (11$\bar{1}$) is read "one one bar one."

It is found in practice that all the indices of faces are small whole numbers, chiefly 1, 2, 3, or 4, larger numbers occurring but rarely. This means that the intercepts which inclined faces make on the axes are always rational multiples of the parametral unit lengths of

---

* These faces are often called the a, b, and c faces respectively, *i.e.* the a face is that which intercepts the a axis only, and so on. Other letters are conventionally used to denote other faces, *e.g.* m and p for prism faces, o for pyramidal faces, and sometimes the Millerian indices are prefixed by the appropriate letter, thus "the a(100) face," etc.

those axes, and really follows from the fact that crystals are built upon a regular space lattice. Hence is formulated the *Law of Rational Indices*, which states that *the ratios of the intercepts on the axes made by any face can always be expressed as rational multiples of the parameters.*

The X-ray investigation of a substance reveals the absolute lengths of the sides of the unit cell. The ratios in which these lengths stand to one another are sometimes not the same as the parameters determined from goniometrical measurements on the same substance, but are always simply and rationally related to them. This discrepancy arises because the plane selected as the parametral plane may not be the (111) plane of the fundamental cell.

The Symmetry of Crystals. An examination of well-developed crystals shows that the faces are commonly arranged symmetrically

Hexagonal (a)   Tetragonal (b)   Trigonal (c)   Digonal (d)

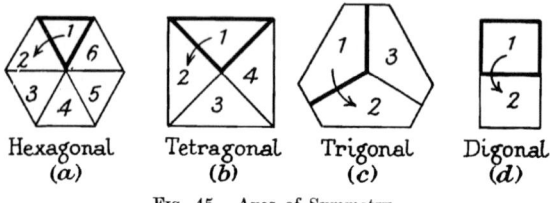

FIG. 45.—Axes of Symmetry.

around a line which can be passed through the centre of the crystal, or they may be so placed that each face is balanced by a similar one opposite to it. The symmetry of crystals is often masked by the unequal development of the faces, but will be revealed if drawings and projections based solely on the interfacial angles are made. In the following treatment of symmetry, ideally formed crystals are considered. The symmetrical disposition of crystal faces can be described by means of the following *elements of symmetry* : (i) *simple rotation axes*, (ii) *planes of symmetry*, and (iii) *rotation-inversion axes*.

A *simple rotation axis* is such that repetition of parts of the crystal takes place by rotation around it. For example, if the eye is placed above the vertical axis of a hexagonal crystal, a rotation of 60° about this axis will cause the crystal to display an exactly similar aspect. This is shown in Fig. 45, a, the heavily lined segment moving from position 1 to position 2. Six such movements are possible in 360°, and the vertical axis of the crystal is therefore

THE SYMMETRY OF CRYSTALS

said to be a *hexagonal*, or *six-fold*, axis of symmetry. Only three other kinds of such axes of symmetry exist in crystals, namely, *tetragonal* or *four-fold*, *trigonal* or *three-fold*, and *digonal* or *two-fold*, the angles of rotation necessary to repeat a part in these cases being 90°, 120°, and 180° respectively.

The line joining the centres of two opposite faces of a cube is a tetragonal axis, as is also the vertical axis of a tetragonal crystal (Fig. 45, b); the vertical axis of a trigonal crystal (Fig. 45, c), and also the line joining two diagonally opposite corners of a cube (Fig. 46) are trigonal; the lines joining the middles of two diagonally opposite edges of a cube (Fig. 46), and the b axis of a monoclinic crystal (Fig. 47) are digonal axes.

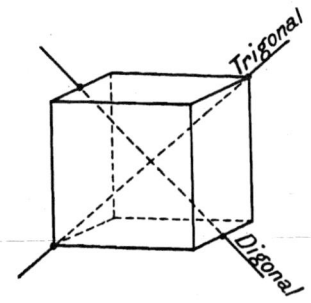

FIG. 46.—Digonal and Trigonal Axes of Symmetry of the Cube.

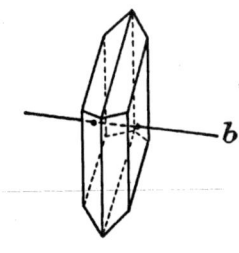

FIG. 47.—Digonal Axis of Symmetry of a Monoclinic Crystal.

A *plane of symmetry* can be described as a plane which divides the crystal into two parts, so that one part is the mirror image of the other, each plane, edge, or solid angle having its counterpart on the opposite side of the symmetry plane. The highest number of planes of symmetry displayed by each crystal system is given below.

Crystal System	Highest Number of Planes of Symmetry
Triclinic	None
Monoclinic	One
Orthorhombic and Trigonal	Three
Tetragonal	Five
Hexagonal	Seven
Cubic	Nine

Fig. 48 shows examples of such planes. A plane of symmetry

is always parallel to a possible face of the crystal and at right angles to a possible edge.

An *inversion-rotation axis* is such that a crystal face is turned about an axis through 120°, 90°, or 60°, and then inverted through the centre of rotation. Only these three types of inversion-rotation axes exist. A four-fold inversion axis is shown in Fig. 59a. If the (111) face is turned through 90° in a clockwise direction about the vertical axis and then inverted through the centre, the ($\bar{1}1\bar{1}$) face of the tetrahedron results. It will be seen from the figure that a four-fold inversion axis is equivalent *morphologically* to a two-fold simple rotation axis.

An axis of symmetry, whether of the simple rotation or inversion rotation type, may or may not be a crystallographic axis, but is

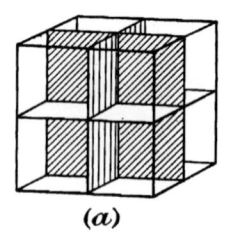

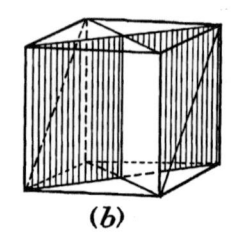

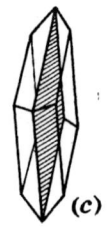

FIG. 48.—Planes of Symmetry.

always in the direction of a possible edge and at right angles to a possible face of the crystal.

A crystal in which every face is accompanied by a similar parallel face, so that a line passed through the centre passes through exactly equivalent points on each side of the crystal, is said to be *centro-symmetrical*, and to possess a *centre of symmetry*. Centro-symmetry results when the operation of rotation by 180° is followed by reflection across a plane of symmetry normal to the rotation axis. This is equivalent to inversion through the centre. Thus, if the case of the centro-symmetrical octahedron of the cubic system is taken (Fig. 59, a, p. 63), the upper right front face (111), may be turned through 180° around the vertical axis, and then reflected across a horizontal plane of symmetry to give its opposite lower left back face ($\bar{1}\bar{1}\bar{1}$).

The following additional terms are often used with which to describe certain types of symmetry:

THE SYMMETRY OF CRYSTALS

Equatorial symmetry occurs when an n-gonal axis is normal to a plane of symmetry.

Polar symmetry is present when an n-gonal axis is perpendicular neither to a plane of symmetry nor to an axis of even symmetry.

Holoaxial symmetry occurs when an n-gonal axis exists alone, or when it is accompanied by digonal axes normal to it, without planes of symmetry.

Tesseral symmetry is a term applied to the symmetry of the cubic system, which is characterised by four trigonal axes equally inclined.

A statement of the number and kinds of elements of symmetry possessed by a crystal is only a convenient way of expressing the manner in which its similar parts are spatially related, the term "similar parts" being taken to mean not merely outward similarity

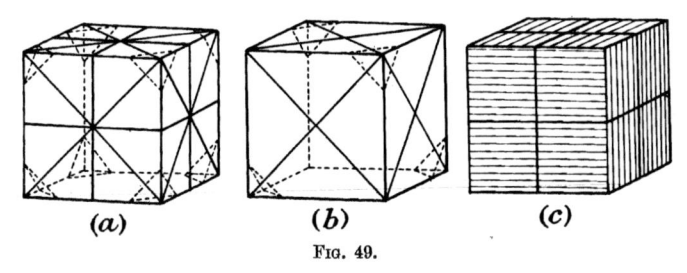

(a)   (b)   (c)

FIG. 49.

of form, but also likeness of orientation with respect to the internal structure. The law which governs crystal growth states that *similar parts of a crystal are similarly modified*. These statements may be illustrated by reference to Fig. 49, *a* and *b*, in which two cubic crystals are shown, the first (*a*) possessing all the planes of symmetry of the holohedral class of the cubic system, namely nine, and the other (*b*) possessing only the six diagonal planes of symmetry belonging to the tetrahedral class of this system. Should one corner of the crystal in Fig. 49, *a*, be modified by truncation parallel to the octahedron, then every other corner must be similarly modified to satisfy the symmetry of the form. It will easily be seen that reflection across the planes of symmetry requires this. In the second case, however, (Fig. 49, *b*) modification of any corner in the same way would only mean modification of *four* similar corners as shown, giving a combination of the cube and tetrahedron. Thus the eight corners of such a cube must be made up of two dissimilar groups of four. In contrast, if a cube of iron pyrites is examined (Fig. 49, c), striations upon its faces will reveal that in this case the diagonal planes of symmetry have gone, and only the three axial ones remain.

The substance calcite, $CaCO_3$, which sometimes displays an hexagonal prism outwardly similar to that of a truly hexagonal substance (in which each such face is crystallographically identical) reveals the *trigonal* nature of its vertical axis, if the faces are etched with dilute acid, when etch pits are developed as shown in Fig. 50, orientated in opposite directions on adjacent faces. This shows that the six faces are divided crystallographically into two groups of three.

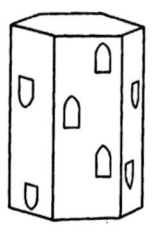

FIG. 50. — Etch Pits on Prism Faces of Calcite.

The different classes of symmetry which exist in each crystal system originate in the different types of symmetry which may be exhibited by the building units of crystals. As illustrating the manner in which the symmetry of these units operates, the three imaginary structures in Figs. 51–53 (reduced for simplicity to two dimensions) may be considered. In these figures, the outlines of the building units are indicated by broken lines, and directions of

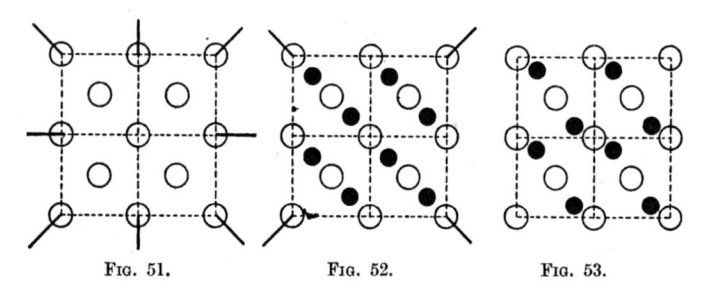

FIG. 51.   FIG. 52.   FIG. 53.

planes of symmetry normal to the paper by heavy lines. All three structures belong to the same system. The structure in Fig. 51 possesses four planes, and a tetragonal axis of symmetry normal to the paper, whilst those in Figs. 52 and 53, which are made up of two different kinds of atoms, possess diagonal planes and no planes respectively, but both have a digonal axis.

Point Group and Space Group Nomenclature. Before discussing in greater detail the relationships of point groups and space

POINT AND SPACE GROUP SYMMETRY 57

groups it is convenient here to describe the symbols by which the various symmetry operations are referred to, and to show how by combining them the various crystal classes may be defined unambiguously.

A plane of symmetry is referred to by the letter m, and simple rotation axes by the numbers 2, 3, 4, or 6, according to whether they are two-, three-, four-, or six-fold. Inversion axes are of three-, four-, or six-fold character and are symbolised by $\bar{3}$, $\bar{4}$, and $\bar{6}$ respectively. A centre of symmetry (*i.e.* a 360° rotation with subsequent inversion) is denoted by $\bar{1}$ and asymmetry by the figure 1, that is, a complete rotation of a point around a centre, which brings it back to its original position. The symbols are combined in the following manner. The symbol denoting the symmetry of the principal axis is given first. For example in the hexagonal system, 6 or $\bar{6}$; in the trigonal system 3 or $\bar{3}$; and in the tetragonal system by 4 or $\bar{4}$. If the principal axis is normal to a plane of symmetry, the fact is indicated by placing the symbol m after the axis symbol as follows : $4/m$.

The symbols for secondary axes follow, and lastly the remaining symmetry planes if they exist, placing those parallel to the principal axis first. Only sufficient symbols are used rigidly to define the point group, since the presence of all the other elements of symmetry proper to the crystal class described inevitably accompany those designated in the point group symbol. For example, in the holohedral class of the cubic system, represented most simply by a cube, the point group symbol may be built up on the lines given above. The vertical axis is four-fold and is normal to a symmetry plane. There are four three-fold axes, but only one is necessary in the symbol as it will be repeated by the four-fold vertical axis. A vertical plane of symmetry comes next, and so the symbol becomes $4/m3m$, or more simply still, $m3m$. A little experimenting with a diagram such as that shown in Fig. 54, will demonstrate that all the other symmetry elements proper to this crystal class are implied by this symbol.

Space group symmetry includes operations additional to those of the point group. These are (*i*) *glide planes*, and (*ii*) *screw axes*. (These, together with the point group operations, comprise the " coincidence movements " of the early mathematical workers referred to on page 41.) Glide planes are such that reflection of a unit takes place across the plane as in an ordinary plane of symmetry, but is followed by a simple translation parallel to the plane (Fig. 57). Screw axes effect a repetition of a point by means of a translation in the direction of the axis, coupled with a rotation

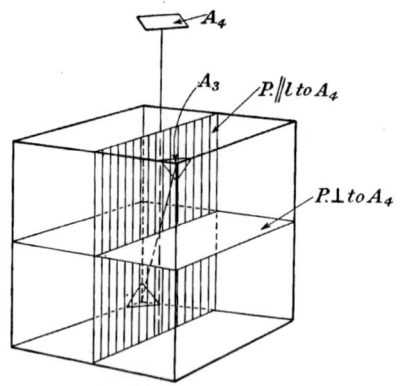

FIG. 54.—Illustration of Point Group $4/m3m$.

of 120°, 90°, or 60°, around it. To complete the space group nomenclature a symbol is used as a prefix to describe in general the type of cell upon which the crystal structure is based. The symbols are explained below.

Cell Type	Symbol	No. in Fig. 29
Simple	P	1, 4, 10, 13, 14
Body centred	I	2, 5, 11
Side centred	A, B, or C*	8 and 12
Face centred	F	3 and 9

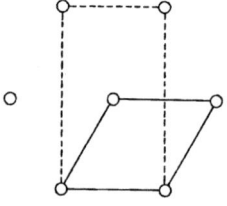

FIG. 55.—Relationship of the C type of hexagonal cell to the Bravais cell, No. 6.

The rhombohedral cell (Fig. 29, No. 7) is given the symbol R, and

* According to whether the front, side, or base is regarded as centred, this depending upon the orientation of the cell chosen: (Cf. Fig. 31.)

the Bravais hexagonal cell (Fig. 29, No. 6) is sometimes described in terms of a face-centred cell, C, shown in Fig. 55 by broken lines. Glide planes are indicated by letters as follows:

Symbol	Translation
a	$a/2$ ⎫ $i.e.$ a translation along one side
b	$b/2$ ⎬ of the unit cell a, b, or c equal
c	$c/2$ ⎭ to $\tfrac{1}{2}$ of its length
n	$a/2 + b/2$, $b/2 + c/2$, or $a/2 + c/2$ (which equals a translation from the corner to the centre of the face parallel to the plane)
d	$a/4 + b/4$, $b/4 + c/4$, or $a/4 + c/4$ (which equals a translation of one quarter of a face diagonal)

Screw axes are indicated by subscripts to the number denoting the type of rotation thus: 6_1. The subscript gives the distance along the axis in terms of the lattice cell dimensions in which a point or unit of the crystal structure will reach a position similar to that which it occupied before, in an adjacent cell. Hence, the symbol 6_1 means that the operation of such a six-fold screw axis is equivalent to a translation of the lattice by the length of one cell. Similarly, a rotation screw axis 4_2 signifies that each turn is one of $90°$ and each accompanying translation is $2c/4$ where c is the cell length in the direction of the axis.

The space group symbol is arranged as follows * : first comes the letter which indicates the type of cell; then the number which describes the principal axis of the crystal, and, followed as in point group nomenclature, by any plane of symmetry or glide plane normal to this axis. Next come the secondary axes and planes of symmetry, or glide planes parallel to the axes. For example, $F\bar{4}3m$ indicates a structure based upon a face centred lattice, with a four-fold inversion axis, a secondary axis of three-fold character (therefore cubic), and a plane of symmetry parallel to the inversion axis. This means that the substance belongs to the tetrahedral class of the cubic system ($\bar{4}3m$).

Examples of the Relationship between Point Group and Space Group Symmetry. It is instructive to compare the structure of zinc blende, ZnS, which belongs to the space group

* The system of notation used is that of Hermann and Mauguin. See F. C. Phillips, *An Introduction to Crystallography* (Longmans, 1946), pp. 271, 289.

just mentioned, with that of diamond and sodium chloride. Each of these substances has a structure based upon a face-centred lattice, but the detailed structure of each is different, and depends upon the way in which more than one face-centred lattice interpenetrate. Hence each belongs to a different space group. Although all the structures and space groups are different, diamond and sodium chloride belong to the same crystal class (point group) and zinc blende to another.

Substance	Crystal Class	Space Group
Diamond }	$m3m$	$Fd3m$
Sodium chloride }		$Fm3m$
Zinc blende	$\bar{4}3m$	$F\bar{4}3m$

Figs. 56–58 show these structures in perspective and in plan. The structure of NaCl is readily seen to be made up of two face-centred cubic lattices one of sodium ions and the other of chloride ions, and these interpenetrate in such a way that one lattice is related to the other by a simple translation of half the length of the unit cell along any one of the cell axes. It is clear that even though two kinds of ions are present the full symmetry of the holohedral class of the cubic system is present (Fig. 56).

In the diamond structure the relationship of the atoms is that of two face-centred cubic lattices, one of which is displaced relative to the other by a translation of one-quarter of a face diagonal $(a/4 + c/4)$ after reflection across a glide plane. One cell of the structure is shown in Fig. 57 and it will be seen that the four atoms situated inside the cell lie at the centres of small tetrahedra, one of which is outlined in the figure. In the plan, the atoms marked 1, 2, 3, and 4, lie on successive lower planes, those marked 2 lying at the centres of tetrahedra at the corners of which atoms marked 1 and 3 lie. It is clear that simple planes of symmetry do not occur in the *structure* but glide planes such as that shown in the plan do. By the operation of such a plane, atoms 1 are reflected across the plane, and then by a diagonal translation carried to occupy positions 2. The positions of two screw axes are also shown in the diagram, the operations of the screw axes being illustrated in the perspective drawing, and in the plan by tracing the positions of atoms 1 to 4 to successively lower layers. A morphological examination of diamond shows that it belongs to the holohedral class of the cubic system. It is clear, therefore, that each set of

C.P.M. 61 C*

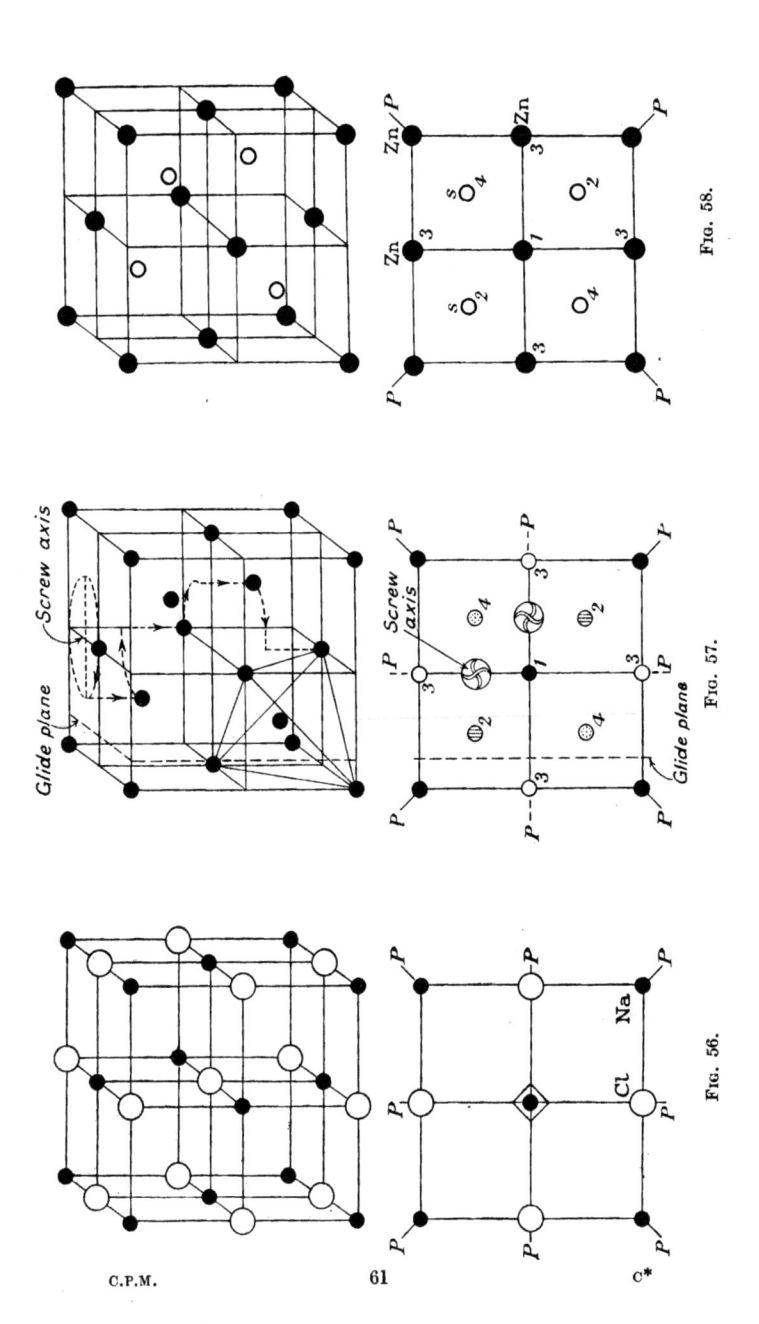

Fig. 58.

Fig. 57.

Fig. 56.

parallel glide planes of the space group are equivalent to a plane of symmetry in the point group, and that each set of four-fold screw axes similarly become simple four-fold rotation axes in the point group.

In Fig. 58 the structure of zinc blende is shown. The actual distribution of the points in the interpenetrating lattices is identical with that of diamond, but in this case the atoms are of two kinds. The sulphur atoms are shown as clear circles within the unit cell. It will now be clear that the glide planes of the diamond structure no longer exist because of this difference in the nature of the atoms. Moreover, the screw axes of the diamond structure also disappear. The tetrahedral symmetry of the space group therefore persists in the point group, and so zinc blende belongs to a class of lower symmetry than either sodium chloride or diamond.

Forms. It will be seen that the faces occurring in a crystal group themselves into sets, in each of which the members are similarly orientated to the crystal axes. Each of these sets of crystallographically similar faces, having the same Millerian indices (except as regards sign), is known as a *form*. Three kinds of form occur in crystals as follows :

Forms consisting simply of two parallel faces cutting only one axis are known as *pinacoids*.

Forms consisting of 3, 4, 6 or more faces parallel to the vertical axis are known as *prisms*.

Forms parallel to a horizontal axis, and cutting the other two, are known as *domes*. (These are really non-vertical prisms.)

Forms cutting three axes are known as *pyramids*. Pyramids may produce solid figures, or *closed forms*, by themselves, but prisms, domes, and pinacoids require the presence of other forms with which to complete the crystal ; they are therefore known as *open forms*. The number of possible faces belonging to a form varies according to the crystal system and class to which the crystal belongs. For example, in the triclinic system the highest number of faces in a form is two, in the orthorhombic system eight, and in the cubic system forty-eight. Forms are conveniently referred to by taking the indices of the positive face, if one occurs, and enclosing it in braces thus : $\{111\}$. The cube is therefore $\{100\}$, and the octahedron $\{111\}$.

In systems of low symmetry, forms are named, where necessary,

HOLOHEDRISM, HEMIHEDRISM, AND TETARTOHEDRISM

after axes to which they are parallel, or in some other distinguishing way. Thus in the orthorhombic system, the form $\{100\}$ is parallel to the *macro-* (longer) horizontal axis, *b*, and is therefore called the *macro-pinacoid*; the form $\{010\}$ is parallel to the *brachy-* (shorter) horizontal axis, *a*, and is therefore called the *brachy-pinacoid*. The form $\{001\}$ which intersects the vertical axis, *c*, is termed the *basal pinacoid*. The same terms are used in the triclinic system.

In the monoclinic system, the axes other than *c* are the *clino-* (or *a* axis) and *ortho-* (or *b* axis). The pinacoid $\{010\}$ is thus named the *clino-pinacoid*, and the $\{100\}$ the *ortho-pinacoid*.

Holohedrism, Hemihedrism, and Tetartohedrism. Crystals which exhibit forms possessing the highest symmetry of the system to which they belong are said to be *holohedral*. Some crystals,

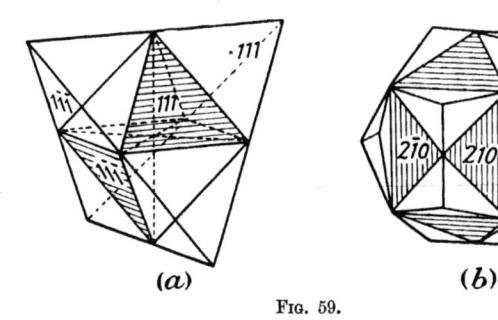

FIG. 59.

however, show forms in which only one half or one quarter of the faces of the corresponding holohedral forms are developed. Such forms are termed *hemihedral* and *tetartohedral* respectively. An example is furnished by the two cubes shown in Figs. 49, *a* and *b*. In the first, the highest symmetry of the cubic system is exhibited, and in particular, the form composed of faces which truncate the corners affects all eight corners; in other words, the form consists of eight faces. In the second cube, however, the corresponding form consists of four faces only, and is therefore hemihedral. It used to be thought that hemihedrism and tetartohedrism were merely further manifestations of the influence of environment during growth on the habit of the crystal (see p. 47), but it is now recognised that this is not the case, the true facts being that crystals showing such forms possess a smaller number of elements of symmetry than those belonging to the holohedral class of the system. In further illustration of the above, Figs. 59, *a* and *b*, show the relation of the octahedron and the tetrahexahedron to the tetrahedron and the pyritohedron respectively. The faces of the holohedral form developed in the hemihedral crystal are shaded. It will be noticed that the vertical simple four-fold axis in each case becomes a digonal one ($= \bar{4}$ in Fig. 59, (*a*) in the form of lower symmetry, and that to each holohedral form there correspond two hemihedral forms termed positive and negative, which taken together present the form of the holohedral crystal.

Hemimorphism. In certain classes, crystals occur which are terminated differently at opposite ends of an axis which is both one of symmetry and a crystallographic axis. These crystals may be termed *uniterminal* or *hemimorphic*. Some examples representative of all the hemimorphic classes are : tartaric acid, $(CH.OH.COOH)_2$, monoclinic, class 2 (see below); hemimorphite, $2ZnO,SiO_2,H_2O$, and ammonium magnesium phosphate, $NH_4MgPO_4,6H_2O$, orthorhombic, class 2*mm* ; iodosuccinimide, $(CH_2CO.CHICO)NH$, tetragonal, class 4*m* ; wulfenite, $PbMoO_4$, tetragonal, class 4 ; tourmaline, $H_6Na_2Fe_4B_6Al_3Si_{12}O_{63}$, class 3*m*, and sodium periodate, $NaIO_4,3H_2O$, class 3, both trigonal ; and in the hexagonal system, greenockite, CdS, class 6*m*, and strontium antimonyl tartrate, $Sr(SbO)_2(C_4H_4O_6)_2$, class 6. Specimens of this type are figured on pp. 68–9, and 72, Figs. 68, 75, and 94.

When two ends of an axis of symmetry are dissimilar, or *polar*, the opposite ends show different physical properties. For example, tourmaline on heating develops a positive electric charge at one end of its trigonal axis, and a negative charge at the other. This phenomenon is known as *pyroelectricity*. Tartaric acid and quartz are other good examples, the latter showing three horizontal electric axes, the six vertical edges of the crystal becoming positive and negative alternately.

Enantiomorphism. Eleven of the 32 classes of crystals have either no symmetry (class 1), or only axes of symmetry (classes 2, 22, 4, 42, 3, 32, 6, 62, 23, and 43). In these classes *enantiomorphous* forms may occur. Enantiomorphous crystals are those of which two types are possible in one class, a right-handed or *dextro-* variety and a left-handed or *lævo-* variety. (Such crystals have already been mentioned from the standpoint of their internal structure on p. 43.) A complete crystal of one sort is the mirror image of the other, and neither can be turned about so as to occupy the same position in space as the other. Fig. 60 shows crystals of right-handed and left-handed quartz and tartaric acid. Enantiomorphous crystals have the property of rotating the plane of polarisation of light, the dextro- and lævo- varieties acting in opposite directions in this respect. This property will be dealt with more fully in subsequent chapters.

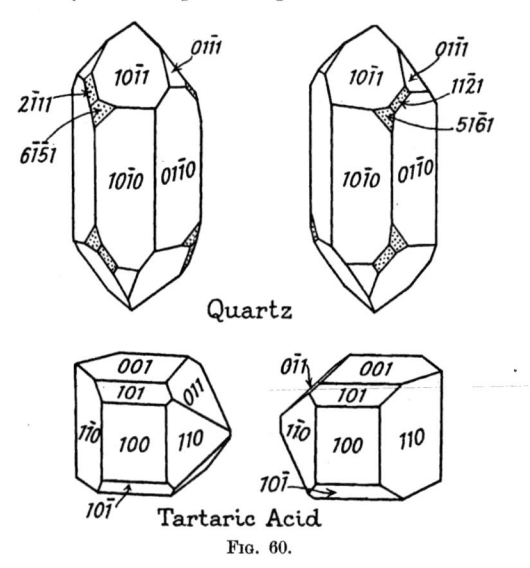

Fig. 60.

The 32 Crystal Classes.

Crystals may be classified according to the number and kinds of elements of symmetry they possess. One class, namely class 1 of the triclinic system, possesses no symmetry at all, and is known as the asymmetric class. Then there are the 10 classes given above in which only axes of symmetry occur. As already stated, these 11 classes in all are those in which enantiomorphous forms may occur, this being due to the absence of planes of symmetry. Next may be taken 17 classes in which planes of symmetry occur, among which is one class (the monoclinic class m), having no axes of symmetry. Lastly there are four classes ($\bar{1}$, $\bar{4}$, $\bar{3}$ and $\bar{6}$) having only inversion axes of symmetry.

The 32 classes of crystals thus broadly outlined are divided into

7 systems (by some authors, into 6). The general characters of each of these systems, and the symmetry of the classes belonging to them are set out below. Figures are given of crystals typical of various classes, the forms developed being indicated by the Millerian indices marked on the crystal faces. With reference to these figures, it may be recalled that pinacoids cut only one axis and have general Millerian indices of the type $\{h00\}$, prisms in general and domes cut two axes and have the type of symbol $\{hk0\}$, whilst pyramids cut three axes and so are represented by symbols of the type $\{hkl\}$.

Triclinic System.

FIG. 61.—Triclinic Axes and three Pinacoids.
$\alpha \neq \beta \neq \gamma$.

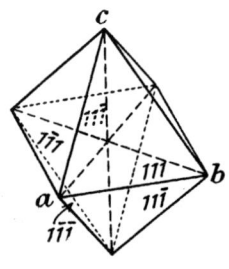

FIG. 62.—Triclinic Pyramid.

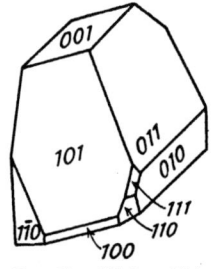

FIG. 63.—Calcium Thiosulphate (after Tutton). Class 1. Each face comprises one *form*.

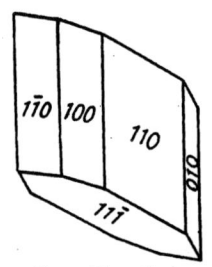

FIG. 64. — Copper Sulphate. Class I. Each face accompanied by a parallel face.

Triclinic, or Anorthic System. *Referred to three unequal axes at unequal angles, α, β, and γ, other than 90° to one another. Any one of the axes may be taken as the vertical one, c, and of the other two, b, the longer and lateral one, is called the* macro-axis, *and a, the shorter, the* brachy-axis *(Fig. 61). Characterised by no symmetry or only centro-symmetry.*

Two classes occur in this system:

Class 1. No symmetry.
  EXAMPLE: Calcium thiosulphate, $CaS_2O_3,6H_2O$.

Class $\bar{1}$. A centre of symmetry, C.*
  EXAMPLE: Copper sulphate, $CuSO_4,5H_2O$.

Under the microscope, no ortho- or right-angle shapes will be seen with crystals of this system.

Monoclinic System. *Referred to three unequal axes, a, b, and c, of which b and c are at right angles, and a is inclined to c at an angle not 90° in a plane normal to that containing b and c. a is called the* clino-axis *and b the* ortho-axis. *The angle between a and c is referred to as β (Fig. 65). Characterised by having a plane of symmetry with a digonal axis of symmetry normal to it, or by only one of these elements.*

Three classes occur in this system:

Class m. 1 P.
  EXAMPLE: Potassium tetrathionate, $K_2S_4O_6$.

Class 2. 1 AII, polar.
  EXAMPLE: Tartaric acid, $(CH.OH.COOH)_2$, enantiomorphous, in two forms, dextro- and lævo-.

Class 2/m. 1 AII at right angles to P; C.
  EXAMPLE: Potassium magnesium sulphate ("schönite"), $K_2SO_4,MgSO_4,6H_2O$.

Orthorhombic System. *Referred to three unequal axes at right angles, of which any one may be selected as the vertical one, c. Of the horizontal axes, the longer, b, is called the* macro-axis, *and the shorter, a, the* brachy-axis. *Characterised by possessing only digonal*

---

* A, followed by II, III, IV, or VI indicates an axis of digonal, trigonal tetragonal, or hexagonal symmetry respectively. Ā indicates an inversion axis; P indicates a plane of symmetry, and C a centre of symmetry. In the description of each crystal class, the point group symbol is followed by a short summary of the complete symmetry.

Monoclinic System.

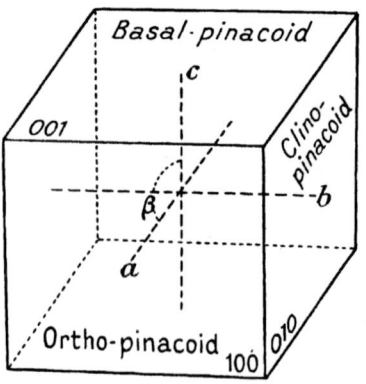

Fig. 65.—Monoclinic Axes and three Pinacoids.

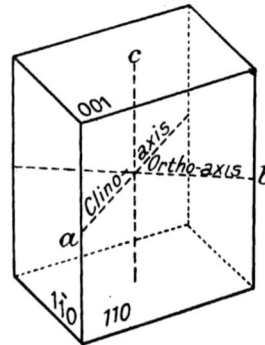

Fig. 66.—Combination of Monoclinic Unit Prism and Basal Pinacoid.

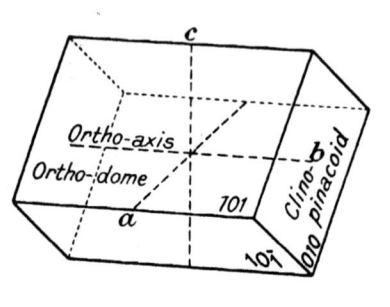

Fig. 67.—Combination of Monoclinic Ortho-dome and Clino-pinacoid.

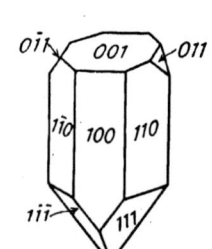

Fig. 68.—Crystal of Potassium Tetrathionate, Class *m*.

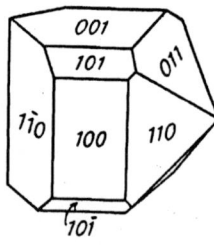

Fig. 69.—Tartaric Acid (Dextro variety), Class 2. Compare Fig. 60.

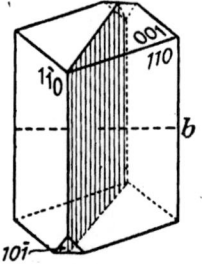

Fig. 70.—Ferrous Sulphate. Class $2/m$. Showing symmetry plane and the digonal axis of symmetry.

Orthorhombic System.

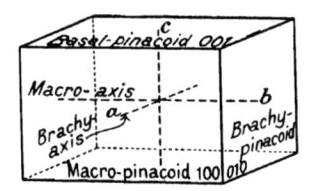

FIG. 71.—Orthorhombic Axes and three Pinacoids.

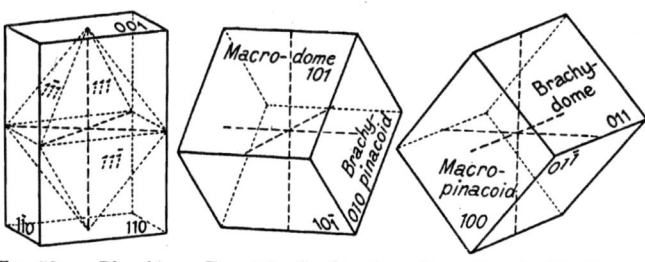

FIG. 72.—Rhombic Unit Prism and Bipyramid.

FIG. 73.—Combination of Macro-dome and Brachy-pinacoid.

FIG. 74.—Combination of Macro-pinacoid and Brachy-dome.

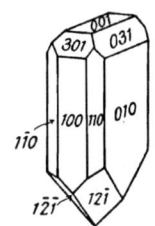

FIG. 75.—Hemimorphite (Zinc Silicate) $(OH)_2Zn_4Si_2O_7,H_2O$, Class mm.

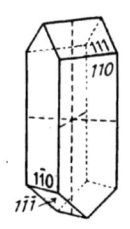

FIG. 76.—Epsomite (Magnesium Sulphate), $MgSO_4,7H_2O$, Class 222.

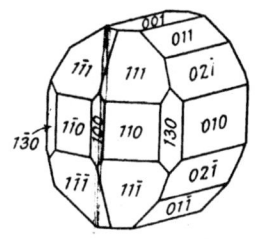

FIG. 77.—Ammonium Sulphate (after Tutton). Class mmm. Combination of pinacoids, brachy-domes, prisms and pyramid.

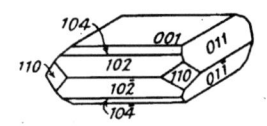

FIG. 78.—Celestine ($SrSO_4$), Class mmm.

axes of symmetry, of which c is one, and a and b may be the other two. Two symmetry planes may intersect in the axis c. In the holohedral class, therefore, three planes of symmetry intersect in the three crystal axes, which are digonal axes of symmetry.

Three classes occur in this system:

Class 222. 3 AII.
EXAMPLE: Magnesium sulphate, $MgSO_4.7H_2O$.
Class mm. 1 AII and 2 P.
EXAMPLE: Ammonium magnesium phosphate,
$$NH_4MgPO_4.6H_2O.$$
Class mmm. 3 AII and 3 P; C.
EXAMPLE: Potassium sulphate, K_2SO_4.

Tetragonal System. Referred to three axes a_1, a_2, and c, all at right angles, the vertical axis, c, being unequal to the other two, which are equal in length. Characterised by one four-fold axis of symmetry parallel to c, or by a four-fold inversion axis.

There are seven classes in this system:

Class 4. · 1 AIV.
EXAMPLE: Wulfenite, $PbMoO_4$.
Class $\bar{4}$. 1 $\bar{A}$IV (= AII morphologically).
EXAMPLE: $2CaO.Al_2O_3.SiO_2$.
Class 42. 1 AIV and 4 AII.
EXAMPLE: Strychnine sulphate, $C_{21}H_{22}N_2O_2.H_2SO_4.6H_2O$.
Class 4/m. 1 AIV normal to P; C.
EXAMPLE: Scheelite, $CaWO_4$.
Class 4mm. 1 AIV and 4 P.
EXAMPLE: Iodosuccinimide, $(CH_2CO.CHICO)NH$.
Class $\bar{4}2m$. 1 $\bar{A}$IV; 2 AII and 2 P.
EXAMPLE: Potassium dihydrogen phosphate, KH_2PO_4.
Class 4/mmm. 1 AIV normal to P; 4 AII and 4 P; C.
EXAMPLE: Zircon, $ZrSiO_4$.

The indices of a face in the tetragonal system refer to the axes a_1, a_2, and c in that order. In drawings, the a_2 axis occupies the same position as the b axis in drawings of crystals of the triclinic, monoclinic, and orthorhombic systems, namely, pointing to the right.

Tetragonal System.

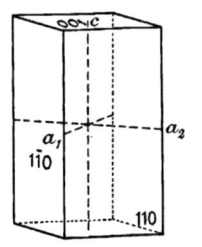

Fig. 79.—Tetragonal Prism and Basal Pinacoid.

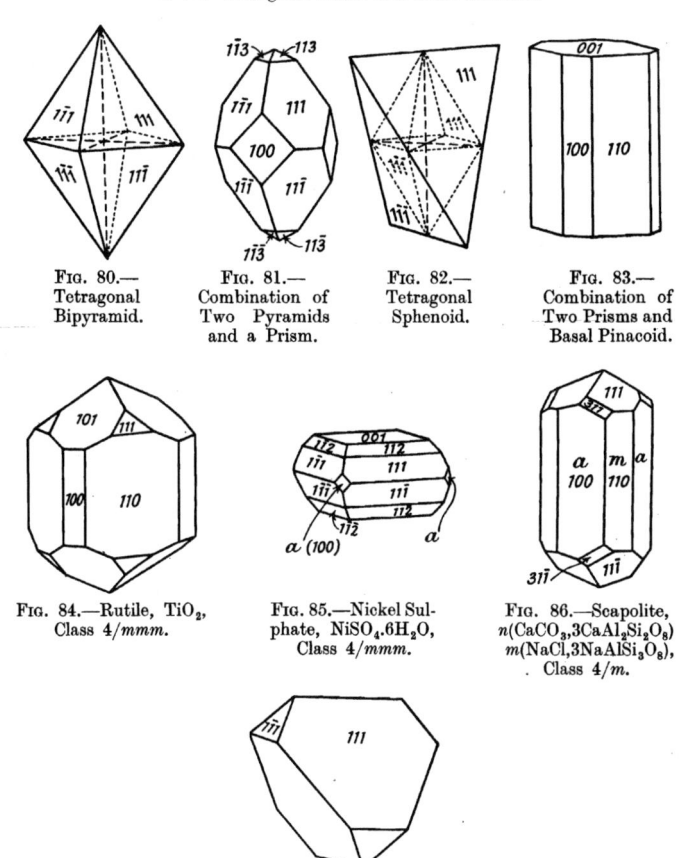

Fig. 80.—Tetragonal Bipyramid.

Fig. 81.—Combination of Two Pyramids and a Prism.

Fig. 82.—Tetragonal Sphenoid.

Fig. 83.—Combination of Two Prisms and Basal Pinacoid.

Fig. 84.—Rutile, TiO_2, Class $4/mmm$.

Fig. 85.—Nickel Sulphate, $NiSO_4.6H_2O$, Class $4/mmm$.

Fig. 86.—Scapolite, $n(CaCO_3,3CaAl_2Si_2O_8)$ $m(NaCl,3NaAlSi_3O_8)$, Class $4/m$.

Fig. 87.—Copper Pyrites, $CuFeS_2$, Class $\bar{4}2m$.

THE MORPHOLOGY OF CRYSTALS

Trigonal System.

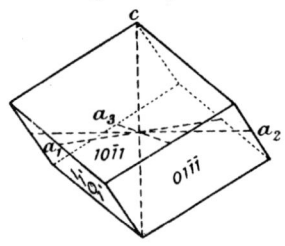

FIG. 88.—A Rhombohedron and Trigonal Axes.

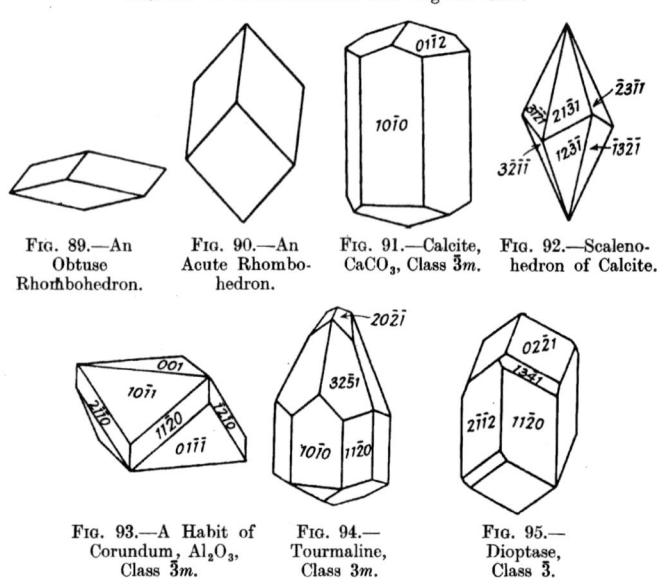

FIG. 89.—An Obtuse Rhombohedron.

FIG. 90.—An Acute Rhombohedron.

FIG. 91.—Calcite, $CaCO_3$, Class $\bar{3}m$.

FIG. 92.—Scalenohedron of Calcite.

FIG. 93.—A Habit of Corundum, Al_2O_3, Class $\bar{3}m$.

FIG. 94.—Tourmaline, Class $3m$.

FIG. 95.—Dioptase, Class $\bar{3}$.

Trigonal, or Rhombohedral System. *Referred by some authors to three axes of equal length, a_1, a_2, and a_3, meeting at equal angles other than $90°$, and by others to four axes, three horizontal, a_1, a_2, and a_3, at angles of $120°$, and a vertical axis, c, of different length from the others. Characterised by the possession of a vertical trigonal axis of symmetry.*

There are five classes in this system:

Class 3. 1 AIII.

EXAMPLE: Sodium periodate, $NaIO_4, 3H_2O$.

HEXAGONAL SYSTEM 73

Class $\bar{3}$. 1 AIII and C.
    EXAMPLE : Dioptase, CuH_2SiO_4.
Class 32. 1 AIII normal to 3 AII.
    EXAMPLE : Quartz, SiO_2.
Class 3*m*. 1 AIII and 3 P.
    EXAMPLE : Tourmaline.
Class $\bar{3}m$. 1 AIII normal to 3 AII ; 3 P ; C.
    EXAMPLE : Calcite, $CaCO_3$.

In the first system of axes (above), the indices of a face refer to the axes a_1, a_2, and a_3 in that order. In the second system (Bravais-Miller), the indices of a face refer to the axes a_1, a_2, a_3, and c in that order. In drawings, the a_1 axis points to the left front, the a_2 axis to the right, and the a_3 axis to the left rear. The negative portion of the a_3 axis is therefore in the front of the drawing. This system is used in the figures opposite.

Hexagonal System. *Referred to four axes, three horizontal ones, a_1, a_2, and a_3, of equal length, inclined to one another at 120°, and a vertical axis, c, different in length from the others, and at right angles to them. Characterised by a vertical axis of hexagonal symmetry or by a hexagonal inversion axis.*

There are seven classes in this system :

Class 6. 1 AVI.
    EXAMPLE : Strontium antimonyl tartrate, $Sr(SbO)_2(C_4H_4O_6)_2$.
Class $\bar{6}$. 1 $\bar{A}$VI (= morphological 3-fold axis normal to P).
    EXAMPLE : None known.
Class 62. 1 AVI normal to 6 AII.
    EXAMPLE : Double salt of barium antimonyl tartrate and potassium nitrate, $Ba(SbO)_2(C_4H_4O_6)_2,KNO_3$.
Class 6/*m*. 1 AVI normal to P ; C.
    EXAMPLE : Apatite, $Ca_5F(PO_4)_3$.
Class 6*mm*. Dihexagonal-pyramidal class. 1 AVI and 6 P.
    EXAMPLE : Greenockite, CdS.
Class $\bar{6}2m$. 1 $\bar{A}$VI (= 3-fold axis normal to P) ; 3 AII and 3 P.
    EXAMPLE : Benitoite, $BaTiSi_3O_9$.
Class 6/*mmm*. 1 AVI normal to P ; 6 AII and 6 P ; C.
    EXAMPLE : Beryl, $Be_3Al_2(SiO_3)_6$.

THE MORPHOLOGY OF CRYSTALS

Hexagonal System.

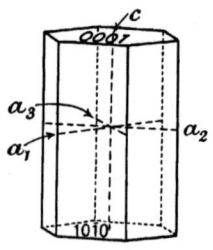

FIG. 96.—Hexagonal Prism with Basal Pinacoid showing Axes.

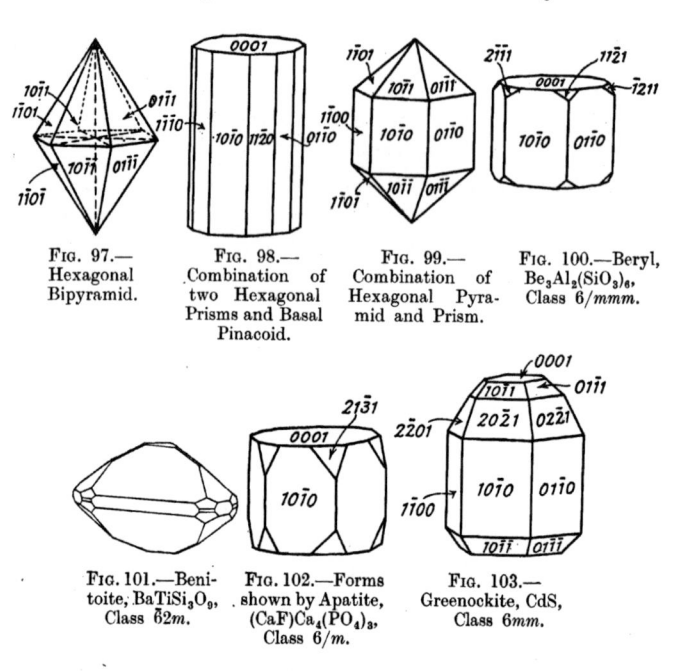

FIG. 97.—Hexagonal Bipyramid.

FIG. 98.—Combination of two Hexagonal Prisms and Basal Pinacoid.

FIG. 99.—Combination of Hexagonal Pyramid and Prism.

FIG. 100.—Beryl, $Be_3Al_2(SiO_3)_6$, Class $6/mmm$.

FIG. 101.—Benitoite, $BaTiSi_3O_9$, Class $\bar{6}2m$.

FIG. 102.—Forms shown by Apatite, $(CaF)Ca_4(PO_4)_3$, Class $6/m$.

FIG. 103.—Greenockite, CdS, Class $6mm$.

Cubic System. *Referred to three equal axes, a_1, a_2, and a_3, all at right angles. (In drawings, these axes are orientated respectively like the a, b, and c axes in the orthorhombic system.) Characterised by possessing four trigonal axes of symmetry, and three rectangular axes of symmetry which may be tetragonal or digonal ($= \bar{4}$ in class $\bar{4}3m$).*

CUBIC SYSTEM

Cubic System.

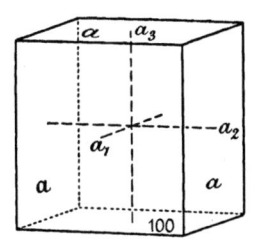

FIG. 104.—Cube and Cubic Axes.

FIG. 105.—Octahedron.

FIG. 106.—Rhombic Dodecahedron.

FIG. 107.—Tetrahexahedron.

FIG. 108.—Icositetrahedron.

FIG. 109.—Trisoctahedron.

FIG. 110.—Hexakis-octahedron.

FIG. 111.—Pyritohedron, Class $m3$.

FIG. 112.—Tetrahedron, Class $\overline{4}3m$.

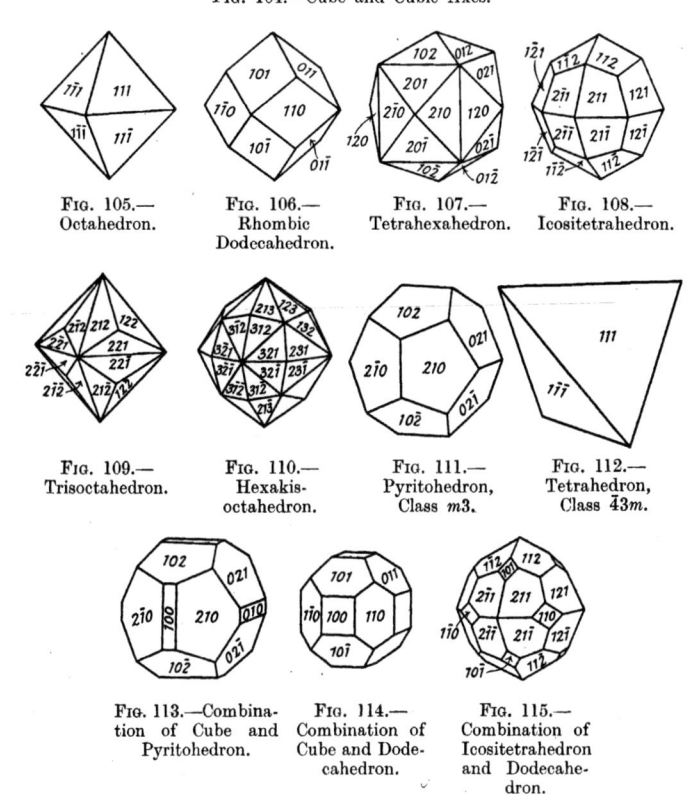

FIG. 113.—Combination of Cube and Pyritohedron.

FIG. 114.—Combination of Cube and Dodecahedron.

FIG. 115.—Combination of Icositetrahedron and Dodecahedron.

FIGS. 105–110 inclusive are simple forms of holohedral symmetry—Class $m3m$.
The reader is referred to Fig. 42, p. 47, for combinations of the cube and octahedron.

There are five classes in this system:

Class 23. 4 AIII and 3 AII.
  EXAMPLE: Barium nitrate, $Ba(NO_3)_2$.
Class 43. 4 AIII, 3 AIV, and 6 AII.
  EXAMPLE: Cuprite, Cu_2O.
Class $m3$. 4 AIII, 3 AII, and 3 P, and C.
  EXAMPLE: Iron pyrites, FeS_2.
Class $\bar{4}3m$. 4 AIII, 3 AII, and 6 P.
  EXAMPLE: Tetrahedrite, Cu_3SbS_3.
Class $m3m$. 4 AIII, 3 AIV, 6 AII, 9 P, and C.
  EXAMPLE: Fluorspar, CaF_2.

Cleavage. Many crystals show a marked tendency to break in certain definite directions, the planes thus produced often presenting bright surfaces which have a pearly or vitreous lustre. This *cleavage*, as it is called, is always parallel to a possible crystal face.

It is readily seen that crystals with layer structures will give rise to a good cleavage by breaking parallel to the widely spaced layers crowded with atoms, the interatomic attraction within the layers being stronger than the inter-layer forces. Examples of this type are graphite, molybdenite (MoS_2), cadmium iodide (CdI_2), gypsum ($CaSO_4,2H_2O$), and members of the mica family of minerals.

It is more difficult to understand why two structures based on the same type of lattice do not necessarily give rise to similar cleavage. Diamond and zinc blende, for example, have structures which are geometrically similar (Figs. 57 and 58, p. 61), but the former has octahedral {111}, and the latter dodecahedral {110} cleavage. This case has been discussed by A. F. Wells (*op. cit.*, second paper) who explains it as follows. A {111} cleavage would break the smallest number of bonds per unit area, and might therefore be expected to occur in both cases. In zinc blende, however, this cleavage would mean the separation of oppositely charged surfaces (owing to the appreciable ionic character of the Zn—S linkage), whilst in diamond cleavage in any direction results in two neutral surfaces. The {110} cleavage in zinc blende, on the other hand, leaves surfaces each containing equal numbers of zinc and sulphur atoms, all with the same residual charge, so that the surfaces are electrically neutral.

CLEAVAGE 77

In a number of the silicates, the main structure is determined by parallel silicon-oxygen chains of different types, the cleavage being developed parallel to the chains, which are stronger than the forces holding them together. The well-known mineral families of the pyroxenes and the amphiboles provide typical examples of these structures. In Fig. 116 the silicon-oxygen chains are illustrated, those of the pyroxenes being formed of single rows of linked SiO_4 tetrahedra. They are seen in plan, and in side (end) and front elevations respectively at a, b, and c. The cross-section of the chain is outlined. The amphibole chain is a double one, being composed of two pyroxene chains linked together forming a series of rings, and the cross-section is therefore more elongated than the pyroxene one. This is illustrated in Fig. 116, d and e. In the crystal structure of both families the chains are packed parallel to the c axis as shown at f and g respectively, $\{110\}$ cleavages being developed as shown to the right of the figure. It must be remembered that the cleavage surface shown in the diagram as a zig-zag line is microscopically a plane surface.

In the felspar family, silicon-oxygen chains of different shape from those mentioned above lie parallel to the *clino*-axis, and cleavage is developed along them parallel to $\{010\}$ and $\{001\}$.*

The intimate connection between cleavage and the outward form of a crystal makes the study of the degree of perfection, and direction of cleavage important, cleavage fragments often being as useful objects of study as complete crystals in the identification of a substance under the microscope.

Cleavage may be developed in various directions in a crystal, but a set of parallel cleavage planes will only be the exact equivalent of another set if both are developed parallel to crystallographically similar faces. For example, fluorite, CaF_2 (cubic), may be cleaved parallel to the octahedron, $\{111\}$, and therefore has four directions of equivalent cleavage. Galena, PbS, and sodium chloride (both cubic) cleave parallel to the cube faces, $\{100\}$, and therefore possess three equivalent cleavage directions. Calcite, $CaCO_3$, develops three equivalent cleavages parallel to the faces of the fundamental rhombohedron.

Cleavage is defined in terms of *direction* (*i.e.* the face or form to which it is parallel—cubic, octahedral, prismatic, pinacoidal,

* See W. L. Bragg, *Atomic Structure of Minerals* (Cornell University Press, 1937), pp. 197 and 232.

Fig. 116.

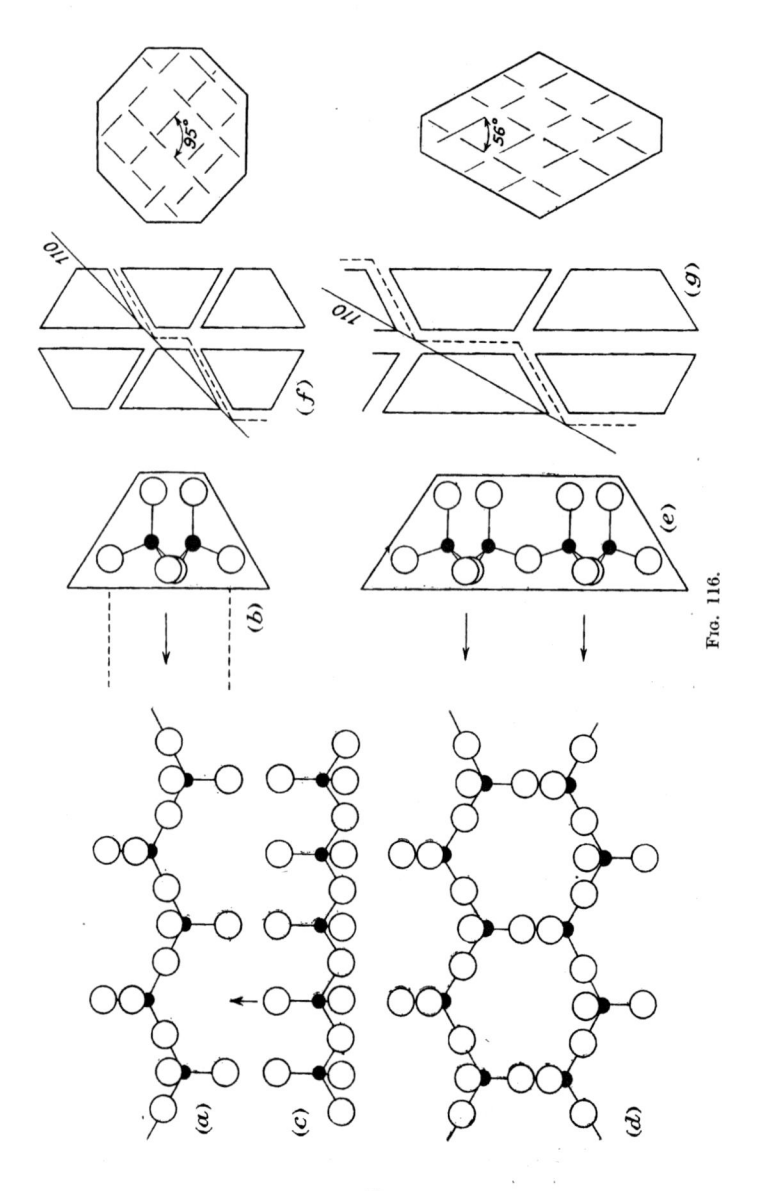

78

basal, etc.), and *quality* (*i.e. perfect* with lustrous surfaces, *imperfect, difficult*, etc.).

In the following table the types of cleavage to be found in the crystal systems are summarised, the rarer ones being italicised. It should be noted that in the case of prismatic or pyramidal cleavages, the indices given in the table may not be the ones actually found in the crystal; the idea is simply to list the general types of cleavage which occur. The number of exactly equivalent sets of cleavage planes possible is inserted before the Millerian indices denoting the direction.

System	Cleavage Commonly Developed		
Cubic	cubic	3	{100}
	octahedral	4	{111}
	dodecahedral	6	{101}
Tetragonal	basal	1	{001}
	prismatic	2	{100}
		2	{110}
	pyramidal	4	{111}
Hexagonal	basal	1	{0001}
	prismatic	3	{10$\bar{1}$0}
	pyramidal	6	{10$\bar{1}$1}
Trigonal	as for hexagonal;		
	also rhombohedral	3	{10$\bar{1}$1}
Orthorhombic . . .	pinacoidal	1	{001}
		1	{010}
		1	{100}
	prismatic	2	{110}
	(domal)	2	{011}
		2	{101}
Monoclinic	pinacoidal		
	clino-	1	{010}
	ortho-	1	{100}
	basal	1	{001}
	prismatic	2	{110}
	hemi-pyramidal	2	{111}
	(inclined prism)	2	{11$\bar{1}$}
Triclinic	the crystal is usually orientated so that a prominent cleavage is that of a pinacoid, *i.e.* {100}, {010}, {001}.		

Fig. 117 shows an orthorhombic crystal with some possible equivalent cleavage directions distinguished by similar types of line.

Different types of *fracture* surfaces may be developed upon crystals with or without cleavage. Fracture is termed *conchoidal*

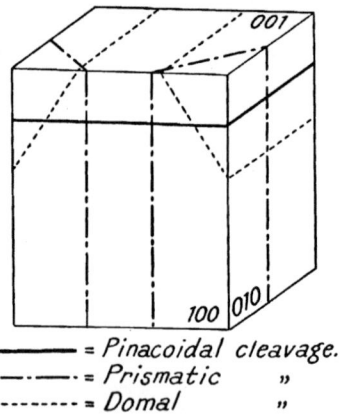

FIG. 117.—Some Types of Orthorhombic Cleavage.

——— = *Pinacoidal cleavage.*
—·—·— = *Prismatic* ,,
- - - - - - = *Domal* ,,

when it is curved like a shell, *e.g.* as in quartz or glass; *splintery* when it is like that of wood or other fibrous substances; or simply *uneven* when it has neither of the foregoing characteristics.

Twinned Crystals. Regular crystal groups composed of two or more distinct individuals often occur. These groups may be *parallel growths*, in which each crystal is orientated similarly, or *twinned crystals*, in which case the two or more coalescent individuals of the group may be related in various ways, and only a part of their similar edges and faces may be orientated in a parallel manner. Twinned crystals usually reveal themselves by the presence of re-entrant angles, or elbow or cruciform shapes, but sometimes examination under the microscope in polarised light is necessary to reveal the true character of the structure. Figs. 118 to 123 show typical twinned crystals. Those which are joined upon one face, which is common to both parts, are said to be *juxtaposition*, or *contact* twins (Figs. 118–121), whilst those shown in Figs. 122 and 123 are known as *interpenetration* twins.

The component parts of a twinned crystal are usually related to one another in one of the following ways:

(*i*) by reflection of one part across a plane which is common to both parts, or (*ii*) by rotation of one part through 180° about an axis common to both parts, or

(*iii*) by both the above operations taken simultaneously.

In centro-symmetrical crystals the effects of reflection of a part across a plane and rotation about an axis at right angles to. that plane are equivalent, but in crystals which are not centro-symmetrical, these two operations produce different results.

The plane across which reflection is imagined to have taken place, or the plane normal to the axis of rotation, is called the *twinning plane*, and the axis, the *twinning axis*. No symmetry plane of the untwinned crystal can become a twinning plane, nor can an axis of even symmetry (*i.e.* digonal, tetragonal, or hexagonal) become a twinning axis, but a symmetry plane of a higher class may be the twinning plane of a crystal belonging to a class of lower symmetry, which does not possess the plane of symmetry in question. A twinning axis is always a possible crystal line, *i.e.* a crystal axis, or the normal to a possible crystal face, and the twinning plane is most often parallel to a fundamental crystal face which is not a symmetry plane, as stated above.

The common plane of contact between two individuals of a twinned crystal is known as the *composition plane*, but in the case of interpenetrant twins, optical examination reveals that a true plane of contact does not exist, the surface between the two parts being step-like, or entirely irregular.

The terms reflection and rotation, used in connection with twinning planes and axes, are used only to describe the relative positions of different units of twinned crystals, and not the actual methods by which twinning has been brought about. The composition plane is often the same as the twinning plane, but not necessarily so, and, when not coincident, they are often at right angles to one another.

Fig. 118 shows a regular octahedron twinned on a plane parallel to the octahedron face (111), Fig. 119 shows a hexagonal prism twinned on a plane parallel to a pyramid face (10$\bar{1}$1), and in Fig. 120 a gypsum crystal is shown twinned on a plane parallel to (100). In each case the position of the parts can be described by imagining one part to have been rotated 180° about an axis normal to the twinning plane, or by supposing that one part is the mirror image of the other across the twinning plane, which now becomes a symmetry plane of the twinned crystal. It will be noticed that both the twinning plane and the twinning axis bear the same relation to both parts of the twin.

In certain cases different positions of twinning planes and axes produce the same result. For example, in orthoclase, $KAlSi_3O_8$,

Fig. 118.—"Spinel Twin" Octahedron, twinned on (111).

Fig. 119.—Hexagonal Prism twinned on a Pyramid Face.

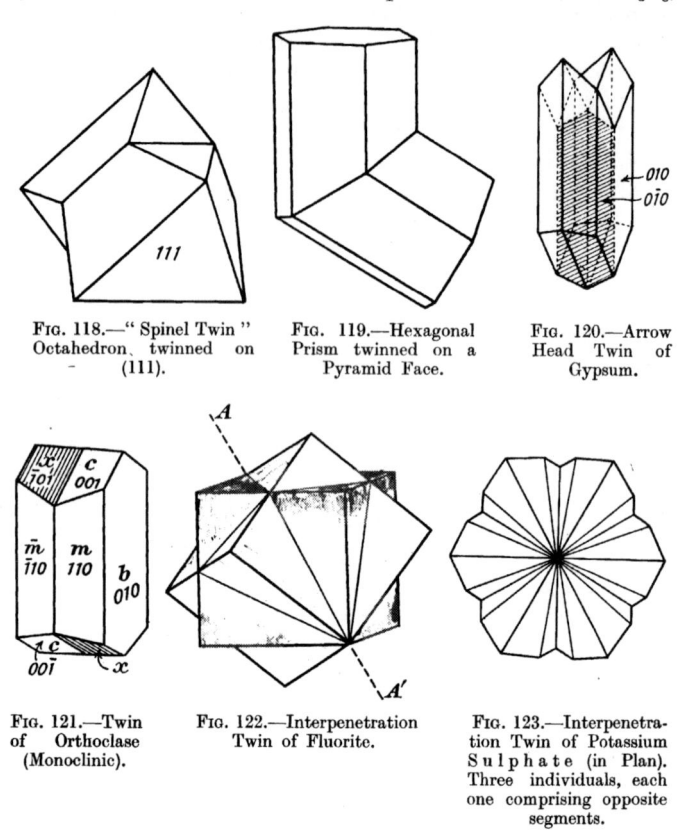

Fig. 120.—Arrow Head Twin of Gypsum.

Fig. 121.—Twin of Orthoclase (Monoclinic).

Fig. 122.—Interpenetration Twin of Fluorite.

Fig. 123.—Interpenetration Twin of Potassium S u l p h a t e (in Plan). Three individuals, each one comprising opposite segments.

(Fig. 121) the composition plane is parallel to $b(010)$, but the disposition of the parts may be described by supposing the twinning axis to be either normal to $a(100)$, or parallel to the vertical axis, the twinning plane being $a(100)$ in the former case, and perpendicular to the vertical axis in the latter. In both cases the twinning plane is perpendicular to the composition plane.

MULTIPLE TWINNING AND MIMETIC TWINS

When interpenetrant twinning takes place, the parts obey the same laws as apply in contact twins, some common axis of rotation or common reflection plane existing as in the cases described above. This fact can often be seen more clearly if it is imagined that the two parts are entirely separated, while retaining the appropriate orientation. The interpenetrant cubes of Fig. 122 are related by reflection across an octahedral plane, or by rotation about the axis AA'.

Twinning might be defined as an attempt to attain a higher degree of symmetry than the separate individuals of the twin possess, for further planes and axes of symmetry are thereby introduced. When two individuals having less than holohedral symmetry interpenetrate so as to produce a crystal having outwardly the holosymmetry of the system to which the individuals belong, the twinning is called *supplementary twinning*. An example is furnished by the orthorhombic mineral hemimorphite, $(OH)_2Zn_4Si_2O_7,H_2O$, belonging to Class mm (Fig. 124). When twinned on (001) and completely interpenetrant, outwardly the full orthorhombic holohedral symmetry is exhibited. Left- and right-handed enantiomorphous forms may also be twinned by interpenetration, giving apparent holosymmetry.

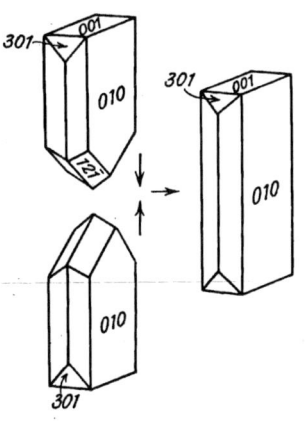

Fig. 124.—Supplementary Twinning in Hemimorphite.

Multiple Twinning and Mimetic Twins. Sometimes more than two individuals are twinned together. When the twinning plane is parallel to a pinacoid, every alternate member of the series is similarly orientated, but where the twinning plane is inclined as in the case of a prism or pyramid face, a cyclical grouping results (Fig. 125). A very fine example of the former is furnished by the triclinic mineral albite, $NaAlSi_3O_8$, and also by γ o-nitroaniline (see Expt. 13, Chap. X, p. 380). A simple twin of albite is shown in Fig. 126, where it will be seen that the twinning and composition plane is $b(010)$, and that a re-entrant angle is formed between the (001) face of the first crystal and the (00$\bar{1}$) face of the second. This kind of twinning is typically repeated many times, each part being an extremely thin plate, or lamina, of microscopic thickness. To the eye, the outward form of such a *polysynthetic* crystal would appear as in Fig. 127, the shaded faces being finely striated. If the lamellæ are of equal thickness and sufficiently fine, the substance may simulate monoclinic symmetry by the apparent elimination of the re-entrant angle between the successive basal planes. Under the polarising microscope, however, the twin lamellæ are easily revealed in thin slices or small fragments of the substance.

The forms resulting from such polysynthetic or multiple twinning are sometimes termed *mimetic twins,* since they simulate a higher symmetry than actually corresponds to the internal structure of the

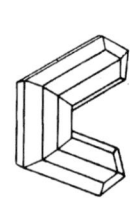

Fig. 125.—Cyclical Twinning in Rutile.

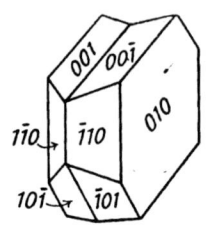

Fig. 126.—Simple Twin of Albite.

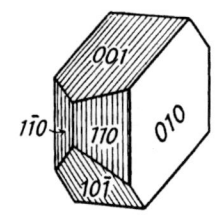

Fig. 127.—Multiple Twin of Albite.

substance. They may readily be distinguished from true polymorphic forms (see Chap. I), because their physical properties are the same as those of the simple untwinned crystals of which they are composed, whereas the different polymorphic forms of a substance have different physical properties. The case of the two potash felspars, microcline and orthoclase, is interesting. They have the same chemical composition, $KAlSi_3O_8$, the former being triclinic and composed of two sets of polysynthetic twin laminæ parallel to the face $b(010)$ and the twin axis b, whilst the latter is monoclinic and appears homogeneous under the microscope. The physical properties of both are similar, and their principal refractive indices (as measured on natural material) almost identical, thus :

	α	β	γ
Orthoclase (adularia)	1·519	1·523	1·525
Microcline	1·518	1·522	1·525

SUMMARY

It is therefore possible that orthoclase is really a triclinic substance *ultramicroscopically* twinned, thus imitating monoclinic symmetry.*

Many other substances show mimetic twinning. For example, isopropyl-ammonium chloroplatinate, $(C_3H_7NH_3)_2PtCl_6$, is apparently orthorhombic, but microscopic examination proves it to be composed of monoclinic laminæ twinned on (100). Above 32° C. the crystals change to a truly orthorhombic form.

Boracite, $Mg_7Cl_2B_{16}O_{30}$, is pseudo-cubic, being built up of orthorhombic lamellæ.

Another interesting example is potassium sulphate, K_2SO_4,† which is orthorhombic, having prism angles closely approximating to those of the hexagonal system, namely:

$$a\,(100) : p\,(110) = 29°\ 48'$$
$$p\,(110) : p'(130) = 30°\ 0'$$
$$p'(130) : b\,(010) = 30°\ 12'.$$

A common tendency in this substance is to crystallise in triplets (Fig. 123) outwardly simulating hexagonal symmetry still more closely.

When it is remembered that twinning often appears to be an unsuccessful attempt to attain a higher symmetry, it is interesting to note that under appropriate conditions many mimetic forms change to the actual forms they imitate. The example of isopropyl-ammonium chloroplatinate has already been quoted; potassium sulphate also becomes truly hexagonal, changing from biaxial to uniaxial (see next chapter), between 600 and 650° C.; and the pseudo-cubic mineral leucite, $KAl(SiO_3)_2$, which shows polysynthetic twinning at normal temperatures, becomes truly cubic between 500 and 600° C.

From what has been said it will be realised that it is important to look for evidences of twinning when examining substances under the microscope.

SUMMARY

The building units of a crystal (which may be atoms, ions, molecules, or groups of molecules) are arranged in an orderly manner, characteristic of the substance of which it is composed. The

---

* See Bragg, *Atomic Structure of Minerals* (1937), pp. 238–42, for a discussion of this problem from the point of view of X-ray results.

† Tutton, *Crystalline Form and Chemical Constitution* (1926), p. 172.

starting-point in the classification of the different types of arrangement that are possible is the conception of the *space lattice*.

A space lattice is a homogeneous structure of points in three dimensions, such that the environment of any one point is similar to that of any other. There are *14 types of space lattice*, and these can be classified into 7 groups, which correspond with the *7 systems* into which crystals have been divided on the basis of their outward form.

The 7 systems can be subdivided into *32 crystal classes* which show varying degrees of symmetry, or in one case, none at all. The symmetry of each class is defined by a set of symmetry elements known as a *point group* operating around a centre. The underlying cause of these variations is to be found in the way in which the atoms of the building units are arranged around the points of the space lattice. This arrangement can be described in terms of a set of symmetry operations which is known as a *space group*. There are 230 possible space groups, each one resulting in the point group symmetry of one of the 32 crystal classes.

The superficial appearance of a crystal depends upon the conditions which attended its growth. When growth is impeded mechanically, the outward form may be quite irregular, but otherwise the crystal will normally develop plane faces, each of which corresponds to a plane of atoms (commonly a closely packed one) in its structure. The particular combination of faces developed, *i.e.* its *crystal habit*, depends (very markedly in some cases) upon its environment during growth, *e.g.* upon the composition of the mother liquor.

The positions of crystal faces are referred to three imaginary *crystallographic axes*, a, b, and c, which are determined by the intersections of three prominent faces at right angles, or as nearly at right angles as possible. For each substance a face is selected, inclined to all three axes, and making intercepts on them, which are as nearly equal as possible. This face is called the *parametral plane*, and its intercepts, expressed in terms of the intercept on b as unity, are called the *parameters*, or the *relative lengths of the crystal axes*. The ratios of these intercepts are called the *axial ratios*, and these together with the *interaxial angles* are called the *crystal elements*.

By means of the intercepts which it makes on the axes (the intercepts being expressed in terms of the parameters), the position

SUMMARY 87

of any other face may be denoted. It is found that *the intercepts on the axes made by any face can always be expressed as rational multiples of the parameters.* This is the *Law of Rational Indices* and really follows from the fact that crystals are built up according to a regular plan.

The symmetry of crystals can be described with reference to (i) *axes of symmetry*, (ii) *planes of symmetry*, and (iii) *inversion-rotation axes*.

The symmetry of crystals is not governed solely by geometrical considerations. For instance, two corners of a crystal may be geometrically identical and similarly related to the axes, but have different physical properties, owing to the building units of the crystal not being completely symmetrical; they are then crystallographically dissimilar, and if, by a change in the environment of the growing crystal, a new face forms across one corner, one need not be formed across the other.

Crystals which show the full symmetry of the system are termed *holohedral* or *holosymmetric*. Crystals which show only one half, or one quarter, of the faces developed by the holohedral form are termed *hemihedral*, and *tetartohedral* respectively; crystals which develop different sets of faces at opposite ends of an axis are termed *hemimorphic*; whilst those capable of existence in two different forms, that are mirror images of one another, are said to be *enantiomorphic*.

The classification of crystals into 7 systems is done on the basis of the angular relationships of the crystallographic axes and their relative lengths. The distinguishing features of each system have already been clearly given in the body of the chapter, and it is unnecessary to repeat them here. Crystals belonging to the cubic system stand out from others, because their interfacial angles are characteristic of the system and not of the substances of which they are composed. With other crystals, some, at least, of the interfacial angles are characteristic of the substance.

Many crystals show a marked tendency to break along certain definite planes, which are always parallel to faces, or possible faces. This property is known as *cleavage*.

Regular crystal groups composed of two or more coalescent individuals often occur. These groups may be *parallel growths*, in which each crystal is orientated similarly, or *twinned crystals*, in which case the two or more individuals of the group are orientated differently. Twinned crystals usually show re-entrant angles, or elbow or cruciform shapes.

When the individuals in a twinned crystal group are very numerous and of microscopic thickness, the result is to simulate outwardly a higher symmetry than actually corresponds to the internal structure of the substance. This is known as *polysynthetic*, or *mimetic* twinning.

RECOMMENDED FOR FURTHER READING

Crystallography and Practical Crystal Measurement, A. E. H. Tutton (Macmillan, 1911, or Vol. I of the 1928 edition).

Mineralogy, H. A. Miers (Macmillan, 1929).

An Introduction to Crystallography, F. C. Phillips (Longmans, 1946).

CHAPTER IV

THE OPTICAL PROPERTIES OF CRYSTALS

The essential difference between the optical properties of non-cubic (anisotropic) crystals and those of cubic crystals, glasses, liquids, and gases (isotropic substances) has already been mentioned in Chapter I. It consists mainly in that the velocity of light in non-cubic crystals varies with the direction of propagation, whereas in isotropic media it is independent of the direction. Solid isotropic media when in a state of strain, however, acquire to some extent the optical properties of anisotropic crystals.

Double Refraction. An allied property of anisotropic crystals is that of *double refraction*, or *birefringence*. A ray of light entering such a crystal is, in general, divided into two rays, which travel through the crystal with different velocities, and usually in different directions. *These two rays are polarised in planes at right angles to one another.* These important facts form the basis of the optical method of studying crystals. In crystals of the tetragonal, trigonal, and hexagonal systems there is one direction, and one only, in which double refraction does not occur. This direction, which is the same as that of the *c* crystallographic axis, is called the *optic axis*, and such crystals are termed *uniaxial*. Crystals of the orthorhombic, monoclinic, and triclinic systems possess *two* directions, which for most practical purposes may be regarded as corresponding to the optic axis in uniaxial crystals; they are therefore called *biaxial*.

The first recorded observation of the phenomenon of double refraction was made by Erasmus Bartolinius in 1669 on the transparent variety of calcite known as Iceland Spar, the same chemical compound, curiously enough, as that which was to be the starting-point of Haüy's systematisation of crystal morphology about a century later. The phenomenon was studied by Huygens (1690), who was successful in accounting for all but that part of it requiring a knowledge of polarisation. The complete explanation was given by Fresnel, and by Arago, in 1811, following the discovery of polarised light by Malus in 1808.

If a small black spot marked on a piece of white paper is covered with a cleavage rhombohedron of Iceland Spar, half an inch thick or more, two virtual images of the spot are seen through the spar. One of the images is due to rays which obey the ordinary laws of refraction, and are therefore called *ordinary* rays. When the line of vision is normal to the paper this image appears immediately above the spot itself, and remains stationary when the crystal is rotated in a horizontal plane. The other image is due to rays which do not obey the ordinary laws of refraction, and are there-

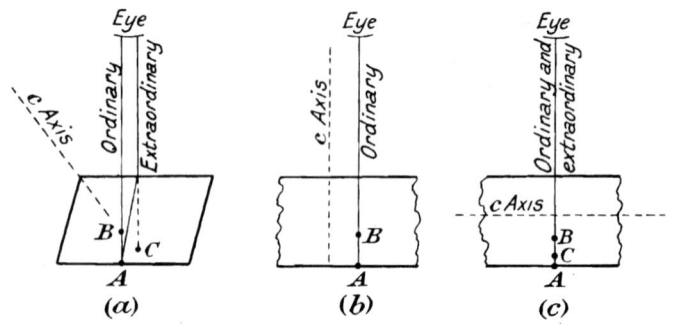

FIG. 128.—Passage of Light through Calcite.
A is the black spot, B the ordinary image, and C the extraordinary image.

fore called *extraordinary* rays. When the line of vision is normal to the paper, this image appears to one side of the actual spot, and rotates with the crystal when the latter is turned in a horizontal plane. The paths of the two sets of rays (indicated for simplicity by single lines) are shown in Fig. 128, *a*, in which A is the spot, B the ordinary image (that is the image due to the ordinary rays), and C the extraordinary image. B is higher than C because in calcite the extraordinary rays travel more rapidly than the ordinary rays.

This division of the light into two rays is observed no matter what face of the rhombohedron be laid on the paper. But if a plate of the crystal cut perpendicular to the *c* axis were used, only one image of the spot would appear if the line of vision were normal to the paper, and this image would be above the spot, and would remain stationary during rotation of the plate in a horizontal plane (Fig. 128, b). For oblique lines of vision, however, two images would be seen. Finally, if a plate of the crystal cut parallel to the c axis were used, a normal line of vision would yield, apparently, a single image, which, however, could be shown by a suitable focussing device to consist of two images immediately above one another (Fig. 128, c).

Calcite belongs to the trigonal system. Results similar to the above are theoretically obtainable with any trigonal, tetragonal, or hexagonal crystal, but it is rare to find transparent crystals, other than those of calcite, which are sufficiently thick and birefringent to show a clear separation of the two images.

Wave Surfaces. Huygens' explanation of these results was based on a consideration of the propagation of light in all directions from any point inside the crystal. For uniaxial crystals which, like calcite, are completely transparent and not optically active (see later), it has been fully established by later work. It was to the effect that if a point source of monochromatic light be imagined to exist anywhere inside the crystal, the light will be propagated as two independent sets of rays. The rays of one set (ordinary rays) travel with the same velocity in all directions, as they would in an isotropic medium, and their *wave surface*,* *i.e.* the surface formed by the ends of all the rays at any instant, is a *sphere*, having the point source as centre. The wave surface of the other set (extraordinary rays) is an *ellipsoid of rotation*, which touches the spherical wave surface, corresponding to the same instant of time, at the two points where the axis of rotation emerges. The axis of rotation is the optic axis, mentioned above, and in this direction light travels with a single velocity, and does not suffer double refraction. For a given crystal the wave surfaces are the same in every particular, no matter where the point source be situated within it. This is a natural consequence of its regular internal structure, by reason of which any two paths through it which are parallel to one another are optically identical.

The ellipsoid may lie either inside or outside the sphere. Figs. 129 and 130, which depict *principal sections* of the wave surfaces, *i.e.* sections containing the optic axis, illustrate these two possibilities respectively. In calcite, since the extraordinary rays travel

* Called by some authors the *ray surface*, since it is the locus of points attained by rays.

92   THE OPTICAL PROPERTIES OF CRYSTALS

with a greater velocity than the ordinary rays (except along the optic axis where the distinction between the two vanishes), the wave

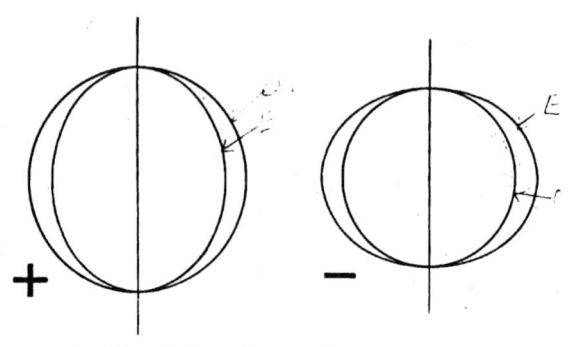

FIG. 129.—Wave Surfaces of
a Positive Uniaxial Crystal.

FIG. 130.—Wave Surfaces of
a Negative Uniaxial Crystal.

surfaces are related as in Fig. 130. Such crystals are said to have *negative* birefringence. Those with wave surfaces related as in Fig. 129 are said to have *positive* birefringence.

Wave Front. The plane tangent to the wave surface at any point is known as the *wave front* corresponding to the ray which meets the surface at that point. The idea of a wave front in connection with a single ray is, like that of a single ray itself, merely a conception which is convenient in a theoretical treatment. Contact with reality is made by supposing that the ray forms one of a beam of parallel rays; the tangent plane referred to above, or, more precisely, that part of it intersected by the beam, is then the wave front of the beam, because it is the common tangent plane to the wave surfaces corresponding to all the rays.

The direction normal to this plane is called the *wave-front direction*, *wave direction*, or *wave normal*, and this is the direction in which the wave front as distinct from the ray is to be regarded as moving. For ordinary rays, since the wave surface is a sphere, the tangent plane is always normal to the ray, and the wave normal is therefore coincident with the direction of the ray itself. But for extraordinary rays this is, in general, not the case. Consider, for example, the ray ORR' in Fig. 131, which shows an elliptic section through the centre of an extraordinary wave surface. The tangent plane to the surface at R is PRP', and the normal to this, namely ON, is not coincident with the ray direction ORR'. The same is true of all other rays in the section, with the exception of those travelling along

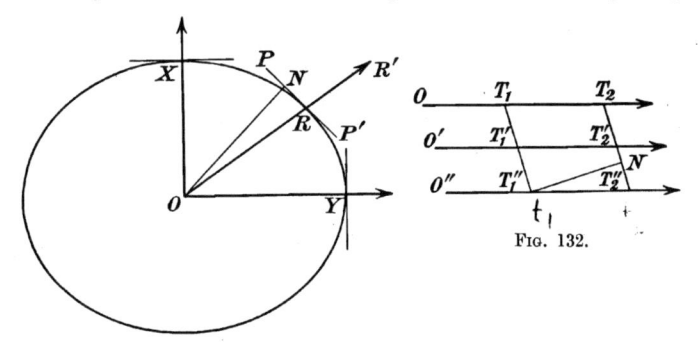

Fig. 131.—Wave Fronts of Extraordinary Rays.

Fig. 132.

its axes, for example the rays OX and OY. In these cases the ray and wave directions coincide.

Wave Velocity and Ray Velocity. Consider a beam of parallel extraordinary rays, $OT_2, O'T_2', O''T_2''$ (Fig. 132), and let their wave front at the time t_1 be $T_1 T_1' T_1''$, and at the time t_2 be $T_2 T_2' T_2''$. The distances travelled by the rays in the time $t_2 - t_1$ are $T_1 T_2$, $T_1' T_2', T_1'' T_2''$, which are equal to one another, and the *ray velocity* may therefore be expressed as

$$\frac{T_1 T_2}{t_2 - t_1}$$

The *wave-front velocity*, or *wave velocity*, is, however, measured by the perpendicular distance between the wave fronts, by $T_1'' N$ for example, and may therefore be expressed as

$$\frac{T_1'' N}{t_2 - t_1}$$

Huygens' Construction. Once the orientation and form of the wave surfaces in a crystal are known, the directions into which light will be refracted on entering any face or artificial plate of the crystal can easily be determined by means of a geometrical construction due to Huygens. This construction for the case of light passing from one isotropic medium to another, *e.g.* passing from air into a piece of glass or a cubic crystal, is given in Fig. 133.

Let ABDC represent a section of a parallel beam of monochromatic light entering from the air the plane surface, BD, of a piece of glass. Further, let ABDC be that section of the beam which lies in a plane perpendicular to the surface of the glass.

Draw DE, perpendicular to AB. Then DE represents the wave front of the beam section, when the ray CD has reached the surface of the glass.

Let v and v' be the velocities of the light in air, and in the glass

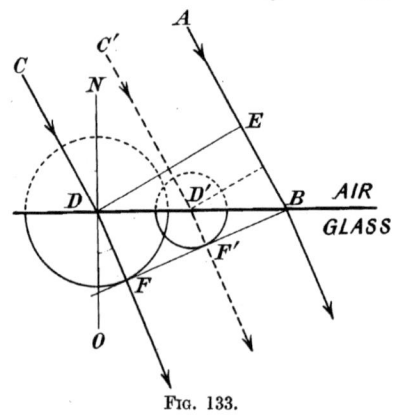

Fig. 133.

respectively. With D as centre, describe a sphere of radius r, such that
$$\frac{BE}{r} = \frac{v}{v'}$$

Let BF represent the plane drawn from B, which is perpendicular to ABDC and tangent to the sphere inside the crystal at F. Join DF. Then DF is the direction into which the ray CD is refracted. If the construction be repeated for any other ray in the beam, e.g. C'D', it will be found that the plane BF' which is tangent to the sphere in this case is the same as the plane BF, and the direction of refraction, D'F', is therefore parallel to the direction CD. Thus all rays are refracted in parallel directions, and the plane BF is the wave front of the refracted beam, when the ray CD has reached F.

If NO is the normal to the surface of the glass at D, it may easily be proved that
$$\frac{\sin NDC}{\sin FDO} = \frac{BE}{DF} = \frac{v}{v'}$$

The angles NDC and FDO are called the *angles of incidence* and *refraction* respectively. In general, then, when light of a given wave-length passes from one isotropic medium to another, the sine of the angle of incidence divided by the sine of the angle of refraction is a constant quantity (*i.e.* it is independent of the angle of incidence), which is equal to the velocity of the light in the first medium divided by its velocity in the second. If, as is commonly the case, the first medium is air, the sine ratio is called the *refractive index* of the second medium for light of the particular wave-length in question. The refractive index is thus equal to the reciprocal of the velocity of the light in the medium, its velocity in air being taken as unity. Strictly, this is the refractive index *with respect to air*, the *absolute* refractive index being the sine ratio when the first medium is a vacuum. Since, however, the ratio of the velocity of light in air to that in a vacuum is very nearly equal to unity (1·00029 for the D line), the distinction is of no consequence for most practical purposes. The refractive index is a property both of the medium and of the rays of the particular wave-length of light concerned. Therefore, instead of speaking, as above, of the refractive index of a medium for light of a given wave-length, we may with equal correctness speak of the refractive index of a particular ray for a certain medium.

A further feature of refraction in isotropic media is that the refracted rays lie in the plane of incidence. Thus, for example, in Fig. 133 the refracted rays in the area DBF are in the same plane as the incident rays in the area ABDC.

Consider now the application of the Huygens construction to the case of a parallel beam of monochromatic light entering a plate of calcite, cut so that its surface is perpendicular to a principal section through the wave surfaces.

In Fig. 134 the lines AB, CD, DE, and NO have the same significance as in Fig. 133. BD is the section of the crystal surface. With D as centre, describe the wave surfaces of the crystal plate, namely the sphere and ellipsoid of rotation, on such a scale that they bear the same relation to the velocities of the light in the crystal as BE does to its velocity in the air.

Let BF and BG represent the planes drawn from B, which are perpendicular to ABDC and tangent to the sphere and ellipsoid at F and G respectively. Then DF and DG are the directions of the ordinary and extraordinary beams in the crystal respectively, BF and BG being the respective wave fronts.

96   THE OPTICAL PROPERTIES OF CRYSTALS

When, as in this case, the crystal plate is perpendicular to a principal section of the wave surfaces, it is a geometrical consequence of the construction that the extraordinary rays are in the plane of incidence. Thus the rays in the area DBG are in the same plane as the incident rays in the area ABDC. But for plates perpendicular to other sections of the wave surfaces, this only happens when the optic axis is in the plane of the plate. On the other hand, the ordinary refracted rays are always in the plane of incidence as in the case of isotropic media.

The extraordinary beam also deviates from the law of sines, and in two respects. Not only does the refractive index depend

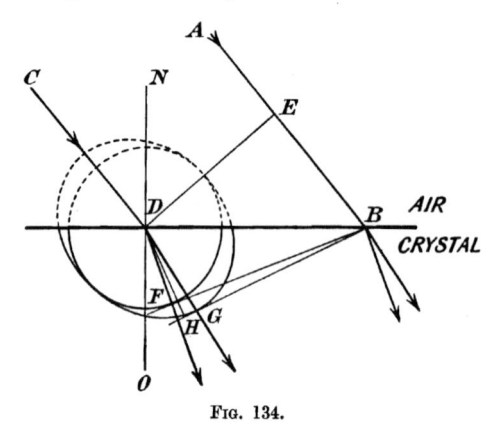

FIG. 134.

on the direction of the beam, but the sine of the angle of incidence divided by the sine of the angle of refraction does not, in general, give the refractive index, $i.e.$ is not equal to the reciprocal of the velocity of the beam in the crystal. Thus in Fig. 134

$$\frac{\sin NDC}{\sin GDO} \neq \frac{BE}{DG}$$

If, however, instead of considering the beam direction, we consider the *wave-front direction*, DH, where DH is perpendicular to the plane BG, then it may easily be proved that

$$\frac{\sin NDC}{\sin HDO} = \frac{BE}{DH}$$

Thus for an extraordinary ray, the sine of the angle of incidence *divided by the sine of the angle of refraction of the wave front is equal to the velocity of the ray in air divided by the velocity of its wave front in the crystal.* This sine ratio is the extraordinary refractive index of the crystal for the particular direction through it, along which the ray travels.

It should be pointed out that this definition of refractive index holds for all rays, whether ordinary or extraordinary. For with ordinary rays the wave-front direction and ray direction coincide, so that it is immaterial which term is used.

The wave-front direction, or wave normal, of a refracted ray lies in the plane of incidence in all media, whether isotropic, uniaxial, or biaxial, because the wave front is always perpendicular to the plane of incidence. In uniaxial crystals, the wave normal of an extraordinary ray lies in the same principal section of the wave surface as the ray itself. This is a geometrical consequence of the wave surface being an ellipsoid of *rotation*.

In Fig. 134 the ordinary ray is refracted more than the extraordinary ray, and it might be supposed that this is necessarily the case, since in all directions the velocity of the extraordinary ray is the greater. That this is by no means true will readily be seen, if the construction be repeated with the wave-surface section turned through about 90° in a clockwise direction. The extraordinary ray will then be found to be bent out of the course of the incident light more than the ordinary ray. The wave normal of the former, however, will still be less refracted than the ordinary ray.

An important case, with which we shall frequently have to deal in practice, occurs when the incident light is perpendicular to the crystal surface. The Huygens construction for this case is given in Fig. 135. ABDC is the incident beam section, the rays AB, CD, and all the other rays of which it is composed being perpendicular to the surface of the crystal, BD.

Describe wave surfaces for the crystal about the points B and D, and on the same scale. Let EF represent the plane parallel to the crystal surface, and tangent to the two spheres inside the crystal at E and F respectively, and let GH represent the plane parallel to the crystal surface and tangent to the two ellipsoids. Then BEFD is the beam section of the ordinary refracted rays, its wave front being EF, and BGHD is the beam section, corresponding to the same instant of time, for the extraordinary rays. If the crystal surface is perpendicular to a principal section of the wave surfaces,

98   THE OPTICAL PROPERTIES OF CRYSTALS

then the extraordinary rays are in the plane of incidence. If the crystal surface is not perpendicular to a principal section, the extraordinary rays are not necessarily in the plane of incidence. In both cases, however, the wave normals, BJ, DK, etc., of the extraordinary rays travel in the same direction as the ordinary rays, *i.e.* normally to the surface of the crystal. In the case of biaxial crystals also, when the incident light is perpendicular to the crystal surface, the wave normals of both rays travel normally to the surface.

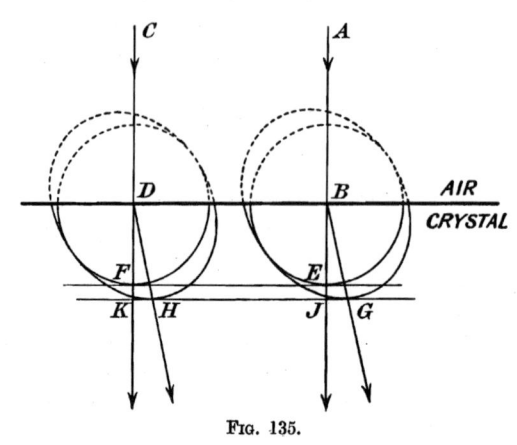

FIG. 135.

There are two cases of perpendicular incidence in uniaxial crystals which should receive special consideration. In the first, Fig. 136, *a*, the crystal surface is perpendicular to the optic axis. Here, it is clear, there is no double refraction whatever, and all rays pass through the crystal with the same velocity. In the second, Fig. 136, *b*, the crystal surface contains the optic axis. The directions of the extraordinary and ordinary rays coincide, and neither deviates from the course of the incident light. The light, however, travels through the crystal as two distinct sets of rays having different velocities, and gives rise to two images of any object viewed in this direction, the one image being immediately above the other. These two cases were discussed in a different way on p. 90, Figs. 128, *b* and *c*.

When an extraordinary ray passes out of the crystal into air, its refraction is governed by the same principles as those which

THE POLARISATION OF LIGHT 99

determined its refraction on entering the crystal. Accordingly, after passing through a parallel-sided crystal plate, it regains the

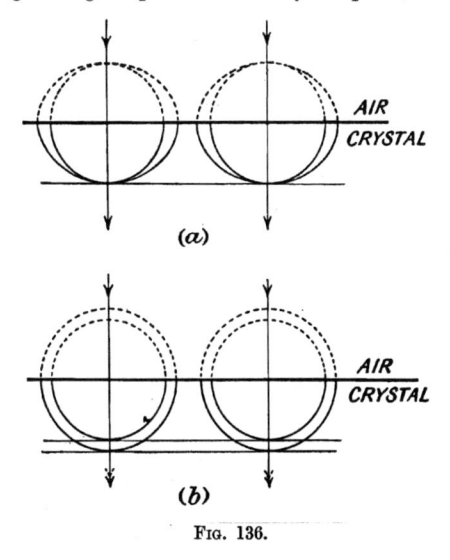

FIG. 136.

original direction of the incident light, just as the ordinary ray does, and therefore travels along a path which is parallel to that of the latter.

The Polarisation of Light by Crystals.* If the rays emerging from a crystal of Iceland Spar be allowed to enter another crystal of spar, they will, in general, undergo a second double refraction, so that an object viewed through the two crystals will give four images. For certain positions of the two crystals relative to one another, however (not including those in which the rays pass along the optic axes), this does not occur, and only two images, as with a single crystal, are seen. This shows that the rays which emerge from the first crystal are of a different nature from rays of ordinary light, for otherwise they would undergo double refraction at all positions of the second crystal. These facts were known to Huygens, but he was not able to find an explanation, and the problem remained unsolved until an accidental observation by Malus in 1808 gave

* In this section, no treatment is given of the *rotation* of the plane of polarisation by the so-called optically active substances.

the clue. Malus, who was engaged in a study of double refraction, happened to examine the sunlight reflected from the windows of the Luxembourg Palace through a prism of quartz (like calcite, a doubly refracting uniaxial substance), and was surprised to find that, when he rotated the quartz about the line of direction of the light, the two images of the sun disappeared alternately every 90°. In the hands of Young, Fresnel, and Arago, this observation led to the theory of the polarisation of light, which may be briefly stated as follows.

Light from ordinary luminous sources, such as the sun or a lamp, consists of rays, the vibrations of which take place in all possible directions perpendicular to the direction of the rays. By direction of vibration is meant the direction of electric intensity of the field associated with the passage of the light, or, more shortly, the direction of the electric vector. If the light suffers reflection or refraction by isotropic media, or double refraction by anisotropic crystals, however, part or all of the light becomes *polarised* (or, strictly, *plane polarised*), which means that, while still vibrating perpendicularly to its line of propagation, it does so *in one plane*, instead of in all directions.* It should be mentioned, that in the case of extraordinary rays, there is some doubt whether the vibrations take place perpendicularly to the ray direction or to the wave direction. Although the question is of little moment from the practical standpoint, we shall assume in what follows, in conformity with general opinion, that the vibrations take place perpendicularly to the wave direction, *i.e.* parallel to the wave front.

The maximum amount of polarisation is produced by an isotropic medium when the reflected and refracted rays are at right angles to one another. Thus in Fig. 137, let AO be a ray striking the surface of a piece of glass at O, and at such an angle that the angle between the reflected ray, OB, and the refracted ray, OC, is 90°. Under such circumstances, it is found that the reflected ray is completely, the refracted ray partially, polarised. It is generally assumed that the reflected ray vibrates perpendicularly to the plane of incidence (this vibration is represented in the figure by the dots along OB). The polarised part of the refracted ray vibrates

* According to the electromagnetic theory of light, in a ray of polarised light an electrical disturbance occurs in one plane, and a magnetic disturbance in the plane at right angles to it. Both of these planes therefore may be thought of as "planes of vibration", but the term is always confined to the former. The latter is referred to as the *plane of polarisation*.

THE POLARISATION OF LIGHT 101

in a plane perpendicular to the vibration plane of the reflected ray, i.e. according to the above assumption, it vibrates in the plane of incidence (this vibration is represented in the figure by strokes drawn across OC). Some polarisation occurs at all angles of incidence, but the reflected ray is only completely polarised when the above condition holds. The refracted ray is never completely polarised.

When a ray of ordinary light enters an anisotropic crystal so as to travel through it in any direction other than an optic axis, the two rays into which it is divided are always completely polarised, and, moreover, vibrate in planes at right angles to one another. The orientation of these planes depends on the directions in which the rays are travelling through the crystal. If the incident ray is

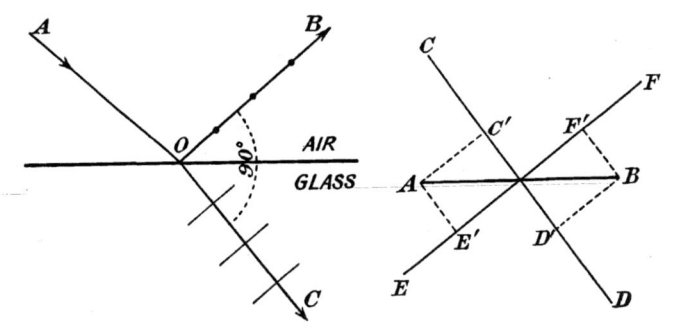

FIG. 137.—Polarisation by Reflection.   FIG. 138.

already polarised, its behaviour on entering the crystal depends on the orientation of its plane of vibration with reference to the planes of vibration offered by the crystal. If its plane of vibration happens to be perpendicular to one of these, then no light is transmitted along the path corresponding to that plane ; all the light travels along the other path, and consequently emerges on the other side of the crystal vibrating in one plane only. For other cases, the light distributes itself between the two paths according to the principle of the parallelogram of forces. This is illustrated in Fig. 138. Let AB be the trace of the vibration plane of a ray of polarised light travelling normally to the surface of the paper, which we will take to represent the surface of a crystal. Let the length AB represent the amplitude of vibration of the ray. Now when a ray strikes a crystal surface perpendicularly, the vibration planes offered by the crystal make traces on its surface, which are at right angles to one another. Let these traces be represented by CD and EF. Since we have assumed that the vibrations take place at right angles to the wave directions (see p. 100), CD and EF represent the *vibration directions* of the two rays passing through the crystal, because for perpendicular incidence the wave directions of the rays are also perpendicular (see Fig. 135, p. 98). From A and B drop perpendiculars AC', BD' and AE', BF' on the CD and EF respectively. Then C'D' represents the amplitude of light vibrating along CD, and E'F' represents the amplitude of light vibrating along EF.

From this it can readily be seen that if AB coincides with either CD or EF, the resolved amplitude along the other is zero. Thus if AB coincides with EF, C'D' = 0, and light will only pass along the path corresponding to EF.

We are now in a position to explain Malus' observation. The sunlight must have struck the palace windows at, or near, the angle giving complete polarisation to the reflected rays. Every time their plane of vibration became perpendicular to one of the vibration planes of the quartz crystal, the rays vibrating in that plane were extinguished. Obviously this state of affairs would occur every 90° of rotation of the quartz.

The experiments with two crystals of calcite described at the beginning of this section may also now be explained. When the two crystals are arranged so that, for light passing through both, the vibration directions in the one are parallel to those in the other, only two images are produced, because each ray emerging from the first crystal is vibrating in one of the directions offered by the second one, and so no further resolution occurs. This happens, for example, when the crystallographic orientations of the two crystals in space are identical, the effect being tantamount to that which would be produced by increasing the thickness of the first crystal. In one particular case, one image only is produced. This is when both crystals are of the same thickness, and are orientated in relation to one another as an object and its mirror image. The rays diverge in the first crystal and converge in the second, as the reader may prove by applying Huygens' construction.

In the general case where the vibration directions in the two crystals are not parallel, each of the rays emerging from the first

THE INDICATRIX

undergoes resolution on entering the second, after the manner illustrated by Fig. 138, and so four images are produced.

The Indicatrix. We have already seen how the velocity of light of a given wave-length in different directions in a uniaxial crystal may be represented by the sphere and rotation ellipsoid of Huygens. To give a complete representation of the optical properties of the crystal for light of this wave-length, however, we need to know also the vibration planes for any given direction. With reference to the assumed plane of vibration of reflected light (see p. 100), these crystal vibration planes prove to be related to the wave surfaces very simply as follows. An extraordinary ray always

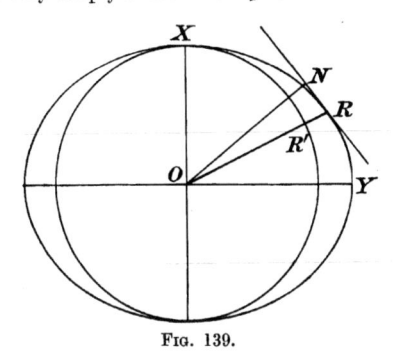

FIG. 139.

vibrates in the principal section of the wave surface in which it lies, *i.e.* in the plane containing the ray and the optic axis. An ordinary ray always vibrates in the plane perpendicular to the principal section in which it lies. This is illustrated in Fig. 139, which represents a principal section through the wave surfaces of a negative uniaxial crystal. OR and OR' are extraordinary and ordinary rays respectively, which lie in this section. The extraordinary ray OR vibrates in the plane of the section, *i.e.* in the plane of the paper, and the ordinary ray OR' vibrates in the plane perpendicular to this, *i.e.* in the plane containing OR' and any line drawn from OR' perpendicular to the plane of the paper. Taken in conjunction with these facts, the wave surfaces provide a complete description of the optical properties of the crystal for light of the wave-length to which they refer. We can, for example, obtain the refractive index of any ray from the wave-surface figure. The refractive index of a ray is, as we have seen, the reciprocal of

the velocity of its wave front with respect to that in a vacuum (p. 97), and so if we make the radii of the wave surfaces represent actual velocities, the refractive index of any ray will be found simply by calculating the reciprocal of the length of the line representing its wave velocity in the wave-surface figure. Thus in Fig. 139 if the length OR represents the actual velocity of the ray travelling in the direction OR, then the length ON will be its wave velocity, where ON is the wave direction, being the perpendicular drawn from O to the tangent plane to the wave surface at R.*
The refractive index of the ray OR is thus equal to 1/ON. The refractive index of an ordinary ray is given simply by the reciprocal of its velocity, for its direction and its wave direction are identical. Thus in the figure, the refractive index of the ray OR′ is 1/OR′. Extraordinary rays which travel in the circular section of the rotation ellipsoid, e.g. OY, are coincident with their wave directions, and so their refractive index is given by the reciprocal of their velocity, as with ordinary rays. Similarly, in the direction of the optic axis, OX, the ray and wave directions coincide, and the refractive index is simply 1/OX.

The optical properties can, however, be more conveniently represented by means of single-surfaced figures, the one most commonly employed being the so-called *optical indicatrix*, due to Fletcher.† For light of a given wave-length in the case of a uniaxial crystal, this is derived from the corresponding extraordinary wave surface as follows. In Fig. 140 let XRY be the extraordinary wave surface (shown for simplicity in principal section as an ellipse), the radii of which represent actual velocities of the light with respect to that in a vacuum. Let OX be the semi-axis of rotation (optic axis), and OY a radius of the circular section. Let the radius OR be an extraordinary ray travelling in any other direction. Then the lengths OY and OR represent respectively the velocities with which the extraordinary rays travel in those directions, and OX represents the velocity with which the light travels along the optic axis, and with which the ordinary rays travel, no matter what their direction.

On the axes OX and OY produced, construct another ellipsoid of rotation, X′R′Y′ (shown in the figure in principal section), OX′

* ON is in the same principal section as OR, because the wave surface is an ellipsoid of *rotation*, and the tangent plane to any ray end is therefore perpendicular to the principal section in which the ray lies.

† *Min. Mag.*, 1891, **9**, 278.

THE INDICATRIX

being the axis of rotation, and the dimensions being $OX' = 1/OY$ and $OY' = 1/OX$. This is Fletcher's indicatrix, which, from its construction, is concentric with, and geometrically similar to, the extraordinary wave surface.

Produce the ray OR to meet the indicatrix at R'. Draw the radius OD in the same principal section of the indicatrix as OR' and conjugate * to it. Complete the parallelogram R'ODE. Draw the wave normal ON of the ray OR and produce it to meet R'E at N'.

Now since the parallelogram R'ODE is formed by conjugate radii, its area is equal to that of the rectangle OX'ZY' (where

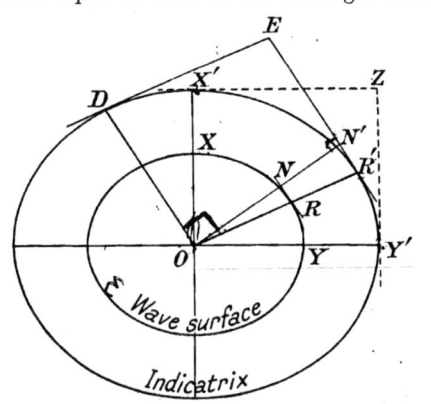

Fig. 140.—Derivation of Indicatrix from Wave Surface.

X'Z and Y'Z are the tangents to the section of the indicatrix at X' and Z' respectively), for OX' and OY' are also conjugate radii.

i.e. area of $R'ODE = ON'.OD = OX'.OY'$,

or
$$OD = \frac{OX'.OY'}{ON'} \quad . \quad . \quad . \quad (i)$$

But since the wave surface and indicatrix are similar,

$$\frac{ON'}{ON} = \frac{OY'}{OY},$$

or
$$ON' = \frac{OY'.ON}{OY}.$$

* Conjugate radii are such that the tangents to the ellipse at the extremities of the radii form a parallelogram with them. For a given ellipse, this parallelogram has a constant area.

Substituting in (i),

$$OD = \frac{OX'.OY}{ON} = \frac{1}{ON} \left(\text{since } OX' = \frac{1}{OY}\right).$$

But 1/ON is the refractive index of the crystal for the ray OR. It follows therefore that the refractive index for any extraordinary ray, the direction of which is represented by a radius of the indicatrix, is given by the length of the conjugate radius lying in the same principal section. In the case under consideration, a ray travelling in the direction OR' has a refractive index equal to OD, the conjugate radius to OR' in the same principal section. It also follows that the refractive index of a wave front travelling in any direction through the centre of the indicatrix is equal to the radius of the indicatrix which is perpendicular to, and in the same principal section as, the line of direction. In the present case, the refractive index of the wave front travelling along ON' is OD, the angle N'OD being a right angle, and OD being in the same principal section of the indicatrix as ON'. Further, the line OD, besides giving by its length the above information, gives by its direction the vibration direction of the ray OR', and of the wave front travelling along ON', for, as already stated, an extraordinary ray vibrates in the plane of the principal section of the wave surface in which it lies, and this section occupies the same plane as the principal section of the indicatrix in which the ray lies.

Consider now how the indicatrix expresses the vibration direction and refractive index of an ordinary ray. It has been shown that the refractive index of such rays is independent of their direction, and in the present case is equal to 1/OX. This in turn is equal to OY', which is a radius of the circular section of the indicatrix. The vibration plane, as already stated, is normal to the principal section in which the ray lies, and the vibration direction may therefore be represented by any line drawn from the ray and perpendicular to this principal section. Such a line is either of the two radii of the circular section, which are perpendicular to the ray, and the choice of either of these has the advantage that its length represents the refractive index of the ray. Thus in the case of an ordinary ray (or wave front) travelling along ON' in Fig. 140, its refractive index and vibration direction are represented by a line drawn from O normal to the plane of the paper, and of length equal to OY'. This is the line OF in Fig. 141, in which the indicatrix is drawn in perspective. The other letters in this figure have the same significance as in Fig. 140.

We have seen that the properties of the extraordinary wave front travelling along ON' are represented by OD, those of the ordinary wave front travelling in the same direction by OF. Now it may be shown that OD and OF are the semi-axes (*i.e.* smallest and greatest radii) of the elliptic section * of the indicatrix which is normal to ON' and passes through O. Since ON' represents any direction, it follows that *the refractive indices and vibration directions of the two wave fronts (ordinary and extraordinary) which can travel*

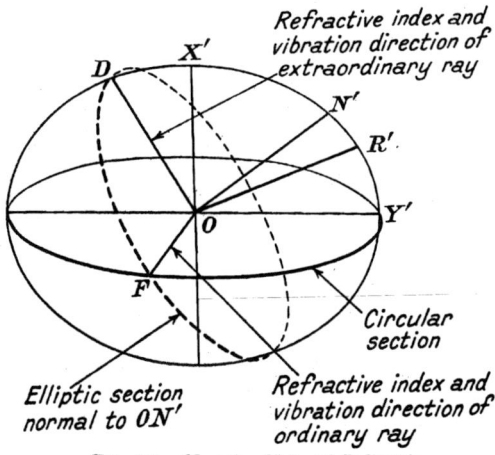

FIG. 141.—Negative Uniaxial Indicatrix.

along any direction in the crystal, are given by the semi-axes of the elliptic section of the indicatrix which passes through its centre, and to which the direction in question is normal. The one semi-axis is always a radius of the circular section, and this is the one which refers to the ordinary ray.

There are two special cases to be considered. The first is that in which the wave fronts travel along OX'. This is the optic axial direction, and the section of the indicatrix normal to this is a circle, which of course has no smallest and greatest radii. This

* All sections of an ellipsoid of rotation which pass through its centre are ellipses, with the exception of the circular section at right angles to the axis of rotation.

represents the fact that the light can vibrate in any direction, and does not suffer double refraction. The second case is that in which the wave fronts travel along OY' (or any other radius of the circular section). The elliptic section of the indicatrix normal to this direction has the special feature that it is that section in which the difference between the semi-axes is a maximum—the semi-axes being in fact those of the indicatrix itself, namely OX' and OY'. In this case, then, the crystal shows its maximum double refraction. These are the two cases which were considered from the standpoint of Huygens' construction in Figs. 136, a and b, p. 99.

The refractive indices corresponding to OX' and OY' are conventionally denoted by ε and ω. Thus, ω is the refractive index of the ordinary ray, and ε is the extreme value of the refractive index of the extraordinary ray, *i.e.* the value which differs most widely from ω. In negative crystals $\varepsilon < \omega$, whilst in positive crystals $\varepsilon > \omega$ (*cf.* Figs. 129 and 130, p. 92). The sign of the double refraction is thus given by $\varepsilon - \omega$.

In Fig. 135, p. 98, the fact is illustrated that when a ray of light enters a plane crystal surface perpendicularly, the two wave fronts travel through the crystal in the same direction as the incident ray, although the two *rays* in the crystal pursue, in general, different paths. In practice, crystals are usually examined under such, or nearly such conditions, and it is for this reason that attention has been mainly drawn to this case in the above treatment of the indicatrix. When, however, the incident ray is not perpendicular to the crystal surface, neither the two rays formed from it inside the crystal, nor their wave fronts, travel in the same direction as each other (see Fig. 134, p. 96). Nevertheless they are always polarised in planes at right angles to one another.

A study of the indicatrix should make it clear that for a given *vibration* direction, there is one refractive index and one only, whereas for a given direction of *propagation*, there are two refractive indices corresponding to the two rays (ordinary and extraordinary) which can travel along that direction.

Circularly and Elliptically Polarised Light. When two rays of the same colour, polarised in different planes, and differing in phase, are caused to travel along the same path, their rectilinear vibrations combine to give a resultant vibration which is in general elliptical. The composite ray is therefore said to be *elliptically polarised*. In the special case where the two planes are perpendicular, the ampli-

BIAXIAL CRYSTALS 109

tudes the same, and the difference one-quarter of a wave-length, the resultant vibration is circular and the light is said to be *circularly polarised*. Another special case is that in which the two planes are perpendicular and the difference of phase is half a wave-length. The resultant vibration is then linear, *i.e.* the light is plane polarised.

Elliptically polarised light is produced when parallel light emerges from a birefringent crystal plate (at least in the region where the two refracted beams overlap on emergence), because, with uniaxial crystals, at each point on the crystal surface an ordinary ray derived from a particular incident ray emerges coincident, but out of phase, with an extraordinary ray derived from some other incident ray, whilst with biaxial crystals, two extraordinary rays, out of phase and vibrating in planes at right angles to one another, are similarly coincident.

Elliptically polarised light may be considered simply from the standpoint of the vibration planes or vibration directions of the two component rays, and will be treated in this way in the sequel.

[Before going farther, the reader may find it helpful to turn to the summary at the end of the chapter for a recapitulation of the main points of the preceding sections. If he has experienced difficulty in forming a clear mental picture of the three-dimensional relationships involved in this subject, he will probably derive great assistance from simple models. A model of the uniaxial indicatrix may be made by cutting a circular section and a principal section out of cardboard, and slotting them so that they may be fitted together at right angles to one another. An inclined elliptical section may also be added (*e.g.* by cutting it into four quadrants, and attaching these to the rest of the model by means of gummed label, sealing-wax, or fine wire), so that the finished model has the appearance of Fig. 141. A model of a biaxial indicatrix (to be described shortly), though more complicated, may be made on similar lines.]

Biaxial Crystals. The treatment which follows in this section refers strictly only to crystals which are completely transparent for the wave-lengths considered, and are not optically active (see p. 126).

As already stated on p. 89, crystals of the orthorhombic, monoclinic, and triclinic systems are *biaxial*, *i.e.* possess two directions which correspond very closely to the optic axis in uniaxial crystals.

In general, when a ray of monochromatic light enters a biaxial crystal, it is divided into two rays polarised in planes at right angles to one another, but neither of these rays obeys the ordinary laws of refraction; in other words, two extraordinary rays are formed. Passing through any point in the crystal, however, there are three planes at right angles to one another, each of which is characterised by the fact that one of the two rays which can travel in any direction in it has a constant refractive index. These three planes are defined by three mutually perpendicular axes which are, respectively, the vibration directions of rays having the maximum, the minimum, and a particular intermediate refractive index. In an orthorhombic crystal these axes run in the same directions as the crystallographic axes. In a monoclinic crystal one of them has the same direction as the b crystallographic axis, and the others lie in the ac plane, i.e. the plane containing the a and c axes, but anywhere in that plane. In a triclinic crystal they have no necessary relation with the crystallographic axes whatever.

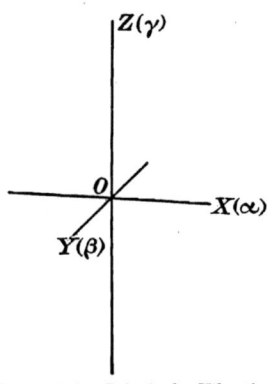

Fig. 142.—Principal Vibration Directions of a Biaxial Crystal.

Let these three axes for a given crystal and a defined wave-length of light be represented by OX, OY, and OZ in Fig. 142, OX being the vibration direction for rays having the smallest refractive index in the crystal, OZ the vibration direction for rays having the greatest refractive index, and OY the vibration direction perpendicular to OX and OZ, and corresponding therefore to a particular intermediate refractive index. These three indices are usually referred to as α, γ, and β* respectively, and these symbols have been included in Fig. 142 adjacent to the axes to which they refer.

Consider now rays travelling in the plane XOY through O. In any such direction, two rays can travel, and one of them vibrates along OZ (i.e. normal to the plane), and has the constant refractive index γ, whilst the other vibrates in the plane, and its refractive

* The N_p, N_g, and N_m of some American authors. The subscripts mean "petty", "greatest", and "mean" respectively.

index depends on its direction as follows : when it is travelling along OY, it vibrates along OX and its refractive index is α ; when it is travelling along OX, it vibrates along OY and its refractive index is β ; for all other directions of travel its vibration direction is intermediate between OX and OY, and its refractive index is intermediate between α and β.

The behaviour of rays passing through O in the planes YOZ and XOZ is exactly analogous to the above. In YOZ one ray always vibrates along OX with refractive index α, whilst the index of the other varies between β and γ, being β when it is travelling along OZ and vibrating along OY, and γ when it is travelling along OY and vibrating along OZ. In the plane XOZ, one ray always vibrates along OY with refractive index β, and that of the other varies

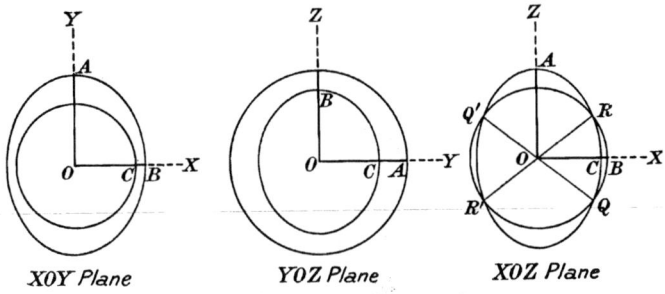

XOY Plane   YOZ Plane   XOZ Plane

FIG. 143.—Principal Sections of Biaxial Wave Surfaces.

between α and γ, being α when it is travelling along OZ and vibrating along OX, and γ when it is travelling along OX and vibrating along OZ.

In each of these three planes, then, one ray behaves as an ordinary ray. Its wave direction is the same as its own direction, and the section of its wave surface made by the plane is a circle. The corresponding wave-surface section for the other ray is an ellipse, the semi-axes of which coincide with the two axes defining the plane (*e.g.* OY and OX in the case of the plane XOY). For this ray, the wave-front direction is coincident with the ray direction when, and only when, it is travelling along these axes.

Fig. 143 gives the wave-surface sections made by these three planes. OA, OB, and OC represent the velocities of rays travelling along the three axes of Fig. 142 with refractive indices α, β, and γ

respectively. Therefore $OA = 1/\alpha$, $OB = 1/\beta$, and $OC = 1/\gamma$, and $OA > OB > OC$. The positions of the three axes of Fig. 142 with respect to the wave-surface sections are indicated by broken lines in Fig. 143. In the section made by the plane XOY, the circle lies entirely inside the ellipse, for the ray which travels with constant refractive index, and therefore constant velocity, has the largest index (γ), and therefore the smallest velocity (OC), offered by the crystal. OC is therefore the radius of the circle, and the major and minor semi-axes of the ellipse are OA and OB respectively. OA is the ray direction for which the refractive index is α; it therefore corresponds to OY in Fig. 142. Similarly OB is the ray direction for which the refractive index is β, and corresponds to

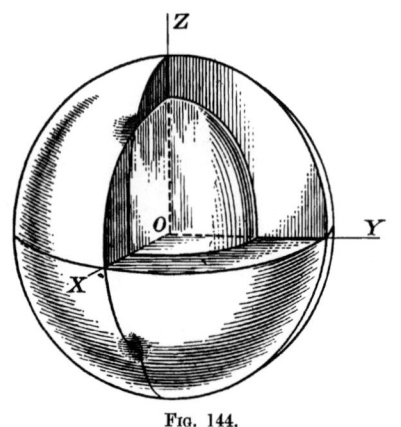

FIG. 144.

OX in Fig. 142. A little consideration will make clear the relations between the wave-surface sections for the planes YOZ and XOZ on the one hand, and the axes OX, OY, and OZ on the other. The wave-surface section for the plane XOZ is especially interesting, because in it the circle and ellipse intersect at four points. This is because the ray of constant velocity has the index of refraction β, the intermediate value, whilst the other changes its index from the minimum, α, to the maximum, γ. Rays which travel from O through any of the points of intersection (e.g. OR) have the same velocity whether they have wave fronts corresponding to the circle or to the ellipse. The point of intersection opposite to R, namely R', is in the same straight line as OR, and similarly QOQ' is a

BIAXIAL CRYSTALS 113

straight line, where Q and Q' are the other two points of intersection. There are thus two directions, ROR' and QOQ', through the crystal along which all rays travel with the same velocity. These two directions are known variously as *secondary optic axes*, *lines of single ray velocity*, or *optic biradials*.
The complete wave surface figure is shown in Fig. 144. For the present purpose it is sufficient to note that it has three planes of symmetry, which are the three sections shown in Fig. 143, and that, as can in a general way be gathered from these sections, it is a double-surfaced figure, the surfaces meeting one another at the four points R, R', Q, and Q', where the secondary optic axes emerge. These points form the bottoms of four funnel-shaped depressions

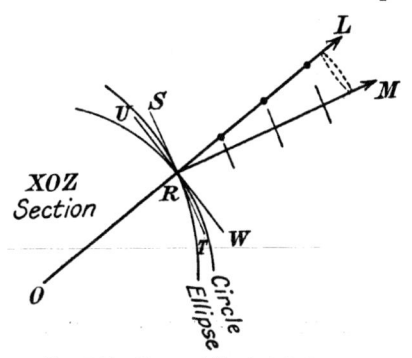

FIG. 145.—External Conical Refraction.

on the outer shell of the figure. In consequence of this, there are innumerable planes tangent to the figure at each of these points. Thus, for example, taking the XOZ section at R (Fig. 145), there is the plane UW, tangent and normal to the circle, and another, ST, tangent and normal to the ellipse. These planes are the wave fronts of the two rays travelling along OR, which belong respectively to the circular and elliptical parts of the XOZ section. They are normal to the section because the latter is a plane of symmetry of the wave-surface figure. For any other section of the figure containing OR, there is a corresponding pair of planes tangent to the two sides of the depression, and these planes are the wave fronts of the two rays which travel in the section along OR.

It follows that a ray which travels along OR can have a wave front corresponding to any one of these tangent planes, the actual one in any particular case being determined by the respective inclinations of the incident ray and OR to the crystal surface. If, for any given inclination of OR to the crystal surface, *all* the incident rays which are refracted along OR are considered, it is found that they lie on the surface of a converging cone ; consequently they emerge on the other side of the crystal as a diverging hollow cone. This emergent cone is represented by MRL in Fig. 145. The ray RL is that which, in the crystal, had the wave front UW. It therefore belongs to the circular part of the XOZ section (see above), and, like all other rays in this part, vibrates at right angles to the section. The ray RM, diametrically opposite to RL on the cone, is that which, in the crystal, had the wave front ST. It therefore belongs to the elliptical part of the XOZ section, and so vibrates in the plane of the section. The vibration directions of

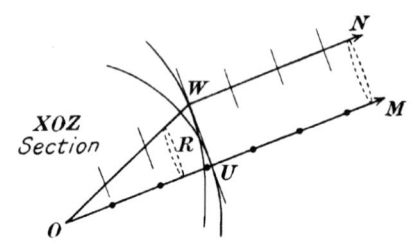

FIG. 146.—Internal Conical Refraction.

these two rays are represented respectively by dots and strokes in the figure.

The other rays in the cone are also polarised, but each ray has a different vibration direction, this property varying continuously as we pass round the cone, and all possible vibration directions being represented. Light which travels along a secondary optic axis, then, may vibrate in any plane.

The phenomenon just considered is known as *external conical refraction*. Actually the cone is not observed in practice, for it must be remembered that the secondary optic axis is indeed a *direction*, and not a single line through the crystal, so that when a real beam of light is passed in this direction, the light emerging is the result of the overlapping of innumerable cones.

Consider again the XOZ section at one of the points of intersection of the circle and ellipse, R (Fig. 146). Let UW be the common tangent to the circle (at U) and to the ellipse (at W). Now the form of the complete wave surface figure is such that the plane which contains UW, and which is normal to the XOZ section, is tangent to the wave surface all round R, the line of tangency being a circle, and U and W being two opposite points on this circle. It follows therefore that all members of the hollow cone of rays travelling from O to points on this circular ring of tangency have the same wave front, *i.e.* their wave-front velocities and, therefore, their refractive indices are identical. The direction in which this common wave front travels is that of the ray OU, for this ray is perpendicular to the wave front. OU is therefore called a *direction of single wave velocity*. OU is also perpendicular to the tangent plane covering the depression in the wave surface opposite to R, namely R' in Fig. 143, and similarly there is a direction which is perpendicular to the tangent planes covering the two depressions Q and Q' (Fig. 143), so that there are two directions of single wave velocity in the crystal. These two directions are also known as *primary optic axes, optic binormals*, or, simply (and usually), *optic axes*.

The ray OU (Fig. 146) belongs to the circular part of the XOZ section, and consequently vibrates in a direction perpendicular to the section. On the opposite side of the cone of rays having the same wave velocity as OU, there is the ray OW which belongs to the elliptical part of the XOZ section, and therefore vibrates in the plane of the section. These vibration directions are shown in the figure by dots and strokes respectively. The other rays in the cone are also polarised, but each ray vibrates in a different direction, which varies continuously as we pass round the cone, just as in the case of the cone of rays which emerge from a secondary optic axis.

If a parallel beam of unpolarised light enters a crystal plate at perpendicular incidence, the surface of the plate being perpendicular to a primary optic axis, Huygens' construction shows (Fig. 147) that each ray as it enters the crystal is refracted into a hollow cone of rays, which, on emerging on the opposite side of the crystal, forms a hollow cylinder. All the rays travelling through the crystal have the same wave velocity and therefore the same refractive index. Thus in the figure the rays AB and A'B' form the cones BDE and B'D'E' inside the crystal, and the cylinders DEHG and D'E'H'G' outside the crystal, respectively. In similar circumstances, the cone UOW in Fig. 146 would form the cylinder MUWN. Light emerges from the crystal vibrating in all directions, as it was before it entered the crystal, but having undergone what is termed *internal conical* refraction.

If the light is polarised before it enters the crystal, the other circumstances being the same as in the preceding paragraph, the vibration direction of the light is not altered by its passage through the crystal, but its direction of transmission is determined by its vibration direction. Thus in Fig. 147, if the ray AB is vibrating

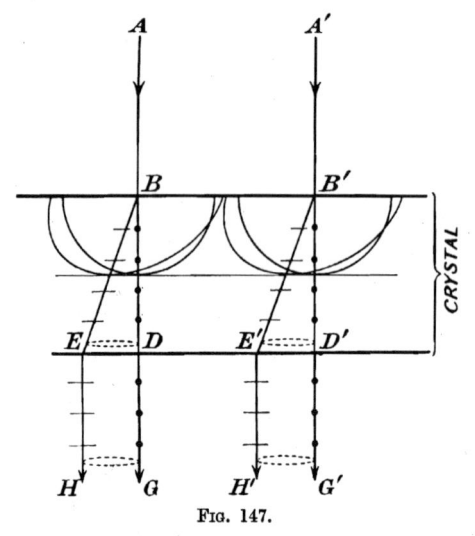

Fig. 147.

in the plane of the paper, *i.e.* in the plane of the XOZ section, it will travel through the crystal in the direction BE. If, on the other hand, it is vibrating in the plane perpendicular to the paper, it will travel along BD, *i.e.* without deviation. For other vibration planes it will travel along the appropriate side of the cone BDE, between BE and BD.

The properties shown by the single optic axis in uniaxial crystals are thus divided between two sets of directions in biaxial crystals. The primary optic axes possess the property of single wave velocity, the secondary optic axes that of single ray velocity. For practical purposes, however, it is only the primary optic axes which are of importance, and it is these directions which behave similarly to the

THE BIAXIAL INDICATRIX

optic axis of uniaxial crystals in ordinary interference phenomena (see, however, p. 250). To illustrate the truth of this statement, let us compare the case considered in Fig. 147 with that in which parallel light at normal incidence enters a crystal cut normal to a secondary optic axis (Fig. 148). Huygens' construction shows that in the latter case two beams polarised in planes at right angles to one another are produced. In other words, the light behaves no differently from the general case of light entering an anisotropic medium.

The angular difference between the primary and secondary optic axes increases with the birefringence of the crystal (*i.e.* the value of $\gamma - \alpha$), and with the value of the acute angle between the second-

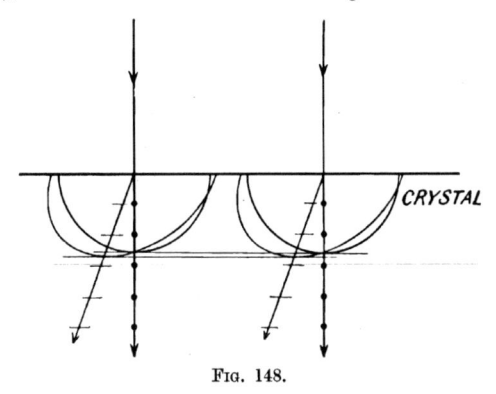

FIG. 148.

ary optic axes. In most inorganic crystals it is considerably less than 1°, an exceptional case being that of orthorhombic sulphur ($\gamma - \alpha = 0.29$) in which it is more than 3°. In the preceding diagrams it has been much exaggerated for the sake of clearness, and, consequently, so have the cones of inner and outer conical refraction.

The Biaxial Indicatrix. By an extension of the general principle by which the indicatrix was derived from the wave surfaces in the case of a uniaxial crystal, the indicatrix for a biaxial crystal can be obtained.* This is a triaxial ellipsoid, constructed on the

---

* In point of fact Fresnel, to whom the pioneer work on biaxial crystals was due, reversed this procedure, and deduced the wave surfaces from a single-surfaced ellipsoid having semi-axes equal to the velocities corresponding to α, β, and γ.

three axes OX, OY, and OZ of Fig. 142, their lengths, for this purpose, being such that they represent the refractive indices to which they respectively refer, when regarded as directions of vibration. Thus, $OX = \alpha$, $OY = \beta$, and $OZ = \gamma$. Fig. 149 shows such an indicatrix. All sections of the indicatrix passing through its centre are, with two exceptions shortly to be dealt with, ellipses, and, just as in the case of the uniaxial indicatrix, *the semi-axes of any such elliptic section give the vibration directions and refractive indices of the two wave fronts which can travel in a direction normal to the section.* Thus in Fig. 149, let ON represent a given direction, and let OL and OM be the semi-axes of the elliptic section (shown

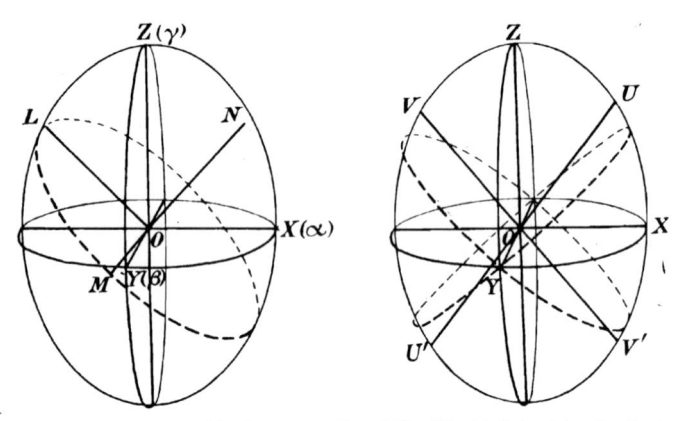

FIG. 149.—Biaxial Indicatrix.   FIG. 150.—Biaxial Indicatrix showing Circular Sections and Optic Axes.

by a broken line) which is normal to ON. Then OL and OM represent the vibration directions and refractive indices of the two wave fronts which can travel along ON.

Consider now all sections of the indicatrix which have OY as a common semi-axis. The value of the other semi-axis in these sections varies from OX to OZ, and somewhere between these limits it must be equal to OY, since OY is intermediate in length between OX and OZ. In this case, then, the two "semi-axes" of the section are equal, and it is in fact a circle. There are two such circular sections, equally inclined to OZ. The two directions normal to these two sections are the primary optic axes, or optic axes, with which we have already dealt. They are represented by

THE BIAXIAL INDICATRIX

UOU' and VOV' in Fig. 150, the circular sections to which they are normal being shown by broken lines. These circular sections represent geometrically the fact that rays, the wave fronts of which travel along the optic axes, can vibrate in any plane, and that they all have the same refractive index, which, since OY is the radius of the circles, is β.

From the above it will be seen that the optic axes necessarily lie in the XOZ, or $\alpha\gamma$ section of the indicatrix. The angles between the optic axes are bisected by the indicatrix axes OX (α) and OZ (γ), the one which bisects the acute angle (commonly called the *optic axial angle*) being termed the *acute bisectrix*, or *first median line*, and the other the *obtuse bisectrix*, or *second median line*. When γ is the acute bisectrix the crystal is said to be *positive*, and when α

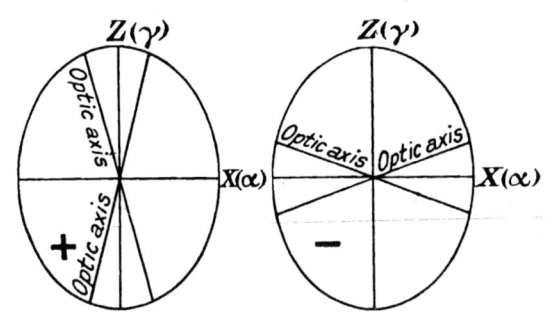

FIG. 151.

is the acute bisectrix the crystal is said to be *negative*. These two cases are distinguished in Fig. 151, which shows XOZ sections of positive and negative indicatrices.

It is a useful fact to remember that (provided the optic axial angle is not near to 90°) the acute bisectrix corresponds to the refractive index (α or γ) which is the more removed in value from β.

It may be noted that the positive *uniaxial* indicatrix (see p. 103) can be regarded as a positive biaxial indicatrix in which the optic axial angle is zero. The former is thus the limiting case of the latter in the direction of decreasing optic axial angle. Similarly the negative uniaxial indicatrix may be regarded as the limiting case of a negative biaxial indicatrix.

The plane in which the optic axes of a biaxial crystal lie is called the *optic axial plane*.

The Orientation and Dispersion of the Indicatrix. The optical properties of a crystal vary with the wave-length of the light used. In all anisotropic crystals, then, there is a different indicatrix for every wave-length. In any direction the refractivity increases as the wave-length diminishes, so that the indicatrix for blue light is larger than that for red light. The orientation and dispersion of the indicatrix are governed by the cardinal principle that the symmetry of the crystal determines the symmetry of the optical, and other physical properties. It should be noted that as far as optical properties are concerned, it is the symmetry of the holohedral class of the system to which the crystal belongs which is determinative, or in other words, the symmetry of the Bravais lattice. This is necessarily so because a beam of light cannot be unsymmetrical about its direction of propagation, so that if, for example, the crystal belongs to a class in which a given plane is not one of symmetry, whereas it is so in the holohedral class, the optical properties will still be symmetrical about this plane. Thus if we construct the indicatrix about the origin of the crystallographic axes, its orientation for the different crystal systems is as follows:

In uniaxial crystals (tetragonal, trigonal, and hexagonal) the optic axis for light of all wave-lengths coincides with the c crystallographic axis.

In orthorhombic crystals the three axes of the indicatrix coincide with the three crystallographic axes, whatever the wave-length.

In monoclinic crystals one axis of the indicatrix, but any one, coincides with the b crystallographic axis, no matter what the wave-length. The other two indicatrix axes therefore lie in the ac plane (*i.e.* the symmetry plane in the holohedral class), but anywhere in that plane, their positions depending on the wave-length.

In triclinic crystals the orientation and dispersion may be in any direction whatever.

The above facts are presented graphically in Figs. 152 and 153, in which the crystals of the various systems are represented by simple prism and pinacoid forms, and also in Fig. 154, in which the orientation of the optic axial plane in biaxial crystals is shown by means of the stereographic projection.

From the foregoing it will be seen that the dispersion of any

ORIENTATION AND DISPERSION OF INDICATRIX   121

anisotropic crystal is made up of one or more of the following elements of dispersion :
  (a) Dispersion of each principal index ;
  (b) Dispersion of the birefringence, i.e. of $(\gamma - \alpha)$ or $(\varepsilon - \omega)$;
  (c) Dispersion of the directions of the indicatrix axes.

For the whole range of the visible spectrum, (a) and (b) commonly

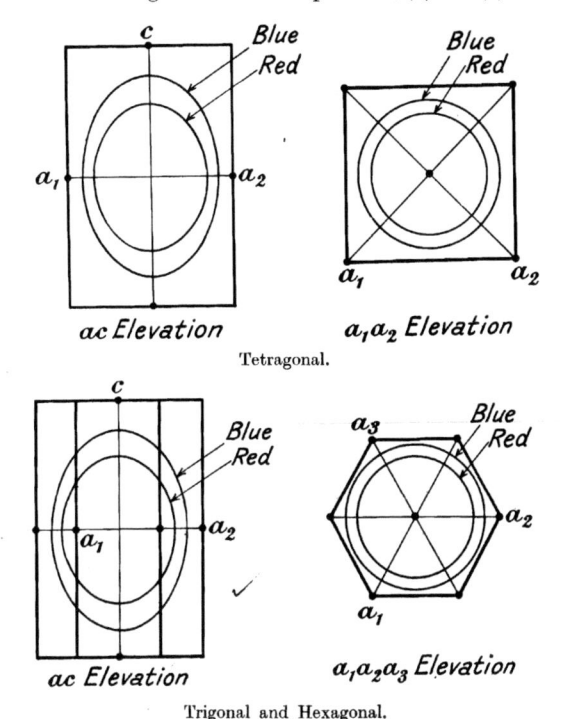

ac Elevation $a_1 a_2$ Elevation
Tetragonal.

ac Elevation $a_1 a_2 a_3$ Elevation
Trigonal and Hexagonal.

FIG. 152.—Orientation and Dispersion of the Indicatrix in Uniaxial Crystals.

affect only the second decimal place (at least in inorganic crystals), and (b) is usually much smaller than (a), whilst (c) only occurs in monoclinic and triclinic crystals. In all biaxial crystals, however, the disproportionate dispersion of α, β, and γ leads to dispersion of the optic axes which, though often very small, amounts in some cases to large angles, and even to a change in the orientation of the optic axial plane, quite apart from any shift of this plane due to (c).

122   THE OPTICAL PROPERTIES OF CRYSTALS

Thus the orthorhombic mineral brookite has optic axial angles which are considerable in both red and blue light, but the optic axial plane is (001) in the first case and (010) in the second. For an intermediate yellowish-green light the optic axial angle is zero, *i.e.* the crystal is uniaxial for this wave-length. This phenomenon

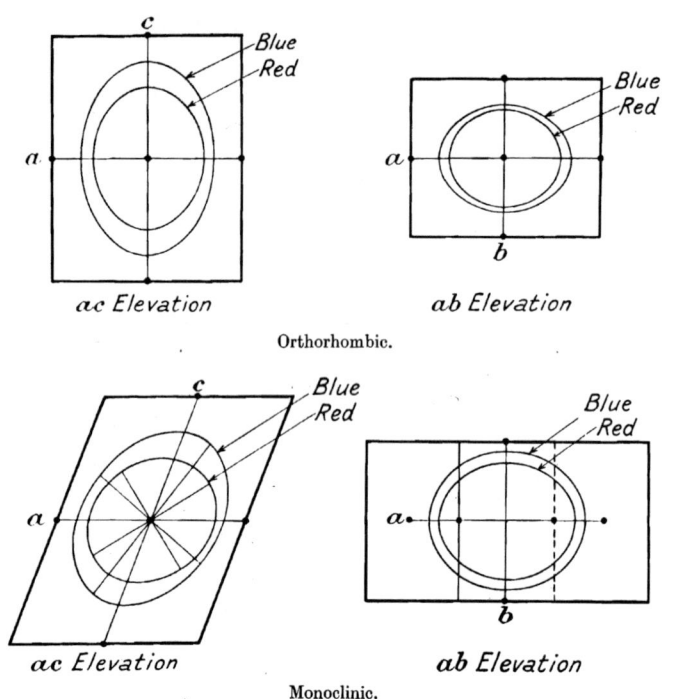

ac Elevation   *ab Elevation*
Orthorhombic.

ac Elevation   *ab Elevation*
Monoclinic.

FIG. 153.—Orientation and Dispersion of the Indicatrix in Biaxial Crystals.
(A diagram for the triclinic case is not given because of the completely unsymmetrical orientation of the indicatrix in crystals of this system.)

is known as *crossed-axial-plane dispersion*. Another effect of dispersion observed in some crystals is *change in the optical sign*. For example, the uniaxial mineral torbernite (air-dried to $8H_2O$), is positive in red light, negative in blue light, whilst for green light (5150 Å) it is isotropic. Another example is afforded by *trans*-stilbene (monoclinic), which is positive for wave-lengths above 4070 Å and negative for shorter wave-lengths.

Alteration in the temperature of a crystal is also accompanied by dispersion of the indicatrix, but this is too small to affect microscopic observations made within the range of laboratory temperatures.

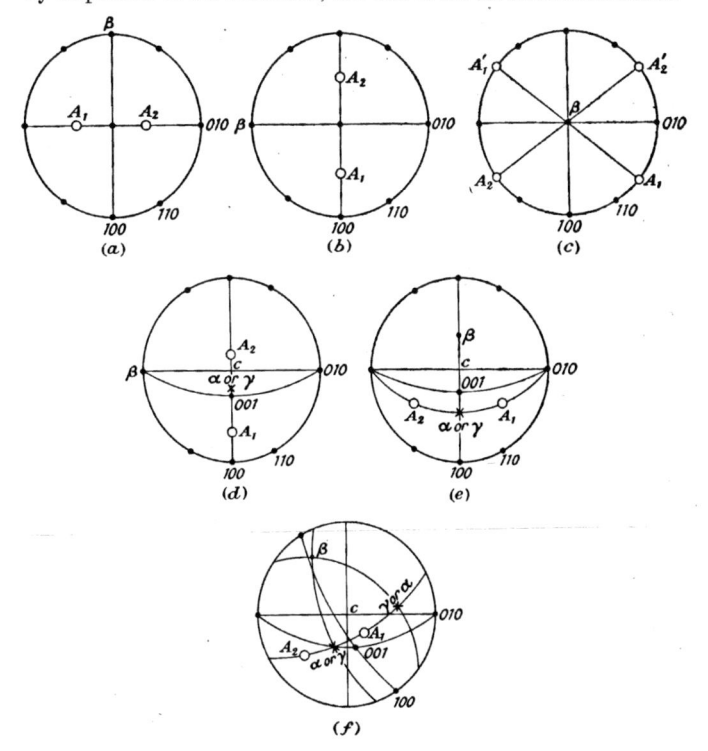

Fig. 154.—Optic Orientation in Biaxial Crystals represented by means of the Stereographic Projection.

(a), (b), (c) Orthorhombic, optic axial plane parallel to (100), (010), and (001) respectively. (d), (e) Monoclinic, optic axial plane parallel to (010) and normal to (010) respectively. (f) Triclinic optic axial plane in random orientation.

Dispersion is treated in further detail in Chapter VIII.

Optical Sections under the Microscope. So far, the optical properties of crystals have been described mainly with reference to the passage of light at perpendicular incidence through parallel-sided crystal plates. These conditions commonly prevail when well-formed crystals are examined under the microscope, because each crystal tends to lie upon a broad face, which is usually opposed by a similar parallel face. Frequently, however, crystalline particles of quite irregular shape have to be examined. It might appear at first as though observations on such material would be valueless, because the light would be deviated irregularly on entering the particles, and the observed phenomena would therefore correspond to no one section of the indicatrix. This situation is avoided, however, by mounting the particles in a liquid of the same, or nearly the same, refractive index (see p. 192, Chap. VI), so that the light passes through them with but little deviation, no matter what their shape. The optical properties of any crystal or crystal fragment mounted in this way may thus be taken as corresponding to the section of its indicatrix which is normal to the path of the light.

In the sequel, when a crystal under the microscope is stated to be, or to present, a certain section, its section normal to the optical axis of the microscope is referred to. The statement does not necessarily imply that the crystal is in the form of a plate parallel to the section named.

Absorbing Crystals. Crystals which appear coloured in transmitted light owe their colour to the fact that they absorb one or more bands of wave-lengths in the visible spectrum (*selective absorption*). The width of such an *absorption band* increases in general with the thickness of the crystal, and its edges are diffuse, *i.e.* the limits of absorption are not sharply defined. Absorption consists in a progressive diminution in the intensity of the light as it travels through the crystal, the light energy which thus disappears being converted into heat. The diminution in intensity for a given wave-length follows *Lambert's exponential law*:

$$I_x = I_0 e^{-\kappa x}$$

where I_0 is the intensity of the light on entering the crystal, I_x is its intensity after passing through a thickness x, and κ is the *absorption coefficient* for the wave-length considered. The meaning of the law is that the fraction of the incident intensity which is absorbed per unit thickness is a constant (κ). Further, since the amplitude, A, of the light vibration is measured by the square root of the intensity, it follows that the decrease in the amplitude is given by the equation:

$$A_x = A_0 e^{-\frac{\kappa x}{2}}$$

where A_x, A_0 are the amplitudes corresponding to I_x, I_0 respectively.

ABSORBING CRYSTALS

The treatment given in the preceding sections of this chapter strictly refers only to the propagation of light in cases in which absorption does not occur, or in other words it is concerned with the behaviour of light of wave-lengths to which the crystal is *transparent*. It has also been assumed that the crystal is not optically active (see p. 26). (Actually no crystal is *completely* transparent for any wave-length, there being always some diminution in intensity as the light penetrates more and more deeply, but for small thicknesses this diminution is negligible for wave-lengths which are not in the neighbourhood of an absorption band.)

The optical properties of coloured crystals for the wave-lengths absorbed can in a limited degree be described by means of an indicatrix consisting of two surfaces, one giving the refractive index and the other the absorption coefficient for different directions of propagation. In biaxial crystals these surfaces cannot be represented geometrically since they are defined by complex quantities (involving $\sqrt{-1}$), but they make geometrically representable traces on the optical symmetry planes, *i.e.* on the *ac*, *ab*, and *bc* planes in an orthorhombic crystal and the *ac* plane in a monoclinic crystal. These traces are, however, not ellipses. The physical meaning of this is that for general directions of propagation, the light is resolved into two elliptically polarised components of equal ellipticity, the major axes of the two ellipses being at right angles to one another. Only beams travelling in an optical symmetry plane are resolved into two plane polarised components as in transparent crystals.

Another difference from transparent biaxial crystals is that light travelling along the optic axes is resolved into two beams vibrating in planes at right angles to one another and having the same refractive index, but different absorption coefficients. Making a (usually) small angle with each of these axes, there are two " rotation " axes along which the light travels as a single circularly polarised beam, the sense of rotation being right-handed along one of these axes, and left-handed along the other.

In uniaxial absorbing crystals, the refractive index and absorption coefficient are represented by rotation surfaces, which are symmetrical about a single optic axis parallel to *c*, like the indicatrix of a non-absorbing uniaxial crystal. These rotation surfaces differ but little from ellipsoids of rotation, and the optic axis is a true axis of no double refraction; light travelling along this direction can vibrate in any plane. For other directions of propagation, the light is resolved into two plane polarised waves, vibrating at right angles to one another, as in a non-absorbing crystal.

Absorbing crystals which transmit light freely in sections of the order of thickness suitable for study under the microscope (say 0·02 to 0·03 mm.) have comparatively small absorption coefficients, and most coloured crystals, other than those of substances which are practically opaque, fall in this category. When the absorption coefficient is small, the shape of the refractive index surface of the indicatrix corresponds very closely to an ellipsoid, and in biaxial crystals the directions of the two " rotation " axes mentioned above merge almost completely into a single direction having the properties of an ordinary optic axis. The two elliptically polarised waves travelling along other directions also become to all intents and purposes plane polarised. As far as their practical examination under the microscope is concerned therefore, such crystals can be treated like non-absorbing crystals as regards most of the optical phenomena which they exhibit. There is, however, one important property in which absorbing crystals differ from non-absorbing ones, which is of practical importance. This is the property of *pleochroism*, or variation of colour with the direction of vibration of the light, which is due to the fact that the selective absorption is different for different directions of vibration. This property will be dealt with in Chapter VII.

Optical Activity. In the preceding pages of this chapter, no account has been taken of the so-called *optically active* substances, which *rotate* the plane of vibration of polarised light. It is not necessary to deal with these substances at any length within the compass of the subject of this book, because their rotatory property has so far proved to be of little importance in ordinary work with the polarising microscope. For the sake of completeness, however, their special characters will be briefly described.

As already stated in Chapter III (p. 64) crystals of these substances belong to the eleven enantiomorphous symmetry classes, in which planes of symmetry are absent. Not all substances crystallising in these classes are optically active, however (for example, barium nitrate, class 23, is not), for in addition to the point group symmetry elements, a necessary condition is that the crystal structure must possess screw axes which are parallel to one another and in the same sense. The screws or spirals may be right-handed or left-handed, and this accounts for the existence of two forms of the crystal, one rotating the plane of vibration to the right and the other to the left. This property of the structure may be related to the fact, originally discovered by Reush, that when plane polarised light is directed normally at a pile of thin mica plates, each plate being turned through a small angle with respect to the one preceding it, the plane of polarisation is rotated in the sense of the twist given to the pile.

Optically active substances are of two types, namely (i) those which are active in the fused or dissolved state as well as in the crystalline state ; (ii) those which are active in the crystalline state only. In substances of type (i) the molecules are themselves devoid of planes of symmetry, and can therefore exist in two different forms that are mirror images of one another. To this property of the molecules is attributed the existence of dextro- and lævo-rotatory forms of the substance in the liquid state and in solution. Such molecules form spiral arrangements in the crystal structure, and so in the crystalline state the substance will have a rotatory power which is the resultant of that due to the molecules and that due to the spiral arrangement. Usually these two factors reinforce one another, the effect of the latter being in most cases the greater. All chemists are familiar with many substances of this type, e.g. tartaric acid, many of the sugars, and some inorganic co-ordination compounds.

In substances of type (ii), the structural units have not the symmetry limitations possessed by those of substances of type (i), and do not of themselves give rise to optical activity. Accordingly as soon as the crystal structure is destroyed by fusion or solution, the activity disappears. Some of the better known examples of this type are given in the following table :

	System	Class	Rotation in Na light for a plate 1 mm. thick
$NaClO_3$	Cubic	23	3° 8'
$KLiSO_4$	Hexagonal	6	3° 26'
α-Quartz	Trigonal	32	21° 43'
$MgSO_4,7H_2O$	Orthorhombic	222	2° 36'

In all the above cases, except the cubic $NaClO_3$, the rotations refer to plates cut perpendicular to an optic axis, since only on such plates can the optical activity of anisotropic crystals be studied uncomplicated by double refraction. In optically active cubic crystals, the rotation is the same, no matter what section be examined.

The rotation produced by a given optically active crystal, liquid, or solution is directly proportional to its thickness in the direction in which the light travels, and also varies considerably with the wave-length of the latter, being greater the shorter the wave-length. For example, the rotations given by a plate of α-quartz 1 mm. thick cut perpendicular to the optic axis are as follows for light of different wave-lengths :

Wave-length (Å)	Rotation
6708	16° 32'
5896, 5890	21° 43'
4861	32° 46'
3969	51° 11'

Between crossed Nicols, optically active crystals give rise to phenomena which it is convenient to deal with here, although the reader may not fully understand them until Chapters V to VIII have been read. Isotropic crystals and sections of anisotropic crystals normal to an optic axis, which normally appear dark between crossed Nicols in parallel light, are illuminated under these circumstances, if they are optically active. If monochromatic light is used, darkness can be achieved by rotating one of the Nicols through the rotation angle for the section, but in white light, owing to the dispersion of the rotation angle (see values for α-quartz above), this will not be possible, and moreover, the section will appear coloured, the tint varying with the thickness of the section, and being composed of those portions of the spectrum which are vibrating in planes most nearly parallel to the vibration plane of the Nicol nearer to the eye. In convergent light, the centre of the interference figure which is given if the substance is anisotropic, will be illuminated in the same way, since it is the focal point of light which has travelled along the optic axis.

The above phenomena are clearly shown by plates of α-quartz cut perpendicular to the optic axis and 3 mm. or more thick. It would appear, however, that most optically active crystals of a size suitable for examination under the polarising microscope do not offer a sufficient thickness to the light to give an appreciable rotation. The authors have examined with this instrument small crystals of sucrose, tartaric acid, magnesium sulphate ($MgSO_4, 7H_2O$), and sodium periodate ($NaIO_4, 3H_2O$), all of which are optically active, and in no case could the activity be recognised with certainty. Even the last-named compound, which in optic axial sections 1 mm. thick has a rotation for sodium light of 23° 18′, gave optic axial interference figures which were apparently quite normal. Nevertheless, the possibility of a substance with a particularly large rotation cropping up in work with the microscope must not be overlooked. Such substances are known, for example cinnabar, HgS, which in optic axial section 1 mm. thick has a rotation of 325° for red light, and a number of inorganic co-ordination compounds of cobalt and chromium.

Fresnel proposed a theory of optical activity, which was based on the elementary principle that any simple harmonic motion along a straight line, such as that associated with a light vibration vector, may be resolved into two circular motions having opposite directions. The theory, which was developed with particular reference to uniaxial crystals like quartz, was to the effect that when a ray of plane polarised light enters the crystal parallel to the optic axis, it is resolved into two circularly polarised rays rotating in opposite directions and travelling at different speeds. The resultant of these two circular motions is equivalent to a plane polarised ray, the direction of vibration of which changes continuously with passage through the crystal, and in the same direction as the direction of rotation of the faster of the two circular components. Since the two components travel at different speeds, they must correspond to different refractive indices, and therefore, according to the theory, the ordinary and extraordinary wave surfaces do not touch at the point of emergence of the optic axis as they do in a non-optically active crystal (see Figs. 129 and 130, p. 92). This has been confirmed by experiment, which shows that near these points the outer surface is slightly bulged, and the inner one slightly flattened, from their normal shapes. The refractive indices for the two circular components in right-handed quartz are given below.

Wave-length (Å).	$n.$	
	Right-handed Component.	Left-handed Component.
3960	1·55810	1·55821
7600	1·53914	1·53920

It will be noted that the difference between the two indices affects only the fourth decimal place.

The difference between the refractive indices (Δn), of the two circular components travelling along the optic axis is related to the rotation angle ϕ by the equation:

$$\phi = \frac{\pi}{\lambda}.d.\Delta n$$

where λ is the wave-length and d the thickness of the section (both in the same units), ϕ being expressed in radians. It is evident that for crystals which have rotation angles of the same order as that shown by quartz, or less, Δn will be extremely small. Applying the equation to the case of cinnabar mentioned above, and taking a mean value of 7000 Å for the wave-length of red light, Δn works out to 0·0013. Thus even in the case of this large rotation the deviation of the wave surfaces from the standard shape is quite small and for most purposes can be ignored.

In optically active biaxial crystals the inner and outer wave surfaces do not quite touch at the singular points where the secondary optic axes emerge (*e.g.* at R in Fig. 143, p. 111), but again the deviation from the normal form is very small.

For further information on the subject of optical rotation in crystals, the reader is referred to the comprehensive works cited at the end of this chapter, and for further treatment of the theory, to advanced treatises on optics. An excellent account is given in the *Fundamentals of Physical Optics* by F. A. Jenkins and H. E. White (McGraw-Hill, 1937).

The Relationship between Optical Properties and Crystal Structure. When a non-conducting medium is between the plates of a charged condenser, the atoms of which it is composed are distorted by the electric field, this effect being termed *polarisation*.*
Their electrons, more particularly the outer ones which are the most weakly held, are attracted towards the positive plate, whilst their nuclei suffer a displacement in the direction of the negative plate. In other words, the " centres of gravity " of the positive and negative components of each atom are displaced in relation to one another, so that it becomes an electrical dipole. The effect of

* It is unfortunate that the same term is used to describe the vectorial properties of light vibrations, but the context will always show the sense in which it is used.

OPTICAL PROPERTIES AND CRYSTAL STRUCTURE

passing a beam of electromagnetic radiation through a transparent medium is very similar, for the atoms are polarised by the field due to the electric vector of the waves. This field differs, however, from that just considered in that it alternates in direction with the frequency of the radiation used. For the visible region of the spectrum this means that the field is changing its direction between 4 and $7 \cdot 5 \times 10^{14}$ times every second, and the atomic nuclei are too heavy to respond to such rapid oscillations; only the very much lighter electrons can do so. Polarisation by visible light therefore consists solely of the distortion of the electron atmospheres of the atoms, and for this reason is called *electron polarisation*.

The polarisation of an atom may be expressed quantitatively by the moment of the dipole induced by the electric field acting on it, *i.e.* by the product $s.q$, where s is the distance of separation of the charges $+ q$ and $- q$ at the two poles. This moment is directly proportional to the field strength E (if this is not too great), and for fields produced by electromagnetic radiation it also depends on the vibration frequency. Thus for a given frequency:

$$s.q = \alpha.E,$$

where α is a constant characteristic of the atom, called its *polarisability*. Inspection of the above equation shows that α has the dimensions of a volume. This constant is numerically equal to the moment of the dipole induced by a field of unit strength. For the case of polarisation by visible light (electron polarisation) it is evidently a measure of the ease with which the electron atmosphere of the atom may be deformed.

In what follows the terms " polarisation " and " polarisability " are to be understood as referring to electron polarisation.

Polarisability is a property which varies widely for atoms of different elements. As a rough generalisation it may be said to increase with increase in the size and looseness of binding of the electron atmosphere. Further, atoms of the same element in different states of combination have different polarisabilities, since the state of combination influences the strength of binding of the electron atmosphere. For example the chloride ion (I) is more polarisable than the chlorine atom in gaseous hydrogen chloride (II) in which the hydrogen-chlorine linkage is mainly covalent.

$$:\!\ddot{\text{Cl}}\!:^{-} \qquad\qquad \text{H}\!:\!\ddot{\text{Cl}}\!:$$

(I) (II)

132   THE OPTICAL PROPERTIES OF CRYSTALS

In (I) the nucleus of the atom has to hold an additional electron (as compared with the neutral atom) and this results in a general loosening of the electron system. In (II), which may be regarded as the result of a combination of (I) with a proton (hydrogen nucleus), the effect of the extra electron has been largely neutralised by the proton, and the electron system is therefore more tightly bound and less polarisable.

The refractive index of a medium is closely related to the polarisabilities of its atoms. The polarising action of light waves upon an atom is accompanied by a reaction of the atom upon the light waves, the result of which is that their velocity of propagation is reduced and the more so the greater the polarisability of the atom.

Consider the simple case of a medium consisting of single atoms, *e.g.* one of the inert gases. Suppose also that the medium is in the liquid or gaseous state, so that the atoms are distributed in a random manner. It may then be shown that the relationship between the refractive index, n, of the medium and the polarisability, α, of its atoms is as follows:

$$(n^2 - 1)\overline{E} = 4\pi N'\alpha E' = P \quad . \quad . \quad . \quad (i)$$

where $\overline{E}$ is the mean intensity of the electric field due to the light waves, N' is the number of atoms in unit volume, E' is the intensity of the *local* field acting on each atom, and P is the *total polarisation* for unit volume. In general E' is greater than $\overline{E}$ because the polarisation of every atom is increased by the field set up by the dipoles of the atoms surrounding it, including those not in its immediate neighbourhood. The magnitude of this effect may be deduced as follows. It is supposed that a sphere is described anywhere in the medium, the sphere being of such a size that it contains a large number of atoms but is small in comparison with the volume of the medium. If now it be imagined that the atoms inside the sphere are removed, the surface of the cavity which is left carries positive and negative charges respectively on the two parts into which it is divided by the diametral plane normal to the electric vector; for the one part of the surface is formed by the positive ends of atomic dipoles, and the other of negative ends. It may be shown that these charges create a field at the centre of the sphere, having an intensity equal to $P/3$, *i.e.* to $(n^2 - 1)\overline{E}/3$. This intensity is independent of the direction of the electric vector, since the atoms are distributed at random. Moreover it may be shown that since the atoms removed from the sphere were also distributed at random,

OPTICAL PROPERTIES AND CRYSTAL STRUCTURE 133

their average effect at the centre was zero. Thus the local field acting on the atom which occupied the centre of the sphere, and therefore on every atom, since the sphere was located anywhere in the medium, is given by:

$$E' = \overline{E} + P/3 = \overline{E} + \frac{(n^2 - 1)\overline{E}}{3}.$$

Substituting this in equation (i) we obtain:

$$(n^2 - 1)\overline{E} = 4\pi N'\alpha \left[\overline{E} + \frac{(n^2 - 1)\overline{E}}{3} \right]$$

whence
$$(n^2 - 1) = 4\pi N'\alpha \left(\frac{n^2 + 2}{3} \right)$$

or
$$\frac{(n^2 - 1)}{(n^2 + 2)} = \frac{4}{3}\pi N'\alpha \quad . \quad . \quad . \quad . \quad (ii)$$

When the atoms in a medium are not distributed at random, but have a particular spatial relationship to one another, it is no longer

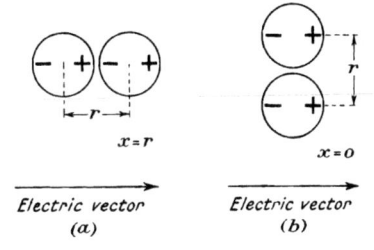

FIG. 155.

necessarily true that the effects on any atom of its nearest neighbours (*i.e.* those inside the sphere in the above argument) cancel out. In general the surrounding dipoles will act so as either to increase or decrease the polarisation of the atom from its "normal" value (that for a random arrangement of neighbours) according to the direction of the electric vector. The reason for this may be understood by considering the simple case (Fig. 155) of two similar atoms in close proximity to one another, and widely separated from other atoms. In (a) the direction of the electric vector is parallel to the line joining the atomic centres, and the positive pole on one atom is adjacent to the negative pole on the other. By induction, therefore, the moment of each dipole will be increased. In (b) the electric vector is at right angles to the line joining the atomic centres, charges of similar sign are adjacent, and so the moments will be reduced. In effect then, the *polarisability* of the atoms is no longer constant, but depends on the direction of the electric vector. Calling this varying polarisability α', it may be shown from electrostatic theory that:

$$\alpha' = \frac{\alpha}{1 - \beta\alpha} \quad . \quad . \quad . \quad (iii)$$

where α is as before the polarisability for a random arrangement of neighbours, and β is given by the expression:

$$\beta = \frac{3x^2 - r^2}{r^5} \quad . \quad . \quad . \quad (iv)$$

r being the distance between the atomic centres, and x the component of this distance in the direction of the electric vector.

The table below shows the result of applying equation (iii) to the cases in Fig. 155 for different values of r, assuming that $\alpha = 1 \times 10^{-24}$ c.c., which corresponds to a medium degree of polarisability.

r (Å).	$\alpha'(\times 10^{24})$.	
	Case (a) Electric Vector Parallel to Line joining Atoms. ($x = r$)	Case (b) Electric Vector Normal to Line joining Atoms. ($x = 0$)
2·0	1·33	0·89
3·0	1·08	0·96
4·0	1·03	0·98
5·0	1·02	0·99

It will be seen that although the effect is very marked at short distances, it falls rapidly as the distance increases and already at $r = 5 \cdot 0$ Å, α' is practically equal to α.

Although in any real medium there would be a number of neighbouring atoms to take into account instead of one only, the simple case just considered will suffice to show that it is the *nearest* neighbours of an atom which have by far the greatest influence on the relationship between its polarisation and the direction of the electric vector. Thus atoms joined by primary bonds in a polyatomic molecule or ion have a much greater effect on one another in this respect than have the atoms of different neighbouring molecules or ions, for these are at appreciably greater distances. Any such atom group therefore which has an elongated shape (like the pair of atoms in Fig. 155, regarded as a diatomic molecule), or in general a non-equant shape, will have as a whole a much greater polarisability when the direction of the electric vector is parallel to a long axis than when it is parallel to a shorter axis.

Before dealing with crystals, it will be convenient to consider liquid and gaseous media in which the molecules are polyatomic. Although, as just shown, the polarisability of such molecules may vary with the direction of the electric vector, this will not result in any corresponding variation in the polarisability of the medium as a whole if it is liquid or gaseous, because the molecules are orientated at random,* and so their directional effects cancel each other. In such a medium therefore, the molecules may be regarded as possessing statistically a mean polarisability $\bar{\alpha}_M$ which is independent of the direction of the electric vector. This may be inserted in equation (ii) in place of α to give the relationship between the molecular polarisability and the refractive index of the medium, N' now being the number of molecules in unit volume. If now both sides of the resulting expression are multiplied by the quotient M/D, where M is the molecular weight and D the density, the factor N' which varies with both temperature and pressure is eliminated, because $N'M/D = N_0$, the Avogadro Number, *i.e.* the number of molecules in a gram molecule. Thus we obtain:

$$\frac{(n^2-1)}{(n^2+2)} \cdot \frac{M}{D} = \frac{4}{3}\pi N_0 \bar{\alpha}_M = [R_M] . \qquad . \qquad (v)$$

where the quantity $[R_M]$ is called the *molecular refractivity*. This expression, known from its originators as the Lorentz-Lorenz equation, holds with considerable accuracy in the sense that for a given frequency of light, $[R_M]$ calculated from the observed values of n and D is independent of temperature. For a given frequency it is a characteristic constant of the substance, and is a measure of

* Modern studies of the liquid state indicate that in many cases regular arrangements of the molecules exist over small elements of volume (containing perhaps a few hundred molecules). The arrangement in the liquid as a whole is, however, random in the sense that along any direction all orientations of these molecular groups will occur with equal frequency.

its mean molecular polarisability for this frequency since it is equal to $\bar{\alpha}_M$ multiplied by the constant quantity $\frac{4}{3}\pi N_0$.

If the molecules each consist of a atoms having a polarisability α_1, b atoms having a polarisability α_2, and so on, then, assuming that the molecular polarisability is a simple additive function of the atomic polarisabilities, we may put equation (v) in the following form:

$$\frac{(n^2-1)}{(n^2+2)} \cdot \frac{M}{D} = \frac{4}{3}\pi N_0 \bar{\alpha}_M$$

$$= \frac{4}{3}\pi N_0 (a\alpha_1 + b\alpha_2 + \ldots)$$

$$= a\left(\frac{4}{3}\pi N_0 \alpha_1\right) + b\left(\frac{4}{3}\pi N_0 \alpha_2\right) + \ldots$$

$$= a[R_1] + b[R_2] + \ldots$$

$$= [R_M] \qquad \qquad \qquad (vi)$$

where $[R_1]$, $[R_2]$, etc. are the *atomic refractivities*, or *refraction equivalents*, of the constituent atoms. It is important to note, however, that these terms represent the refractivities of the atoms *in the particular state of combination in which they exist in the molecule*, and not those which the free atoms would possess.

Equation (vi) holds with a very fair degree of accuracy for liquid organic compounds containing only carbon, hydrogen, oxygen, or the halogens, provided that certain constitutive complications associated with the presence of conjugated double bonds are absent (see below). That is to say, it is possible to attribute constant refraction equivalents to these elements, the simple addition of which gives the molecular refractivities of the compounds to within about 1% of the observed values, if allowances are made for the presence of double and triple bonds and the different states of combination of oxygen. The principles by which these refraction equivalents have been arrived at may be illustrated as follows.

The starting point for the determination of all refraction equivalents has been the value for the CH_2 group, and this may be determined from the difference in molecular refractivity between adjacent members in homologous series of saturated compounds, such as the paraffins, alcohols, or fatty acids (*e.g.* $C_6H_{14} - C_5H_{12} = CH_2$). If now n times the value for the CH_2 group be subtracted from the

OPTICAL PROPERTIES AND CRYSTAL STRUCTURE 137

molecular refractivity of the paraffin having the formula C_nH_{2n+2}, the value for two hydrogen atoms is obtained:

$$C_nH_{2n+2} - n(CH_2) = 2H,$$

and by subtracting this from the equivalent for the CH_2 group, the equivalent for singly linked carbon is obtained. It is easy to see that by an extension of this procedure the refraction equivalents of other elements may be determined, and the following table gives the usually accepted values (D line) for the elements mentioned above (F. Eisenlohr, 1910).

Refraction Equivalents (D line)

Carbon	2·418
Hydrogen	1·100
Oxygen (in CO group)	2·211
Oxygen (in ethers)	1·643
Oxygen (in OH group)	1·525
Chlorine	5·967
Bromine	8·865
Iodine	13·900
Double bond	1·733
Triple bond	2·398

The following examples show the order of the agreement that is obtained between the observed value of the molecular refractivity and that calculated from the above equivalents.

	Benzene, C_6H_6.	Cyclo-hexane, C_6H_{12}.	Ethyl Ether, $(C_2H_5)_2O$.	Acetone, $(CH_3)_2CO$.	Chloroform, $CHCl_3$.
$[R_M]_D$ obs.	26·15	27·71	22·48	16·15	21·40
$[R_M]_D$ calc.	26·31	27·67	22·31	16·07	21·42

Refraction equivalents have also been calculated for nitrogen, but these vary so markedly for different states of combination of this element that about thirty different values are required to cover all possible cases. The same applies to the element sulphur.

Open-chain compounds containing a system of conjugated double bonds, *i.e.* double bonds alternating with single bonds, have molecular refractivities which are very appreciably different from, and usually higher than, the calculated values. For example $[R_M]_D$ for *iso*-diallyl, $CH_3-CH=CH-CH=CH-CH_3$, is 1·76 units higher than the calculated value. There is much chemical evidence that such systems contain loosely bound electrons, and this would account for their abnormally high polarisability. If the conjugated double-bond system forms a closed ring, as in benzene above, the anomaly disappears, but if it is partly in a closed ring and partly in a side chain, as in styrene ⟨◯⟩—CH=CH$_2$, the refractivity is again higher than the calculated value.

Turning now to the case of crystals, we must first consider their broad division into isotropic singly refracting media (crystals of the cubic system), and anisotropic, doubly refracting media (crystals of all other systems). In crystals of the cubic system, the high order of symmetry with which their atoms are arranged results in cancellation of the directional effect of neighbouring dipoles, just as does the random arrangement in a liquid. Such crystals are therefore equally polarisable in all directions, or in other words light of a given frequency passes through them with the same velocity in all directions, and so they are singly refracting. Thus for example, in the sodium chloride structure (Fig. 3, p. 5), each sodium ion is surrounded by six chloride ions situated at the corners of a regular octahedron, and by calculations based on a similar principle to that applied in the case in Fig. 155, it may be shown that for any direction of the electric vector the directional effects of these six chloride ions just cancel one another. Other symmetrical arrangements of the atoms found in crystals of the cubic system (e.g. the regular tetrahedron of atoms around each atom in diamond; Fig. 6, p. 6) may also be shown to result in cancellation of the directional effect.

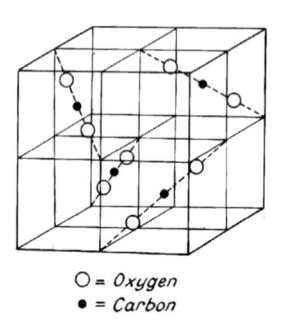

○ = Oxygen
● = Carbon

FIG. 156.—Structure of Solid Carbon Dioxide.

Some cubic crystals are built up of polyatomic molecules or ions which are themselves not equally polarisable in all directions. In such cases the medium as a whole is isotropic because the cubic symmetry requires the anisotropic atom groups to be so orientated with respect to one another that mutual compensation of their directional effects results. For example, in solid carbon dioxide the O=C=O molecules are linear and are thus much more polarisable for light vibrating parallel to the long axis than for light

OPTICAL PROPERTIES AND CRYSTAL STRUCTURE 139

vibrating transversely. They are, however, inclined to one another in the structure, their long axes being parallel to the trigonal axes of the cubic cell as shown in Fig. 156, and the medium as a whole is isotropic.

In anisotropic crystals the spacing of the structural units along at least one of the principal axes of the structure is different from that along the others, and if the units are polyatomic molecules or ions of non-equant shape, these are orientated in such a way that there is a resultant directional effect. The polarisability of such a medium is therefore different in different directions, with this qualification that there must always be one direction of propagation in uniaxial crystals, or two directions of propagation in biaxial crystals, for which the medium is equally polarisable for all directions of vibration. These directions of propagation are the optic axes, and light travelling along them may be thought of as encountering a " cubic " aspect of the structure (*e.g.* consider light travelling along the vertical (and optic) axis in the mercuric iodide structure shown in Fig. 5, Chap. I). If, however, the light travels in any other direction, it is resolved into two polarised components vibrating respectively along the directions of maximum and minimum polarisability for the section of the crystal normal to the path of the light, these two directions being always at right angles to one another. It is in this way that the phenomenon of double refraction arises.

The molecular refractivity of a cubic crystal may be calculated from its refractive index by means of the left-hand side of equation (v), and has the same significance as the molecular refractivity of other isotropic media, namely that it affords a measure of the polarisability or refracting power of the molecules (or the units arbitrarily selected as such*) in the crystal. In the case of anisotropic crystals we may similarly calculate molecular refractivities from each of the principal refractive indices (though the equation is not strictly applicable to such media), *i.e.* from ε and ω for uniaxial crystals, and α, β, and γ for biaxial crystals, and thus obtain constants expressing the refracting power of the molecules for light vibrating along the principal optical directions. The relationship

---

* In ionic crystals or other crystals in which discrete molecules cannot be distinguished, a group of atoms must be arbitrarily chosen as the " molecule ", usually that corresponding to the conventional formula, *e.g.* NaCl for the sodium chloride crystal.

between these crystal refractivities and the refractivities of the constituent atoms, though essentially additive in character, is in general not a simple one. Even for crystals of the cubic system it is not possible to draw up a list of constant refraction equivalents by the mere addition of which we can calculate the refractivities of a wide range of compounds, at least with any approach to accuracy.

Consider for example the case of the alkali halides. These are simple ionic compounds crystallising in the cubic system. They have the NaCl structure (Fig. 3, p. 5), with the exception of the chloride, bromide, and iodide of cæsium, which have a body-centred type of structure in which each ion of a given kind is surrounded by eight of the opposite kind situated at the corners of a cube. In all such structures directional effects cancel out as we have already seen, and we might therefore have expected to find that the molecular refractivity would be a simple additive function of the refractivities of the individual ions. The molecular refractivities of these salts, and of all other true salts, *in dilute aqueous solution* do in fact exhibit such a relationship, that is to say the difference between the refractivities of two salts MX and MY, or MX and NX, remains constant when the nature of the common ion is varied. (The refractivity due to the salt is obtained by subtracting the refractivity of the water, assumed to be the same as that of pure water, from that of the solution.) From such data refraction equivalents can be attributed to the individual ions, if the value for one ion can be determined. Wasastjerna * attempted to solve this problem by assuming that the refractivities of strong acids in dilute solution were the same as the refractivities of their anions, arguing that the simple hydrogen ion, the proton, is unpolarisable. This assumption, however, leads to results for the refractivities of the anion which are too low, because the intense positive field of the proton markedly reduces the polarisability of the water molecules with which it is immediately associated. Therefore the allowance made for the refractivity of the solvent as though it were pure water is too great. Fajans and Joos † attacked the problem later by a more satisfactory method based on certain qualitative relationships between the refractivities of the inert gases (which can be directly determined), and those of ions having the same electronic structures as these gases. By a process of successive approximations, for the details of which the

* *Z. physikal. Chem.*, 1922, **101**, 193.   † *Z. Physik.*, 1924, **23**, 1.

original paper must be consulted, the refractivity of the sodium ion was estimated to be 0·5, and from this as the starting-point refraction equivalents for other ions were calculated.

	F^- $R=2\cdot50$ $r=1\cdot36$	Δ %	Cl^- $R=9\cdot00$ $r=1\cdot81$	Δ %	Br^- $R=12\cdot67$ $r=1\cdot95$	Δ %	I^- $R=19\cdot24$ $r=2\cdot16$	Δ %
Li^+ $R=0\cdot20$ $r=0\cdot6$	⊕⊖ 2·70 *2·337*	−14·6	⊕⊖ 9·20 *7·587*	−21·2	⊕⊖ 12·87 *10·560*	−21·9	⊕⊖ 19·44 *15·978*	−21·7
Na^+ $R=0\cdot50$ $r=0\cdot95$	⊕⊖ 3·00 *3·016*	+0·65	⊕⊖ 9·50 *8·517*	−11·5	⊕⊖ 13·17 *11·560*	−13·9	⊕⊖ 19·74 *17·073*	−15·6
K^+ $R=2\cdot23$ $r=1\cdot33$	⊕⊖ 4·73 *5·162*	+8·3	⊕⊖ 11·23 *10·846*	−3·5	⊕⊖ 14·90 *13·983*	−6·6	⊕⊖ 21·47 *19·754*	−8·7
Rb^+ $R=3\cdot58$ $r=1\cdot48$	⊕⊖ 6·08 *6·740*	+9·8	⊕⊖ 12·58 *12·549*	−0·24	⊕⊖ 16·25 *15·778*	−3·0	⊕⊖ 22·82 *21·708*	−5·1
Cs^+ $R=6\cdot24$ $r=1\cdot69$	⊕⊖ 8·74 *9·507*	+8·1	⊕⊖ 15·24 *15·572*	+2·1	⊕⊖ 18·91 *18·949*	+0·3	⊕⊖ 25·48 *25·143*	−1·35

R = ionic refraction equivalent (Fajans and Joos).
r = effective ionic radius in Å.
Calculated values are shown in plain figures; observed values in *italics*.

Fig. 157.—Molecular Refractivities of Crystals of the Alkali Halides.

In Fig. 157 the observed refractivities of crystals of the alkali halides are compared with those calculated by means of the ionic equivalents of Fajans and Joos. Only in the cases of NaF, KCl, RbCl, CsCl, RbBr, CsBr, and CsI (shaded in the table) is there agreement to within so generous a margin as 5% of the observed value. The other salts show marked discrepancies, both positive and negative, which amount to over 21% for the chloride, bromide, and iodide of lithium.

The explanation is as follows. Since the ionic equivalents are calculated from the refractivities of salts in dilute solution, they refer to ions that are widely separated from one another and are thus unaffected by each others' electric fields. In fact they may be regarded as being approximately the same as the refractivities which the ions would have if they were in the free " gaseous " condition. In the crystal on the other hand the ions are packed closely together, and this affects the binding of their electron atmospheres in the following ways. In the first place, the electron atmosphere of each anion is attracted by the positive fields of the surrounding cations. This tightens the binding of the electrons of the anion and makes it less polarisable. Secondly the negative fields of the anions surrounding each cation loosen the binding of its electrons and make it more polarisable. In most cases the effect of the former of these two opposing tendencies far exceeds that of the latter, so that the medium as a whole has a lower polarisability than that corresponding to the simple sum of the equivalents for the " gaseous " ions. The reasons for this are firstly that the number of electrons in an anion exceeds the number of positive charges on its nucleus, so that the electrons are comparatively weakly held and are easily affected by the positive fields of the surrounding cations. In a cation on the other hand the positive nuclear charges are in excess and so the electrons are firmly held and are not easily affected by the fields of neighbouring anions. A further important result of this strong binding is that in general cations are smaller than anions and their fields are therefore more intense. This too acts so as to enhance the effect of the cations on the anions.

The influence of ionic size is brought out in the figure, which gives the effective radii which the ions have in the crystals, and shows for each salt the relative sizes of cation and anion by means of a small drawing. It will be seen that in those cases in which the size of the cation is approaching that of the anion, there is approximate agreement between the calculated and observed values of the molecular refractivity, $i.e.$ the two opposing tendencies counterbalance one another. As the size of the cation increases further in relation to that of the anion, the second tendency (loosening of the binding of the electrons of the cation) becomes the more important, and where the cation is approximately the same size, or bigger, as it is in RbF and CsF, positive deviations from the calculated refractivities appear. At the other extreme are the lithium salts. As will be seen, the lithium ion is extremely small and thus possesses an intense positive field. The result of this appears in the large negative deviations from the calculated values, particularly when the ion is in partnership with the large and easily deformable chloride, bromide, and iodide ions.

It should be mentioned that it is not strictly correct to compare in this way the chloride, bromide, and iodide of cæsium with the other salts, since, as already stated, they have a different crystal structure. Whereas in these cæsium salts each ion is surrounded by eight of the opposite kind, in the remainder the corresponding number is six, so that the fields acting on the ions in the two cases are not quite comparable. The difference involved, however, is not such as to affect the purely qualitative relationships traced above.

Similar comparisons between the observed molecular refractivities and those calculated from the equivalents for the " gaseous " ions have been made in other series of simple ionic compounds, and the deviations observed can be explained in the same way. For example, the cubic oxides, sulphides, selenides, and tellurides of the metals magnesium, calcium, strontium, and barium [*] mostly show very large negative deviations because not only are all the cations smaller than the anions, but the double charge on the cations results in their positive fields being very intense.

In passing it is of interest to compare the refraction equivalents for Cl^-, Br^-, and I^- given in Fig. 157 with those for the same atoms linked covalently in organic compounds (see table on p. 137). In the ionised state the atoms are far more polarisable, for the reason already given for chlorine at the beginning of this section.

In the case of anisotropic crystals, any attempt to relate quantitatively the molecular refractivities to the refractive properties of the atoms must of course take into account the directional effects of neighbouring dipoles on one another. In 1924, W. L. Bragg [†] showed how this may be done by calculating approximately the refractive indices of calcite and aragonite (the trigonal and orthorhombic modifications respectively of $CaCO_3$), on the basis of their internal structures as determined by X-rays, and the refraction equivalents of Wasastjerna.[‡] The essential features of the structures of these crystals are shown in Fig. 158. In both there are alternate layers of Ca^{++} ions and $CO_3^=$ ions, the latter having an equilateral triangular configuration of oxygen atoms with the carbon at the centre. The planes of these $CO_3^=$

[*] Haase, Z. Krist., 1927, 65, 509; Fajans, ibid., 1928, 66, 325.

[†] Proc. Roy. Soc., 1924, 105, 370. Calculations of refractive indices, based on the crystal structure and the dispersion curves of the rays, were originally made by Born (Dynamik der Kristallgitter, 1915), but Bragg's treatment is simpler and easier to apply.

[‡] loc. cit.

ions are parallel (or in aragonite, nearly parallel) to the layers in which they are situated, as shown in the figure. In calcite the axis normal to these layers (the vertical axis in the figure) is the trigonal axis of the crystal, whilst in the orthorhombic aragonite it is pseudo-hexagonal, the spacing of the structural units along the two horizontal axes, a and b, being very nearly the same. Both crystals

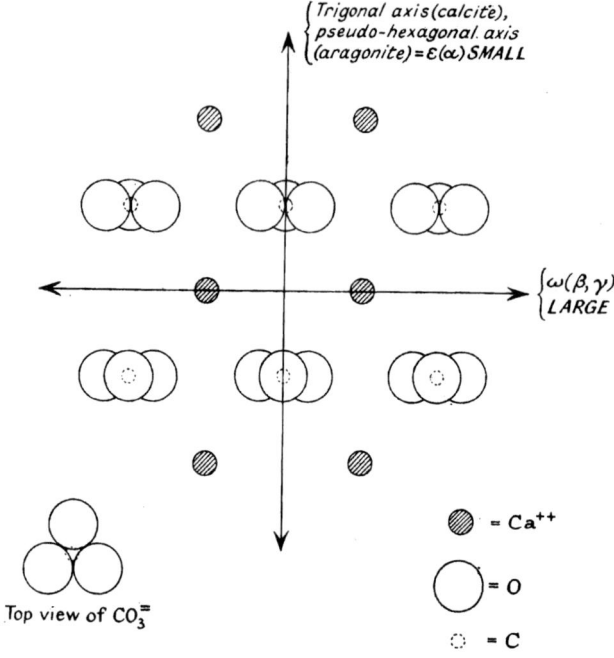

FIG. 158.—Elevation showing General Arrangement of Ions in Calcite and Aragonite (not to scale).

show strong negative birefringence, the largest indices, ω in calcite, and β and γ (which are nearly equal) in aragonite, being those for light vibrating horizontally, i.e. parallel to the planes of the $CO_3^=$ ions (see figure).

The first step in Bragg's calculation was to assign values to the mean refractivities of the calcium ion and the carbon and oxygen atoms, that is to say, the refractivities which these units would have *in the crystal, if directional effects were absent.* There was

OPTICAL PROPERTIES AND CRYSTAL STRUCTURE   145

necessarily some uncertainty about this, because, as shown above this cannot be done accurately even for simple cubic crystals composed of monatomic ions, where the directional effects cancel out. However, the main object of the calculation was to account quantitatively for the *difference* between the principal refractive indices, i.e. the birefringence, and for that purpose the exact values selected for the refractivities were not of great importance. The ones actually chosen were Ca^{++} 1·99; O 3·30; C nil. The first of these was Wasastjerna's estimate for the free calcium ion (probably too high; Fajans and Joos later gave 1·33), and the others were based on a consideration of his values for molecules and ions containing three oxygen atoms joined to a small central atom (Al_2O_3, SO_3, NO_3^-, $CO_3^=$), from which it appeared that the refractivity was due almost entirely to the oxygen atoms; the contribution of the carbon was therefore neglected.

From these values, which are at least of the right order, it will be seen that since there are three oxygen atoms to every calcium ion, the latter contributes only a small fraction of the total refractivity, namely about $\dfrac{1\cdot 99}{1\cdot 99 + 3 \times 3\cdot 30} = 0\cdot 167$. It can also contribute very little to the birefringence, because its nearest neighbours in both calcite and aragonite are oxygen atoms arranged in a roughly equant group around it, so that its polarisation will be practically the same for different directions of the electric vector; in fact Bragg assumed that it behaved as though in an isotropic medium.

The greater part of the refractivity and virtually the whole of the birefringence is therefore to be attributed to the carbonate ions, and their contributions were determined as follows. In the first calculations made it was assumed for simplicity that the polarisation of each oxygen atom was affected only by that of the oxygen atoms in the same ion (these being the nearest exert the strongest influence) and not at all by the polarisation in neighbouring ions. For aragonite this meant that no distinction was made between the β and γ indices, i.e. that the crystal was treated as pseudo-uniaxial with an ordinary index equal to $\dfrac{\beta + \gamma}{2}$, because the assumption ignored the cause of the biaxial character of this modification, namely the difference in the spacing of the $CO_3^=$ ions along the two horizontal axes. It was therefore only necessary to consider the following two cases: (*a*) that for light vibrating normally to the planes of the CO_3^{--} ions, corresponding to the indices ε in calcite and α in aragonite; (b) that for light vibrating parallel to these planes, corresponding to ω in calcite and $\dfrac{\beta+\gamma}{2}$ in aragonite (see Fig. 159). In (a) the moments of the three oxygen atoms are lowered by mutual induction below the value which they would have if the atoms were isolated, and the extent of this lowering was calculated from the relative positions of the atoms and their distance from one another (given by the X-ray analysis of the structures) by an application of the electrostatic principles underlying equations (iii) and (iv) above, extended so as to take account of the effect of

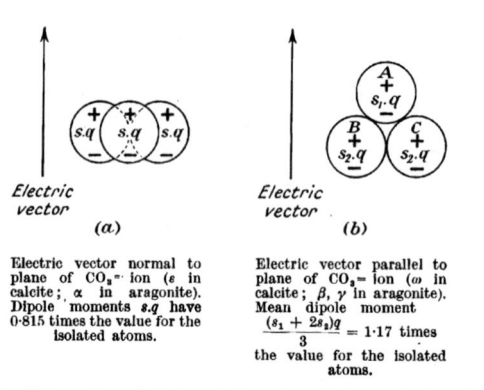

(a) Electric vector normal to plane of CO_3^{--} ion (ε in calcite; α in aragonite). Dipole moments $s.q$ have 0·815 times the value for the isolated atoms.

(b) Electric vector parallel to plane of CO_3^{--} ion (ω in calcite; β, γ in aragonite). Mean dipole moment $\dfrac{(s_1+2s_2)q}{3} = 1\cdot17$ times the value for the isolated atoms.

Fig. 159.—Polarisation of Carbonate Ions (considering only the effect of oxygen atoms in the same ion on one another).

more than one neighbouring dipole. The lowering was found to correspond to a factor of 0·815. In (b), for the direction of the electric vector drawn in the figure, the dipoles of atoms B and C ($s_2.q$) have a depressant effect on each other, but reinforce that of atom A ($s_1.q$), and calculation on the above basis showed that there was a resultant gain in the mean moment of the atoms, $\dfrac{(s_1+2s_2)q}{3}$, amounting to a factor of 1·17. It could also be shown that this same result was obtained for all other directions of the electric vector in the plane of the atoms.

From these two factors and the refraction equivalents of the atoms the principal refractive indices can be calculated by means

OPTICAL PROPERTIES AND CRYSTAL STRUCTURE 147

of equation (vi). Thus for calcite the extraordinary index is given by:

$$\frac{n_e^2 - 1}{n_e^2 + 2} = \frac{D}{M}\{[R_{Ca^{++}}] + (0\cdot815 \times 3[R_O])\}$$

$$= \frac{1}{36\cdot13}\{1\cdot99 + (0\cdot815 \times 3 \times 3\cdot30)\},$$

and the ordinary index by:

$$\frac{n_\omega^2 - 1}{n_\omega^2 + 2} = \frac{1}{36\cdot13}\{1\cdot99 + (1\cdot17 \times 3 \times 3\cdot30)\}.$$

The results of these calculations and the corresponding ones for aragonite (which differ only in the value of $D/M, = 1/34\cdot01$) are compared with the observed values below. It will be seen that they correspond to birefringences of the right order.

		Calc.	Obsd.
Calcite:	ε	1·468	1·486
	ω	1·676	1·658
Aragonite:	α	1·503	1·530
	$\frac{\beta+\gamma}{2}$	1·730	$\begin{cases}\beta & 1\cdot681 \\ \gamma & 1\cdot686\end{cases}$

Bragg also made a more accurate calculation in which account was taken of the effect on each oxygen atom of the dipoles of all other oxygen atoms within a sphere of radius 6 Å drawn around it. In a sphere of this size are included not only the two other atoms in the same ion but 40 more in calcite, and 48 in aragonite, belonging to neighbouring ions.

This refinement permitted the β and γ indices in aragonite to be calculated separately, since it took cognizance of the different spacings of the CO_3^- ions along the two horizontal axes. The result of this treatment was to improve on the whole the agreement between the observed and calculated birefringences as shown below, though it is to be noted that the β value obtained for aragonite was larger than the γ value. Exact agreement was not to be expected since the calculations involved a number of simplifying assumptions, and there was also the doubt already mentioned regarding the values of the atomic refraction equivalents. The agreement is, however, sufficiently striking to show that the theoretical basis of the calculations is essentially sound.

		Calcd.	Obsd.
Calcite :	ε	1·488	1·486
	ω	1·631	1·658
Aragonite :	α	1·538	1·530
	β	1·694	1·681
	γ	1·680	1·686

Other anhydrous carbonates and also some nitrates have structures similar to those of calcite or aragonite, in that they contain parallel planar $CO_3^=$, or NO_3^-, ions, and they also show strong negative birefringence. Examples are $MgCO_3$, $ZnCO_3$, $FeCO_3$, $MnCO_3$, $MgCa(CO_3)_2$, $CdCO_3$, and $NaNO_3$, which are trigonal like calcite, and $SrCO_3$, $PbCO_3$, and KNO_3, which are orthorhombic like aragonite. By a general survey of the refractivities of these compounds, Bragg showed that their birefringences could be explained in the same way, i.e. as due mainly to the effect of the polarised oxygen atoms on one another. Also from the fact that the birefringence of $NaNO_3$ (0·250) is considerably greater than that of calcite (0·172) he inferred that the oxygen atoms are closer together in the NO_3^- ion than in the $CO_3^=$ ion, since this would enhance the inductive effect between them. This has been confirmed by later work, which shows that the O–O distances are 2·10 Å in NO_3^- and 2·18 Å in $CO_3^=$.

Qualitatively the relation, demonstrated by Bragg's work, between the birefringence and optical sign of carbonates and nitrates and the structure and arrangement of their plate-like anions may be extended to all other crystals containing plate-like atom groups (molecules or ions) arranged parallel or approximately parallel to one another, provided that these groups are solely or mainly responsible for the refractivity of the crystal. In all such cases the crystal may be expected to show strong negative birefringence with large indices for light vibrating parallel to the planes of the plates (or a single large one if it is uniaxial), and a small index for light vibrating normally to this direction.

By a further extension of this principle the influence of other shapes and arrangements of atom groups on the optical properties of crystals of which they are sole or most refractive building units may be deduced. Thus for example atom groups which are linear (rod-shaped) are more polarisable for light vibrating parallel to their length than for light vibrating in any direction across the length. If such groups are arranged parallel to one another in a crystal, it will show strong positive birefringence with a large index for light vibrating parallel to the length of the rods and a small index (uniaxial case) or smaller indices (biaxial case) for light vibrating normally to this direction, depending on the cross-sectional shape of the rods and the equality or otherwise of their side-to-side spacing in different directions.

An arrangement of rod-shaped atom groups inclined in different directions but lying in parallel planes is equivalent to one consisting of parallel plate-like groups, and results therefore in negative birefringence. On the other hand, plate-like groups with their planes parallel to a single direction but not to one another constitute an arrangement equivalent to that of parallel rods, and so give rise to positive birefringence.

Finally crystals consisting of atom groups, which are equant in shape or approximately so, or of arrangements of plates or rods inclined in a number of different directions so as wholly or largely to neutralise each other's directional effects, must be isotropic or weakly birefringent.

The above qualitative relationships have been stated more particularly in regard to crystals in which the atom groups are small discrete assemblages. They can, however, be extended to endless structures, as has been done by W. A. Wooster in a survey of the optical data of inorganic substances.* Thus structures consisting of endless chains of atoms arranged parallel to one another (*e.g.* selenium) are positively birefringent, since they are optically equivalent to an arrangement of parallel rods, whilst layer structures (*e.g.* mercuric iodide, Fig. 5, p. 6) are negatively birefringent (except in certain cases when they contain OH groups), since they are equivalent to arrangements of parallel plates. (The presence of OH groups in some cases leads to the formation of hydroxyl bonds between the layers, and this may increase the polarisability for the vibration direction normal to the layers to such a degree that the medium becomes positive, as for example in gibbsite, $Al(OH)_3$.) Lastly structures consisting of a three-dimensional network of symmetrical groups are isotropic or have a weak birefringence. Examples are the various modifications of silica (quartz, cristobalite, and tridymite) which consist of networks of tetrahedral SiO_4 groups in which each oxygen atom is shared between two groups.

* *Z. Krist.*, 1931, **80**, 495.

150   THE OPTICAL PROPERTIES OF CRYSTALS

Table I (pp. 152, 153) summarises all the above conclusions and gives further examples.

So far in this treatment no account has been taken of crystals containing rotating atom groups. As previously stated (Chap. I, p. 12), when the free rotation of a non-equant group occurs, its symmetry is raised statistically, and, as a frequent result, the symmetry of the crystal structure is also raised. This may alter the character of the birefringence or destroy it altogether. For example, taking the cases quoted in Chap. I, in the normal paraffins above the member $C_{20}H_{42}$ the long molecules are packed parallel to one another. At ordinary temperatures the molecules are not rotating, and since their cross-sectional shape is roughly oval (owing to the zig-zag form of the carbon chains), their side-to-side spacing is different in different directions and the crystal has orthorhombic symmetry. At temperatures just below the melting point, the molecules are rotating freely about their long axes, and their cross-sectional shape has therefore become effectively circular. Their side-to-side packing then adopts the regular hexagonal pattern and the crystal becomes uniaxial, though there is of course no alteration in the optical sign which remains positive. In ammonium nitrate above 125°, both the tetrahedral NH_4^+ ions and the planar NO_3^- ions are rotating freely in all directions about their centres, and so are effectively spherical in shape. The crystal therefore adopts cubic symmetry and is isotropic. At lower temperatures the rotation of the ions is not complete, and so their shapes, and in particular that of the planar nitrate ion, compel the crystal structure to take up less symmetrical patterns which are birefringent. In $NaNO_3$ on the other hand, although rotation of the nitrate ions sets in at 180°, it is restricted to rotation about the trigonal axis normal to the planes of the ions, and this has no effect on the symmetry of the structure or the character of the birefringence which remains negative uniaxial.

The relationship between the optical properties of crystals and the shapes and arrangements of their building units plays a most important part in crystal structure analysis. The birefringence and optical sign of a crystal, and the vibration directions corresponding to its maximum and minimum refractive indices, will usually indicate the general plan of arrangement of its building units where these are molecules or polyatomic ions, or will at least enable certain arrangements to be rejected as impossible, thus facilitating the interpretation of the X-ray diffraction pattern. In most cases the purely qualitative relationships as set out in Table I suffice for this purpose, but in some investigations the actual refractive indices to be expected for different possible structures have been calculated approximately and compared with the observed values. The method of Bragg, described above, can be used, though according to St. B. Hendricks and W. E. Deming * it is only appropriate for comparatively simple ionic crystals for which it can be assumed that the ions are approximately spherically symmetrical. For molecular crystals and crystals containing polyatomic ions, these authors consider that the treatment should be based on the polarisability of the molecule or ion as a whole rather than on those of the individual atoms as in Bragg's method. In illustration they show how the refractive indices of a number of oxalate crystals may be accounted for by the anisotropy and orientation of the oxalate group.

A very simple method of estimating approximately the refractive indices in a molecular crystal was used by E. G. Cox † in connexion with a study of the crystal structure of benzene which belongs to the orthorhombic system. The method takes no account of interactions between the molecules, $i.e.$ it is assumed that conclusions drawn about the polarisabilities of the molecules in the liquid and gaseous states are applicable to the solid state—a reasonable approximation to make for a molecular substance, particularly if the molecules are non-polar as in benzene. It depends on the fact that when light is passed through a gaseous medium composed of anisotropic molecules, a small fraction of the light scattered sideways vibrates horizontally ($i.e.$ parallel to the direction of the incident beam), the rest vibrating vertically as it would all do if the molecules were isotropic. The ratio of the intensities of these two polarised components, $I_{horiz.}/I_{vert.}$, called the depolarisation factor, ρ, is related to the principal polarisabilities A and C of an axially symmetrical molecule like benzene as follows:

$$\rho = \frac{I_{horiz.}}{I_{vert.}} = \frac{2(C - A)^2}{4C^2 + 9A^2 + 2AC},$$

A being the polarisability normal to the symmetry axis, $i.e.$ in the plane of the ring for the benzene molecule, and C that parallel to the axis. For benzene vapour, ρ lies between 0·045 and 0·066, whence C/A from the above equation $\approx$ 0·50.

* $Z.\ Krist.$, 1935, **91**, 290.   † Private communication.

TABLE I

RELATION BETWEEN THE OPTICAL PROPERTIES OF CRYSTALS AND THE SHAPES AND ARRANGEMENTS OF THEIR CONSTITUENT ATOM GROUPS (NON-ROTATING)

SHAPE AND ARRANGEMENT OF ATOM GROUPS.	OPTICAL PROPERTIES.	EXAMPLES.

EQUANT. Different arrangements but shape of groups favours cubic or near-cubic lattice. High birefringence cannot arise whatever the arrangement.	ISOTROPIC OR WEAKLY BIREFRINGENT.	*Hexamine*, $(CH_2)_6N_4$. Cubic. Structure of molecule based on a regular tetrahedron of nitrogen atoms.
		Anhydrous salts containing regular tetrahedral anions of type MX_4 *and monatomic cations, e.g.—*
		  K_2SO_4, $B^* = 0.004$; K_2CrO_4, $B = 0.02$; $NaClO_4$, $B = 0.012$; Mg_2SiO_4, $B = 0.015$; $K_2[Zn(CN)_4]$, cubic; $K[BF_4]$, $B = 0.001$.
		Anhydrous salts containing regular octahedral anions of type MX_6 *and monatomic cations, e.g.—*
		  $K_2[PtCl_6]$, cubic; $K_4[CdCl_6]$, $B = 0.0001$.
		Three-dimensional networks of equant atom groups, e.g. forms of silica (tetrahedral SiO_4 groups):
		  quartz, $B = 0.009$; cristobalite, cubic; tridymite, $B = 0.004$.

ROD-SHAPED: (a) All parallel to one another.	MARKED *POSITIVE* BIREFRINGENCE. RODS PARALLEL TO VIBRATION DIRECTION OF GREATEST INDEX.	*Paraffins, normal aliphatic alcohols and acids.* *Sodium azide*, NaN_3. Linear N_3^- ion, $B = 0.36$. *Methyl bixin*, $CH_3O.CO.CH:\{CH.C(CH):CH.CH:\}_4.CH.CO.OCH_3$, $B = 1.1$ ($\gamma_D = 2.6$). (High γ value and birefringence due to conjugated double-bond system.) *Selenium.* Endless spiral chains of Se atoms. $B =$ high. *Natural Fibres.* Long molecules parallel, or approximately so, to fibre axis: *e.g.* ramie, $B = 0.06$.
(b) Inclined in different directions but lying in parallel planes.	LARGE *NEGATIVE* BIREFRINGENCE. PLANES NORMAL TO VIBRATION DIRECTION OF LEAST INDEX.	*Octadecyl ammonium chloride*, $C_{18}H_{37}.H_3NCl$. *Potassium azide*, KN_3. Linear N_3^- ion, $B = 0.14$ approx.

(c) Inclined in different directions and not lying in parallel planes.	ISOTROPIC OR WEAKLY BIREFRINGENT.	Solid carbon dioxide. Cubic. Linear $O=C=O$ molecules arranged parallel to the four trigonal axes of the cubic cell.
PLATE-LIKE: (a) All parallel to one another.	LARGE *NEGATIVE* BIREFRINGENCE, VIBRATION DIRECTION OF LEAST INDEX NORMAL TO PLATES.	Anhydrous salts containing planar, or low pyramidal, anions of type MO_3 arranged parallel to one another, and monatomic cations, e.g.—(Planar anions) $CaCO_3$, B = 0·172 ; $NaNO_3$, B = 0·250. (Low pyramidal anions) $KClO_3$, B = 0·115. *Layer structures, e.g.*—HgI_2 (tetragonal form), B = 0·293. *Organic plate-like molecules, e.g.*—Cyanuric triazide, $C_3N_3(N_3)_3$, B = 0·48 ; s-Triphenylbenzene, $(C_6H_5)_3C_6H_3$, B = 0·35.
(b) Plates parallel to a single direction but not to one another.	LARGE *POSITIVE* BIREFRINGENCE. PLATES PARALLEL TO VIBRATION DIRECTION OF GREATEST INDEX.	*Rhombic sulphur.* S_8 rings parallel to c axis but not to one another. B = 0·29. *Benzene.* C_6H_6 rings parallel to b axis (orthorhombic) but in two sets approximately at right angles to one another, B = high. *Urea.* Planar $(NH_2)_2CO$ molecules parallel to c axis (tetragonal) but in two sets with their planes at right angles. B = 0·12.
(c) Inclined in different directions and not parallel to any single direction.	ISOTROPIC OR WEAKLY BIREFRINGENT.	*Resorcinol, m-$C_6H_4(OH)_2$*, B = 0·049.† $Ba(NO_3)_2$; $Pb(NO_3)_2$. Cubic. Planar NO_3^- ions arranged normally to the four trigonal axes of the cubic cell.

* B = birefringence for Na light, i.e. $(\epsilon \sim \omega)$ or $(\gamma - \alpha)$.
† This is not a particularly low birefringence in comparison with the birefringences of most mineral substances, but it is far lower than the value which would result if the plate-like molecules were arranged parallel to one another (cf. for example the values for cyanuric triazide and triphenylbenzene above).

The molecular refractivity of liquid benzene, $[R_M]$ (eqn. v), is 26·2, and this may be regarded as an average of the components R_C and R_A, corresponding respectively to the refractivities for light vibrating parallel to the C and A directions in the molecule. In the general case where the molecule is not axially symmetrical it has three principal polarisabilities, A, B, and C, and

$$[R_M] = 1/3(R_A + R_B + R_C),$$

and so in the present case where $R_A = R_B$ we have that

$$[R_M] = 1/3(2R_A + R_C).$$

Also since $C/A \approx 0·50$, $R_C \approx 0·50 R_A$. From these relationships and the value of $[R_M]$, the values

$$R_A \approx 31·4 \quad \text{and} \quad R_C \approx 15·7$$

are obtained.

If the molecules were arranged parallel to one another in the crystal, its sign would be negative, and these values would give approximately the refractivities for its principal directions, *i.e.* R_A would be that for light vibrating parallel to the planes of the molecules, and R_C that for light vibrating normally to these planes. We should then have that:

$$\frac{M}{D} \cdot \frac{n_A^2 - 1}{n_A^2 + 2} \approx 31·4$$

and
$$\frac{M}{D} \frac{n_C^2 - 1}{n_C^2 + 2} \approx 15·7,$$

whence
$$n_A = \frac{\beta + \gamma}{2} \approx 1·78$$

and
$$n_C = \alpha \approx 1·34.$$

Actually, however, benzene is strongly positive, and this can only be accounted for by assuming that the molecules are all parallel or nearly parallel to an axis and distributed nearly isotropically about it. This does not effect the polarisability for light vibrating parallel to the planes of the molécules in the direction of this axis, but that for light vibrating at right angles to this direction will be given by the resultant of R_A and R_C for the different orientations of the planes about the axis. For an isotropic distribution, this corresponds to a refractive index of 1·52.* Thus for this case

* Calculated by the ellipse formula (p. 446), taking $\theta = 45°$.

the crystal should have $\gamma \approx 1{\cdot}78$, and $\alpha \approx \beta \approx 1{\cdot}52$. The values observed were:

$$\gamma > 1{\cdot}64$$
$$\beta - \alpha = 0{\cdot}004$$
$$\gamma - \alpha > 0{\cdot}12,$$

and a previous analysis of the structure by X-rays had shown that the molecules were in fact arranged in two sets with their planes approximately at right angles to one another and parallel to the b axis (see Fig. 8, p. 8).*

It should be noted that the preceding treatment in so far as its quantitative aspects are concerned refers only to transparent crystals. The relationship between the pleochroism (see p. 126) and structure of absorbing crystals is dealt with in Chapter VII.

SUMMARY

Crystals of the cubic system, in common with gases, liquids, and unstrained glasses, are *isotropic* to light, *i.e.* the velocity of propagation of light through them is independent of the direction of propagation.

Crystals of all other systems are *anisotropic*, the velocity of propagation of light through them being dependent on the direction of propagation. Also, such crystals are *birefringent*, or doubly refracting, *i.e.* they split up any ray of light which enters them into two rays, except along certain special directions called optic axes. These two rays travel through the crystal with *different velocities*, they are *polarised*, and they vibrate *in planes at right angles to one another*.

Anisotropic crystals are of two kinds, *uniaxial* and *biaxial*.

Uniaxial Crystals. Crystals belonging to the *tetragonal*, *trigonal*, and *hexagonal systems* are uniaxial, which means that they have one direction, and one only, in which double refraction does not occur. This direction is called the *optic axis*, and it is the same as the direction of the c crystallographic axis.

Rays which enter so as to travel through the crystal in any direction other than the optic axis are each split up into an *ordinary*, and an *extraordinary* ray. The velocity of an ordinary ray for a given crystal and wave-length is the same, no matter in what direction

* E. G. Cox, *Proc. Roy. Soc.*, A, 1932, **135**, 491.

the ray travels, and is equal to that of light of this wave-length travelling along the optic axis.

An ordinary ray always vibrates at right angles to the plane containing both the ray and the optic axis. Apart from the fact that it is completely polarised, an ordinary ray is similar to a ray of ordinary light travelling in an isotropic medium ; its wave front is normal to the ray direction, and therefore travels in that direction.

An extraordinary ray always vibrates in the plane containing both the ray and the optic axis. Its velocity varies with its direction, but is the same for all directions that are equally inclined to the optic axis. Its velocity differs most widely from that of the ordinary ray when it travels at right angles to the optic axis. In general, the wave front of an extraordinary ray is not normal to the ray direction, and therefore does not travel in that direction.

The refractive index of any ray of a given wave-length (ordinary or extraordinary) is the reciprocal of the velocity of its wave front, its velocity in a vacuum being taken as unity. For ordinary rays, and for extraordinary rays travelling at right angles to the optic axis, the wave-front velocity is the same as the ray velocity.

The optical properties of a uniaxial crystal for light of any given wave-length may be represented by a three-dimensional figure (the *indicatrix*). This is an ellipsoid of rotation, the axis of rotation coinciding with the optic axis of the crystal, and the length of the radius of its equatorial circular section represents the ordinary refractive index (ω), whilst the length of its semi-axis of rotation represents the extreme extraordinary refractive index (ε), *i.e.* the extraordinary refractive index differing most widely from ω. With the exception of the equatorial circular section, all sections of the indicatrix passing through its centre are ellipses, and *the semi-axes of any such section represent the vibration directions and refractive indices of the two wave fronts which can travel in a direction normal to the section.*

Uniaxial crystals in which $\varepsilon > \omega$ are said to be *positive*, those in which $\varepsilon < \omega$ are said to be *negative*.

Biaxial Crystals. Crystals of the *orthorhombic, monoclinic,* and *triclinic systems* are termed *biaxial*, since they have two directions, or optic axes, which correspond approximately to the optic axis in uniaxial crystals. These *two optic axes are directions of single wave front velocity*, and all rays, the wave fronts of which travel in these

SUMMARY

directions have the same refractive index. Nearly coincident with the optic axes are the directions of single-ray velocity, or secondary optic axes, but these are of little practical importance.

Rays which enter a biaxial crystal so as to travel through it in any direction other than an optic axis are split up into *two extraordinary rays*, which are similar in general properties to the extraordinary rays in uniaxial crystals.

The indicatrix for a biaxial crystal is a *triaxial ellipsoid*, the principal semi-axes of which have lengths corresponding respectively to the smallest refractive index (α), the greatest refractive index (γ), and the refractive index possessed by wave fronts travelling along an optic axis (β). One of the two rays travelling in any direction in the $\alpha\gamma$ plane vibrates normally to the plane with the index β, and the other vibrates in the plane with an index between α and γ, depending on its direction. Analogous properties characterise the $\alpha\beta$ and $\beta\gamma$ planes.

The indicatrix has *two circular sections*, to which the optic axes are normal, and the radius of these sections is β. All other sections passing through the centre are ellipses, and, as in the uniaxial indicatrix, *the semi-axes of any such section give the vibration directions and refractive indices of the two wave fronts which can travel in the direction normal to the section.*

Biaxial crystals are classed as *positive* or *negative* according as the acute angle between the optic axes is bisected by γ or α respectively.

If the biaxial indicatrix be constructed about the origin of the crystallographic axes, then in an orthorhombic crystal, the axes of the indicatrix and the crystallographic axes are coincident; in a monoclinic crystal, one of the axes of the indicatrix coincides with the b axis, the others lying anywhere in the ac plane; and in a triclinic crystal, there is no necessary relationship between the axes of the indicatrix and the crystallographic axes.

Dispersion of the Indicatrix. The size of the indicatrix for a given crystal increases with decrease in the wave-length of the light. In addition to (*a*) *the dispersion of each principal index*, there may be (*b*) *dispersion of the birefringence* (usually small), and (*c*) *dispersion of the directions of the indicatrix axes*. (*c*) occurs only in monoclinic and triclinic crystals.

Absorbing Crystals. The above refers strictly only to crystals which are completely transparent and are not optically active. In coloured anisotropic crystals the indicatrix for a wave-length which is absorbed is not exactly an ellipsoid, though it approximates very closely to one if the absorption is small, and this is so for most crystals which transmit light freely in thin sections.

The selective absorption of light in coloured anisotropic crystals varies in general with the vibration direction of the light, this phenomenon being termed *pleochroism*.

Optical Activity. Crystals which *rotate* the plane of vibration of polarised light have no planes of symmetry and their crystal structures possess screw axes which are parallel to one another and in the same sense. Their building units may or may not be devoid of planes of symmetry. In the former case the units contribute to the activity of the crystal, and the substance is active also in the liquid and dissolved states, whilst in the latter case the activity is a property of the crystalline state only.

The wave surfaces of such crystals differ from those of non-optically active crystals in that the inner and outer surfaces do not quite touch at the points of emergence of the optic axes.

Relation between Optical Properties and Crystal Structure. When light waves pass through a crystal its atoms become deformed (polarised) by the electric vector of the waves, *i.e.* they develop electrical dipoles, the strength of which depends on the intrinsic polarisability of the atoms and the intensity of the local field acting on them. The greater the polarisability of the atoms the more the velocity of the light is reduced.

The local field acting on each atom depends on (i) the mean intensity of the field due to the light waves, (ii) the field set up by the dipoles of surrounding atoms, including those not in its immediate neighbourhood. In a cubic crystal, the polarisation of a given atom is independent of the direction of the electric vector owing to the highly symmetrical arrangement of the surrounding atoms, and the medium is therefore isotropic. In crystals of other systems, the arrangement of the atoms is less symmetrical, and owing primarily to the inductive effects of its nearest neighbours, the polarisation of any atom varies with the direction of the electric vector. Such crystals are therefore anisotropic. In a molecule or polyatomic ion the mutual effect of the atoms composing it is such that its polarisation as a whole will be greater when the light is vibrating parallel to a long axis than when it is vibrating parallel to a shorter one. The optical properties of a crystal built up of such anisotropic

SUMMARY

atom groups will depend on their relative arrangement. Thus for example, rod-shaped molecules packed parallel to one another give a positively birefringent crystal, whilst structures consisting of parallel plate-like molecules are negatively birefringent. Structures composed of molecules of equant shape, or of molecules of non-equant shape if orientated in different directions, are isotropic or weakly birefringent.

RECOMMENDED FOR FURTHER READING

Mikroscopische Physiographie der Mineralien und Gesteine, H. Rosenbusch and E. A. Wülfing (Schweizerbart, 1924), Vol. 1.

Lehrbuch der Mineralogie, P. Niggli, Vol. 1 (Gebruder and Bornträger, 1924).

Manual of Petrographic Methods, A. Johannsen (McGraw-Hill, 1918).

Anleitung zu Optischen Untersuchungen mit der Polarisations Mikroskop, F. Rinne and M. Berek (Leipzig, 1934).

Physical Methods of Organic Chemistry, edited by A. Weissberger, Vol. 1, Chap. XI, by E. E. Jelley (Interscience, New York, 1945). [This chapter includes a useful section on the optics of strongly absorbing crystals.]

CHAPTER V

THE POLARISING MICROSCOPE

The principle of the compound microscope is shown in Fig. 160. Light is reflected by the adjustable *mirror* M through the convex lens, or lens system, C, called the *condenser*, which concentrates it upon the *object* X. The mirror is silvered on both sides, one of which is plane, and the other concave. The concave side is only used in certain special cases to be dealt with later.

A real enlarged image of the object is formed by the *objective* O and the lower or *field lens* of the *eyepiece*, or *ocular* E, in the focal plane F of the upper, or *eye lens* of E. The eye placed at Q where the bundles of parallel rays cross (the so-called *Ramsden disc*), sees an enlarged virtual image of the object, X', apparently at an infinite distance.

The magnification of the instrument is given by the formula—

$$\frac{g \cdot D_v}{f_o \cdot f_e}$$

—where f_o and f_e are the focal lengths of the objective and eyepiece respectively, g is the *optical tube length*, i.e. the distance between the adjacent focal planes of the objective and eyepiece, and D_v is the least distance of distinct vision which is conventionally given as 10" or 250 mm. for normal sight.* Thus for an objective of 6 mm. focal length, an eyepiece of 25 mm. focal length, and a tube length of 160 mm., the magnification will be—

$$\frac{160}{6} \times \frac{250}{25} = 267.$$

* The expression may be separated into two terms to express the respective contributions of the objective and the eyepiece to the total magnification. Two systems are, however, used : English microscope manufacturers usually give the magnifying power of the objective as g/f_o, and that of the eyepiece as D_v/f_e; Continental makers, however, often use D_v/f_o for the objective and g/f_e for the eyepiece.

THE POLARISING MICROSCOPE

This method of obtaining the magnification of a microscope should only be used when an approximate result is sufficient. For accurate work the magnification must be measured directly as described later (p. 224).

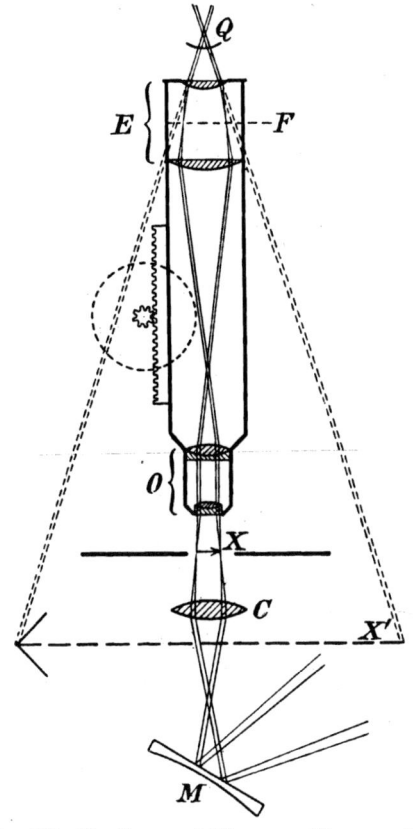

Fig. 160.—The Compound Microscope (diagrammatic).

A polarising microscope is essentially a compound microscope provided with polarising prisms (or discs of "Polaroid") above and below the stage, and some means of altering the orientation of the object (usually in one plane only) with reference to the planes of vibration of the prisms. Provision is also made for the introduction of auxiliary lenses and other accessories into the path of the light.

The arrangement of the main components in the usual type of polarising microscope is shown diagrammatically in Fig. 161. Light is reflected by the mirror M through the polarising or " Nicol " * prism P, whereby it is constrained to vibrate in one plane only. (The action of the Nicol prism will be explained shortly.) P is called the *polariser*, or *lower Nicol* of the instrument. It can be rotated about a vertical axis, the amount of rotation being read from divisions marked upon the tube (not shown) in which it is mounted. In most instruments it is positively located in the " 0 " position by some form of spring catch, and when in this position, its plane of vibration is parallel to one of the cross-wires H in the eyepiece E. When not required, *i.e.* when it is desired to examine an object in ordinary (unpolarised) light, it may be swung to one side as indicated by the arrow in the figure.

After leaving the polariser, the light passes through the condenser C, the purpose of which is to concentrate the light upon the object. To enable this to be done to the best advantage under all circumstances, the condenser is mounted so that it can be moved up and down, this being effected by a rack and pinion in the more expensive instruments. In many polarising microscopes, the condenser is mounted in the same tube as the polariser, so that adjusting the condenser involves moving the polariser too. The condenser can be swung out of the path of the light (see arrow in figure) with or without the polariser according as it is, or is not, attached to the latter.

Above the condenser and immediately below the object may be inserted the *converging lens*, or *converger* N, a plano-convex lens of extremely short focal length, the purpose of which is to converge the light passing through the object into a cone of wide angle. Strictly it is to be regarded as an additional part of the condenser to be brought into use whenever it is necessary to increase the convergence of the light, but in the examination of crystals its function is confined mainly to that of producing suitable illumination for obtaining interference figures (see Chap. VIII), and it is therefore convenient to treat and name it as a separate component. Apart from this it is required for observations of the structural details of an object when using an objective of high numerical aperture, in

* So called because devised by Nicol (1828).

THE POLARISING MICROSCOPE

order to produce a sufficiently wide cone of light to fill the objective and so obtain the maximum resolution of which it is capable (see below).

Some modern polarising microscopes are fitted with " Polaroid " discs instead of Nicol prisms. In those of this type manufactured by Messrs. Cooke, Troughton, & Simms (York, England) the discs are of sufficiently wide aperture to permit the use of a fixed condensing system for all purposes, as will be explained in a later section (p. 182). The " Polaroid " used is of a particularly high quality. It consists of a film of cellulose which has been stretched nearly to breaking point and treated in this condition with a suitable dye. The dye molecules become orientated and the resulting film is intensely pleochroic (see Chap. VII), so that one of the rays resulting from double refraction is almost completely absorbed. The film is mounted between optically worked glass flats. Other types of " Polaroid " which have been made consist of orientated crystals of a very strongly pleochroic substance in a plastic film. Herapathite (quinine trisulphate ditriiodide) is one substance which has been used for this purpose. They do not, however, give a sufficiently high degree of absorption of the one beam for use in a microscope.

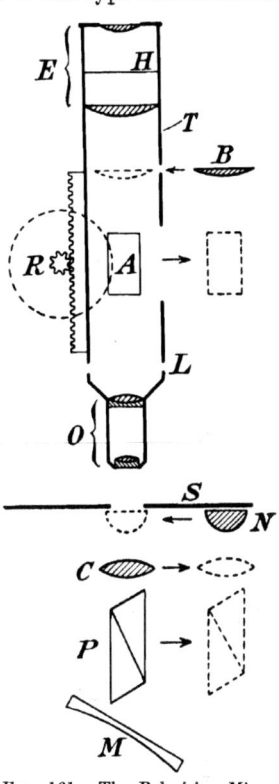

Fig. 161.—The Polarising Microscope (diagrammatic).

" Polaroid " is manufactured by the Polaroid Corporation in America, and the best quality equals or even exceeds the Nicol prism in performance.

The object, mounted on a glass slide, is laid on the *stage* S. This is circular and can rotate in its own plane, the amount of rotation being measurable by the degree divisions marked upon its edge.

Above the *objective* O is the *slot* L through the *body tube* T of the microscope proper. This slot is for the reception of the quartz wedge and other compensators, the purpose of which will appear in the following chapters. Next comes another Nicol prism A, termed the *analyser* or *upper Nicol*. (In some instruments, however, this is placed above the eyepiece, and not in the position shown in the figure; see next section.) Unlike the polariser P, this cannot usually be rotated about its vertical axis, but is set so that its plane of vibration is at right angles to that of the polariser when the latter is in the " 0 " position. When arranged thus in relation to one another, the Nicol prisms are said to be " crossed ", and will not permit light to reach the eyepiece E (so long as the medium between them is entirely isotropic), because the light transmitted by the polariser is completely extinguished by the analyser, according to the principle underlying Malus' experiment (p. 99). When the analyser is not required, it can be drawn to one side as indicated by the arrow in the figure.

Above the analyser, provision is made in most microscopes for inserting the so-called *Bertrand lens* B for the purpose of focussing the interference figures to which reference has already been made. For most other observations this is not required and is drawn to one side. Microscopes not fitted with a Bertrand lens may have for the same purpose another, called the *Becke lens*, which can be fitted above the eyepiece. With this, however, the cross-wires do not remain in view as they do when the Bertrand lens is inserted.

The *eyepiece* E is provided with *cross-wires*, or *cross-hairs* H, which, as has already been indicated, are parallel to the vibration planes of the Nicols when the latter are in the crossed position. The cross-wires are in the focal plane of the eye lens, *i.e.* the plane F in Fig. 160, and are therefore seen at the same time as the object.

The microscope is focussed by moving the tube T up or down by means of the rack and pinion mechanism R. In addition to this movement (*coarse focussing*), a much slower one may be produced by the mechanism for *fine focussing*, which is not shown in the figure.

Some of the important parts of the polarising microscope will now be described in further detail.

The Nicol Prism. Nicol's original design is as follows. A clear flawless cleavage rhombohedron of Iceland Spar (calcite), about three times as long as it is broad, is selected (ABCD, Fig. 162, *a*). By grinding and polishing, the inclination of the end faces,

THE NICOL PRISM

AB and CD, is altered to A'B and C'D, which make an angle of 68° with the edges BC and AD respectively, the angle made by the natural faces being 70° 53'. The prism is then cut diagonally along the plane A'C', perpendicular to the faces A'B and C'D, and parallel to the longer diagonals of these faces, *i.e.* to EF in Fig. 162, *b*, which shows a view of the upper face of the prism. After polishing the cut surfaces, the two halves are cemented together in their original positions with Canada balsam.

When a ray of ordinary light, PQ (Fig. 162, *a*), enters the prism in a direction parallel, or nearly parallel to its length, the extraordinary ray QRS has a refractive index very close to that of the balsam film A'C', and consequently passes through it and emerges at the other end of the prism. The ordinary ray QT, however, has a much higher index than the balsam film, and strikes it at an angle within the critical angle; it is therefore totally reflected at the film and passes out through the side of the prism along TU, where it may be absorbed by mounting the prism in a non-reflecting material, cork painted dead black being commonly used.

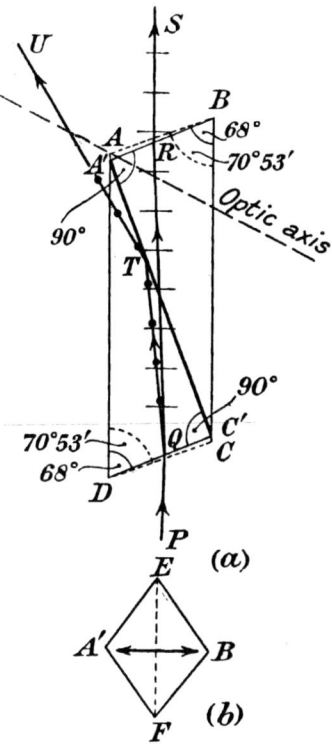

Fig. 162.—The Nicol Prism.

The only light transmitted by the prism, then, is the extraordinary ray, which, as already stated on p. 103, vibrates wholly in the principal section of the wave surface in which it lies, that is to say, in the plane containing both it and the optic axis. This plane (indicated by the strokes across QRS in Fig. 162, *a*, and by the double arrow in Fig. 162, *b*) is parallel to the shorter diagonal of the end face of the Nicol.

The inclination of the end faces in this type of prism results in some loss of light by reflection, and a marked lateral displacement of the image of any object viewed through it. This inclination, in conjunction with the fact that the optic axis of the calcite is not normal to the long axis of the prism (see figure), has the further effect that rays which are inclined to this latter axis (other than those which lie in the plane containing both this and the optic axis) emerge with vibration directions which are slightly different from that of the axial rays. In other words when a narrow cone of light is passed through the prism, the emergent rays are not all polarised in exactly the same direction. These characteristics of the prism are not such as seriously to affect its use as a polariser in the microscope (and it is frequently used for this purpose), but they are less acceptable in an analyser and prisms for this purpose are constructed to one or other of a number of designs (one of which is known as the Glan-Thompson prism) in which the end faces are horizontal and the optic axis is at right angles to the long axis. In such modified prisms loss of light by reflection, and displacement of the image are minimised, whilst axial and inclined rays emerge with almost exactly the same directions of vibration. Their construction involves a much greater wastage of spar than does that of the original design of Nicol, but this disadvantage is somewhat offset by the fact that the analyser need not be so large as the polariser, which must admit a sufficiently wide beam to ensure that objectives of large numerical aperture can be filled with light by the condensing system. In many instruments of high quality, however, the polariser as well as the analyser is of the horizontal-ended type.

All types of calcite polarising prisms are commonly referred to as " Nicols ", whether they are of the original Nicol design or not.

A prism analyser mounted in the body tube of the microscope as shown in Fig. 161 has the following effects on the convergent beam of light travelling from the objective to the ocular : (1) it raises slightly the level of the real image formed by the objective and the field lens of the ocular, and it is therefore necessary to refocus the instrument every time the analyser is inserted or withdrawn ; (2) it introduces astigmatism into the image, that is to say it makes it impossible simultaneously to focus sharply both horizontal and vertical lines in the image. This effect is only observed to a marked degree when using oculars of high power. The first effect may be

THE OBJECTIVE

corrected by a weak convex lens fitted above the analyser, and in the same mounting, so that it moves in and out of the tube with the prism. Both effects can be eliminated by rendering the light beam parallel in the section of the body tube occupied by the analyser. This may be done by mounting in the tube a concave lens of appropriate strength below the analyser, and a convex lens above it to converge the light again. Such corrective lenses to deal with either the first or both effects are usually fitted to the more expensive instruments.

Both the above effects are avoided if the analyser is mounted above the ocular, but the space then occupied by the prism makes it impossible to bring the eye close enough to the ocular to view the whole field simultaneously in comfort, particularly if the ocular is of high power. The body tube position is therefore almost universally adopted in modern microscopes, though an auxiliary analyser mounted above the eyepiece is sometimes required for special determinations, *e.g.* when using the graduated quartz wedge (see p. 263).

The Objective. The most important characteristics of an objective are (1) its *magnifying power*, and (2) its *resolving power*. As already stated, (1) is inversely proportional to the focal length, and is the property usually referred to when speaking of an objective as being of high, medium, or low, power. By (2) is meant the ability to bring out the fine structural details of the object, and it is expressed inversely by the *limit of resolution*, *i.e.* the smallest distance by which two points can be separated in the object and still be recognised as distinct in the image. This distance is given by the following expression:

$$\text{Limit of resolution} = \frac{\lambda}{2n \sin u} = \frac{1}{\text{resolving power}},$$

where λ is the wave-length of the light in air, n is the refractive index of the medium between the object and the objective, and u is half the angle of the widest cone of light that the objective can take up from a point on the object (see Fig. 163 in which O represents the objective, and X a point on the object). The product $n \sin u$ is called the *numerical aperture* (denoted by N.A.). Thus the resolving power is proportional to the numerical aperture and inversely proportional to the wave-length of the light.

With objectives of focal lengths down to about $\frac{1}{10}''$, *i.e.* all but those of the greatest magnifying power, the medium between the object and the objective is air. n is then sensibly equal to unity and so :

$$\text{N.A.} = \sin u.$$

Increase in the magnifying power of objectives must be accompanied by increase in the numerical aperture, if the enlargement of the image is to be of any value. If, for example, the limit of resolution of a particular objective is 0·003 mm., then another objective of higher magnifying power but the same numerical aperture will reveal no more detail than the first, that is to say it will not distinguish between points on the object which are closer than 0·003 mm. to one another ; it will merely give a larger and rather blurred version of the image obtained with the first. A larger numerical aperture at higher magnifications is also needed to give a bright image, without the necessity of changing over to a specially intense source of light.

The following table gives the numerical apertures and other data for a typical range of objectives :

DETAILS OF OBJECTIVES MANUFACTURED BY COOKE, TROUGHTON, & SIMMS

Focal Length.		Magnifying Power.	Working Distance (mm.).*	N.A.	u (degrees).
Inches.	mm.				
$1\frac{1}{3}$	33	3	43·0	0·1	6
1	25	5	17·0	0·15	9
$\frac{2}{3}$	16	10	5·0	0·28	16
$\frac{1}{2}$	12	15	1·5	0·42	25
$\frac{1}{3}$	8	20	2·2	0·45	27
$\frac{1}{6}$	4	40	0·49	0·85	58
$\frac{1}{6}$	4	40	0·71	0·65	40†
$\frac{1}{6}$	4	40	0·83	0·65	40†
$\frac{1}{6}$	4	40	0·37	0·85	58
$\frac{1}{12}$	1·8	95	0·12	1·30	59‡

* The distance between the end lens of the objective and the object when the microscope is in focus.

† In these two objectives, numerical aperture has been sacrificed to obtain a greater working distance—an advantage for some purposes.

‡ Oil immersion objective (see below).

THE OBJECTIVE

In the case of the *oil immersion* objective, shown last in the table, a high numerical aperture is obtained by filling the narrow space between the end lens and the glass slip covering the object with a drop of cedar-wood oil, which has a refractive index of 1·51–1·52. Reference to the expression given above for the limit of resolution shows that the resolving power will be raised in proportion to the value of this index. The reason for using this particular oil is that it has no deleterious effect on the objective, and, what is more to the point, its refractive index is approximately the same as that of glass. Theory shows that this eliminates spherical aberration in the hemispherical end lens of the objective, provided that the object is at a distance r/n from the centre of curvature of this lens, r being the radius of curvature and n the refractive index of the glass. Oil immersion objectives are designed to work with the object at this distance.

In the study of crystals, an objective of high numerical aperture

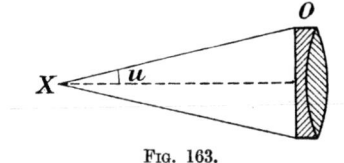

FIG. 163.

is also necessary to obtain satisfactory interference figures. As will be explained more fully later (Chap. VIII), these figures are characteristic patterns of interference colours seen between crossed Nicols when bundles of parallel rays passing in different directions through a crystal section are each brought to a focus in the focal plane of the objective. The wider the angle of the cone of light admitted by the objective, *i.e.* the greater its numerical aperture, the greater the range of directions represented in the figure and the more informative it will be. The result of using an objective with a small aperture would be rather like being asked to identify the Union Jack without being permitted to see more than the centre of St. George's Cross.

In a simple lens, chromatic and spherical aberration increase with increase in the numerical aperture. In order to correct for, or minimise, these aberrations in an objective, it is necessary to divide the refraction between a number of component lenses, the number increasing with the magnifying power and numerical aperture. Thus whereas a 2″ objective may consist simply of an achromatic doublet, a $\frac{1}{6}''$ objective will have five or more components (see Fig. 164). Consequently all high-power objectives are costly and should be treated with the greatest care. The principles underlying the corrections in such lens systems are dealt with in textbooks on optics, and will not be discussed here.

In work with crystals, objectives of very high power are not often needed, and for most purposes a set comprising, say a 2″, a 1″ (or $\frac{2}{3}''$), and a $\frac{1}{4}''$ (or $\frac{1}{6}''$) will suffice. Objectives of higher power than the last, e.g. a $\frac{1}{12}''$ oil immersion, may be of value when studying finely divided material, but it should be noted that crystals small enough to call for a resolving power of this order will probably be too thin to show characteristic interference effects in polarised light, unless their birefringence is very high.

Changes in the magnifying power of a microscope are usually effected by changing the objective, and all but the simplest instru-

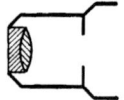

2″ objective (diagrammatic). $\frac{1}{6}''$ objective (diagrammatic).
FIG. 164.—Objectives.

ments are provided with some means of carrying out this operation expeditiously and without risk to the objectives. A simple and (if well made), satisfactory fitment for this purpose is the *revolving nosepiece*, which consists of a revolving mounting screwed to the lower end of the tube, and carrying two, three, or four objectives of different powers, any one of which can be brought into position merely by turning the mounting; some form of stop or catch ensures the correct location of the objective. An objection which is often raised to this device is that its mechanical imperfections, however slight, necessitate the frequent recentring of the objectives (see below), but the authors' experience is that this is by no means the case, at any rate with objectives up to $\frac{1}{6}''$ in power, if the attachment is properly made. The microscopes shown in Figs. 168 and 170 are fitted with a revolving nosepiece.

An extremely satisfactory method of changing objectives is by means of the *objective clutch*. Each objective is provided with a specially shaped collar with a centring adjustment, which fits

THE OBJECTIVE

accurately another collar screwed to the end of the tube of the microscope. The objectives are held in position by a strong spring clutch or catch, which may easily be released. Provided that the collars on the objective and body tube are kept free from dust, an objective which has once been centred should remain centred, no matter how often it is changed.

All polarising microscopes fitted with a revolving stage are provided with means of bringing the axis of rotation of the stage and the optical axis of the instrument into coincidence. This is obviously necessary, particularly when using high-power objectives, if a crystal brought to appear at the intersection of the cross-wires is to maintain that position, or even to remain in the field at all, as the stage is rotated. In almost all instruments the adjustment is effected by two lateral movements of the objective mounting at right angles to one another, actuated by screws with milled heads or with ends shaped to fit a special key.

In a few microscopes, centring is effected by moving the stage instead of the objective. This is not a very satisfactory method, as the centring mechanism reduces the rigidity of the stage assembly, and the adjusting screws are apt to get in the way and be moved accidentally, when the stage is being rotated.

To avoid the necessity for centring objectives, some expensive microscopes are provided with mechanism for rotating the Nicols simultaneously, the stage being kept stationary. Examples are the Swift " Dick " model in which the mechanism consists of toothed gearing, and the Leitz " SY " and the Bausch and Lomb " LD " models in which the Nicols are linked by means of a bar.

Every objective is made to suit a certain microscope tube length, and to obtain the best results it should only be used in microscopes having that tube length. In this connection it should be noted that the addition of a revolving nosepiece to a microscope not originally designed for it will alter the tube length. This will not seriously affect the performance of any but the highest power objectives, but a more important objection is that it may make it impossible to focus interference figures with the Bertrand lens.

A set of objectives to be used with the same microscope should be " parfocal ", *i.e.* they should all focus the image at approximately the same position of the body tube. This makes for convenience and expedition when changing from one objective to another. A particularly undesirable state of affairs is that in which an objective of high power, and therefore small working distance, focusses at a higher position of the body tube than do those of lower power in the set. When changing to this objective there is always the danger of forgetting to rack up the tube first, and thus of bringing the objective into more or less violent contact with the preparation under observation. This is likely to ruin the preparation and may damage the objective.

The Eyepiece. The eyepiece, or ocular, fitted to most microscopes is of the so-called Huygens type, which consists of two plano-convex lenses made of the same glass with their convex sides facing away from the eye (Fig. 165, a). The field lens, *i.e.* the lens farther from the eye, has a focal length which is three times that of the eye lens, and the lenses are separated by a distance equal to one half of the sum of their focal lengths. Under these conditions the

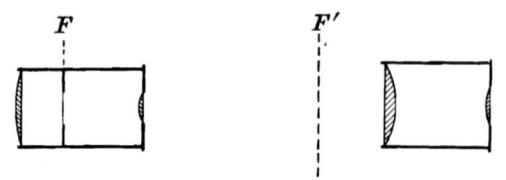

(a) Huygens Ocular (diagrammatic).   (b) Ramsden Ocular (diagrammatic).
Fig. 165.—Oculars.

deviation of the light is shared equally between the two lenses so that spherical aberration is reduced to a minimum, whilst the images formed by different wave-lengths appear to an eye placed close to the eye lens, *i.e.* in the normal position, to be superposed on one another so that chromatic effects are not seen. At the focal plane F of the eye lens, where the real image of the object is formed, are situated the cross-wires as already mentioned on p. 164. The eyepiece as a whole has no external focal plane on the field lens side, and it is therefore said to be a *negative* eyepiece. It is perhaps simpler to think of the field lens as forming part of the objective system, since it combines with the objective proper to form the real image at F.

If it were desired to make linear measurements on the object by means of an eyepiece scale using the Huygens ocular, then it would be necessary to place the scale at F since only here could it be viewed simultaneously with the object. In this position, however, it would be viewed by the eye lens alone, whilst the object image seen is the result of refraction by both field lens and eye lens. Aberrations would therefore affect the scale and object images to different extents and the measurements would be untrustworthy. For such purposes the Ramsden *positive* type of ocular must be used (Fig. 165, b). This consists of two plano-convex lenses of equal focal length with their convex sides facing inwards, and separated by a distance equal approximately to one third of the sum of their focal lengths. The focal plane lies outside the combination at F', and the object image is formed by the objective in this plane. The image of a scale placed here is subject to the same distortions as the object image, since both are viewed through the two lenses of the combination. The same principle is applied in the *micrometer eyepiece*, which affords an alternative means of making linear measurements. This consists of a Ramsden ocular in the focal plane of which a crosshair may be caused to move across the field by the turning of a micrometer screw.

The Becke lens mentioned above as an alternative to the Bertrand lens for viewing interference figures is a Ramsden type of ocular.

In the simple type of Ramsden ocular shown in Fig. 165, the conditions for achromatism are not satisfied, though the resulting defect is not serious. If necessary this defect can be eliminated by using achromatic doublets instead of simple lenses, and some Ramsden oculars are so constructed.

The Stand. Within the meaning of this term may be included those parts of the microscope that support the tube, the stage, the substage (*i.e.* the support or supports on which the polariser and condensing system are mounted), and the mirror. The main parts of the stand (Fig. 168) are the *foot* A connected through the *hinge* B to the *arm* C, which carries at its upper end the tube and its optical parts, and at its lower end the stage, substage, and mirror. It goes without saying that the foot should be massive and well proportioned so that the instrument has no tendency to topple over, whatever the position of the arm. The movement of the latter about the hinge should be smooth, but with sufficient friction to hold the arm in any position. (Usually the friction of the hinge can be easily varied.) The limits of this movement should be definite, and should correspond to the vertical and horizontal positions of the tube respectively. In chemical work the microscope must usually be used with the tube vertical (and the stage therefore horizontal), for on most occasions the substance to be examined is mounted in a liquid medium. Many workers prefer to have the tube inclined when examining dry specimens firmly attached to the microscope slide, but stage clips are then necessary to hold the slide in place. The authors find the vertical position the more satisfactory for all cases, for it enables the stage to be kept clear of clips, and thus permits the slide to be moved about freely and easily. Work with the microscope in this position is not tiring if the table and stool are suitably related in height. The horizontal position is sometimes useful when taking photomicrographs, or when using the microscope for projection purposes.

The coarse focussing mechanism D (Fig. 168) is of the rack and pinion type, the teeth being often cut helically to give greater smoothness of action. Numerous types of mechanism are in use for the fine focussing adjustment (E), but it is unnecessary to describe them in detail here. Some depend on the direct action of a fine threaded screw, whilst in others the motion is transmitted to the body tube through the medium of a lever and a cam. The fine focussing mechanism facilitates the critical focussing of high-power objectives, and if, as in many microscopes, it be graduated so as to read the vertical displacement of the tube corresponding to a given amount of rotation of the operating screw, it may be used to measure the thickness of crystals and other objects placed upon the stage (see p. 224). For accurate measurements of this kind it is important that the mechanism be free from "back-lash" or lost motion.

Diaphragms. In addition to the fixed diaphragms, or stops, in the ocular and objectives, most microscopes are fitted with an adjustable diaphragm below the stage, attached either above or below the polariser. This is called a substage diaphragm, and it is used for cutting out the outer (and most oblique) rays from the cone of light which is focussed on the object by the condenser. Two examples of the use of the substage diaphragm are : (i) to limit the convergence of the light to that just sufficient to fill the objective, and so eliminate glare (see p. 183) ; (ii) to produce the weakly convergent light used in the Becke refractive index test (see p. 226).

Another adjustable diaphragm is often fitted at some place in the tube for the purpose of isolating the interference figures of small crystals. When the crystals examined are so small that several of them appear in the field of view, even when a high-power objective is used, the interference figure given by one crystal can be isolated if the light which has passed through surrounding crystals is prevented from reaching the eye by means of a diaphragm. Otherwise a confused image due to the mixing of several interference figures results. It has been pointed out by J. W. Evans * that a diaphragm for this purpose must be placed where a real image of the object is formed. Various positions fulfilling this condition will be given in Chapter VIII where the methods of observing interference figures are dealt with. Meanwhile it may be noted that the diaphragm fitted *immediately* above or below the Bertrand lens in many microscopes is not satisfactory for the purpose (though frequently claimed to be so); its value is confined to improving the definition of the figures.

Adjustable diaphragms may be of the iris type as fitted to photographic lenses, or may consist of a *slider* or a rotating disc perforated with a series of holes of different sizes. Iris diaphragms have the advantage of giving, within their limits, any desired aperture, and are convenient to operate, but they cannot be closed sufficiently to give extremely small apertures. Slider diaphragms have the merits of simplicity and cheapness, and their apertures can be made as small as is required. Also, when fitted below the stage, such diaphragms afford a ready means of producing oblique illumination, for this can be obtained merely by adjusting the slider so that one of the apertures is a little to one side of the centred position.

Two examples of special diaphragms are : (1) Slawson's objective diaphragm for measuring optic axial angles (p. 315), and (2) Saylor's diaphragms for securing critical oblique illumination for refractive index determinations (p. 228). These are not normal fittings of the microscope, and the descriptions of them can more conveniently be given later.

Additions to Stage Equipment. *The Mechanical Stage.* This extremely useful addition to the microscope is a device for varying the position of the slide on the stage. It consists of a spring frame which grips the slide firmly, and which can be moved in two directions at right angles by racks and pinions, the amount of movement in each direction being given by a scale and vernier. The whole contrivance is clamped on top of the ordinary stage. An example (by J. Swift & Son, Ltd.) is shown in Fig. 166, which is self-explanatory. The advantages of the mechanical stage are

* *Min. Mag.*, 1916, 18, 49.

that the movement of the slide can be controlled to a nicety, *e.g.* when making a preliminary survey of its contents, or when bringing a crystal to the intersection of the cross-wires, and that any particular part of the slide, for which the stage readings are noted, can be rapidly found on subsequent occasions.

Rotation Apparatus. Many instruments have been described for changing the orientation of crystals under observation so that their optical characters may be determined for other positions relative to the path of the light than those which they would naturally adopt on the ordinary stage. Most of these devices involve the mounting of a single crystal on a rotatable support, but others consist of mechanism for tilting a microscope slide, on which the crystals are mounted in the usual way, into any desired position. The

FIG. 166.—Mechanical Stage.

more elaborate examples of the former type are provided with lateral and tilting adjustments for bringing any axis of the crystal into exact coincidence with the axis of rotation, as in a reflecting goniometer. A good example is the *stage goniometer* of Miers, shown in Fig. 167. This is best suited to a microscope with a fixed stage and rotating Nicols, such as the " Dick " model of Messrs. Swift referred to above, and is usually clamped to the stage with the needle support pointing to the observer. It can, however, be fitted to rotating stages. With this apparatus a variety of operations can be carried out, such as the measurement of maximum extinction angles, and optic axial angles. By arranging the microscope so that the needle support points downwards, the crystal can be kept immersed during the observations in a glass cell containing a liquid of known refractive index.

TYPICAL MICROSCOPES 177

Simpler forms of apparatus for the rotation of single crystals are described on pp. 213–16.

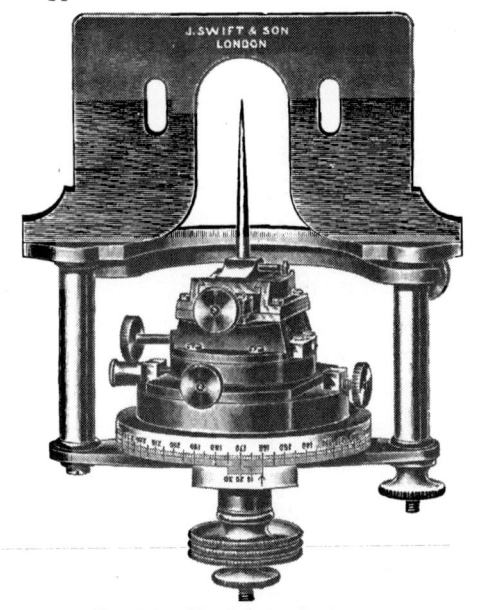

FIG. 167.—Miers' Stage Goniometer.

The outstanding example of rotation apparatus of the other type is the *universal stage* of von Fedorov, the use of which is described in Chapter XI.

Some Typical Polarising Microscopes. The microscopes described below are typical examples of modern designs.

The "Lapidex" Model (J. Swift & Son, Ltd., London). A feature of this instrument (Fig. 168) is its simple and robust construction. It has a heavy and well-proportioned stand, and is provided with both coarse and fine focussing mechanisms. The polariser is mounted so that it can be moved up and down for focussing the condenser which is attached to its upper end, or swung out of the path of the light when unpolarised light is required. The up-and-down movement is direct, *i.e.* without rack and pinion, but works smoothly though with sufficient friction to hold the condenser in any position. The converger is conveniently mounted on a metal slider working in a channel in the stage, so that the

178   THE POLARISING MICROSCOPE

change from parallel to convergent light can be made instantly without disturbing the glass object slide. The objective mounting is provided with centring screws with milled heads, one of which can be clearly seen in the figure. Immediately above these is the

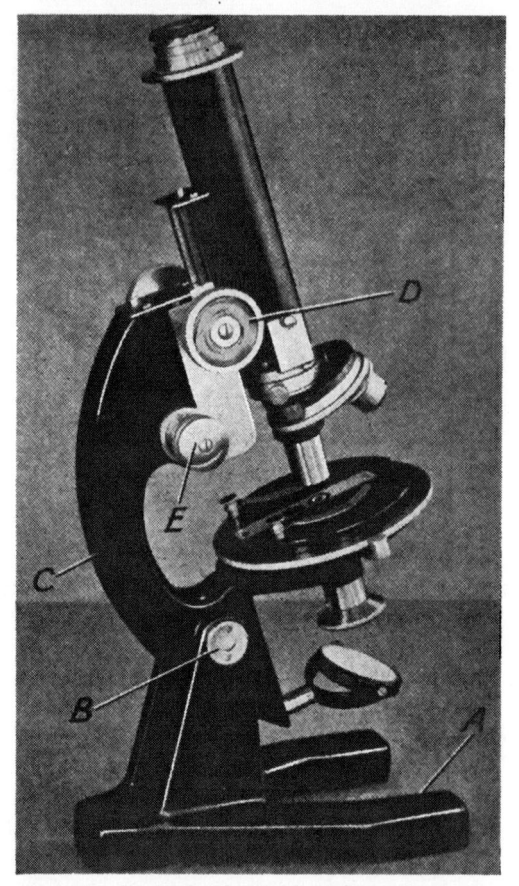

FIG. 168.—The Swift " Lapidex " Polarising Microscope.

slot for the introduction of the quartz wedge and other compensating plates. The slot is cut at 45° to the cross-hairs in the ocular, and is provided with a dust cover.

The analyser is mounted in a brass box which slides in and out

TYPICAL MICROSCOPES

of the tube. No Bertrand lens is fitted, but interference figures can easily be viewed by removing the eyepiece, or by fitting a Becke lens, supplied with the instrument, over the eyepiece. Additional accessories which may be fitted are an iris diaphragm

FIG. 169.—The Swift " Survey " Polarising Microscope.

at the lower end of the polariser and a pinhole stop in place of the eyepiece, for isolating the interference figures of small crystals.

The " Survey " Model (J. Swift & Son, Ltd., London). This instrument (Fig. 169) is of a more advanced type than the

"Lapidex" model. It is provided with a Bertrand lens in a focussing mount, and the analyser can be rotated through 90°. The substage may be focussed by means of a rack and pinion movement, and is provided with centring screws to enable the condenser to be brought into perfect collimation with any objective in use. It also carries an iris diaphragm.

Cooke Polarising Microscopes (Messrs. Cooke, Troughton, & Simms, York, England). Figs. 170 and 171 show respectively the

FIG. 170.—The Cooke "Elementary" Polarising Microscope.

"Elementary" and "Research" models. In these microscopes, "Polaroid" discs are used in place of Nicol prisms. The use of "Polaroid" for this purpose has been discussed by A. F. Hallimond.* Several advantages are gained by this modification. In the first place, the disc used as the analyser introduces no astigmatism into the image when it is mounted in the body tube, as in the model shown in Fig. 171, and the difference in the position of focus of the body tube resulting from its insertion can be com-

* *Nature*, 1944, **154**, 369.

TYPICAL MICROSCOPES 181

pensated quite simply by a glass disc of the same thickness mounted in the same sliding holder, so that the light passes through the glass disc when the analyser is not in use.

Owing to its small thickness, a " Polaroid " analyser can also be fitted above the ocular, as in the " Elementary " microscope (Fig. 170), without the inconvenience, mentioned above, which is associated with a Nicol prism in that position.

The main advantages are, however, connected with the polariser

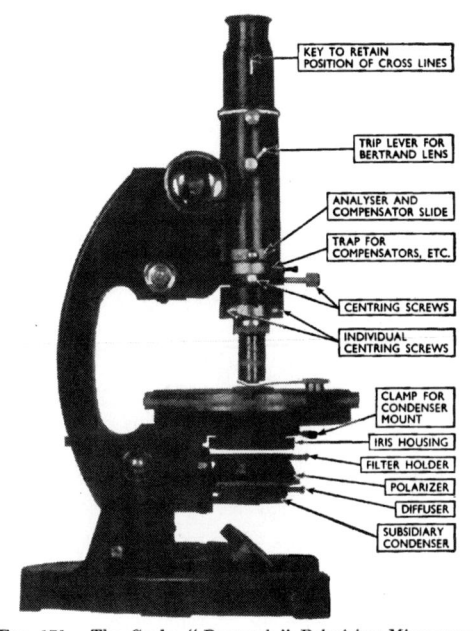

Fig. 171.—The Cooke " Research " Polarising Microscope.

and the condensing system. The use of " Polaroid " permits the passage of a much wider beam of light than is possible in the case of a microscope fitted with calcite prisms. This is because large pieces of optical calcite are practically unobtainable nowadays, and this limits the diameter of the prisms that can be made at a reasonable cost to about 10 mm. " Polaroid " of optical quality can be readily prepared in discs of considerably greater diameter than this. The polariser in the Cooke microscopes has a diameter

of 27 mm. and this means that the condenser can be made wide
enough to fill both high- and low-power objectives with light
without any change in its lens combination. In order to make the
best use of the narrow beam passed by a calcite polariser, it is
necessary, as we have seen, to provide a removable component, the
converger, for use with objectives of high numerical aperture. If
the converger is left in on changing to an objective of low numerical
aperture, only the centre of the field is properly illuminated. The
condenser fitted to the Cooke microscope illuminates the field fully
with all objectives up to a focal length of $1\frac{1}{2}''$ (N.A. about 0·1).
The only adjustment necessary on changing objectives is to open
or close the substage diaphragm, so as to "trim" the cone of
light to suit the numerical aperture
of the new objective (see section on
illumination below).

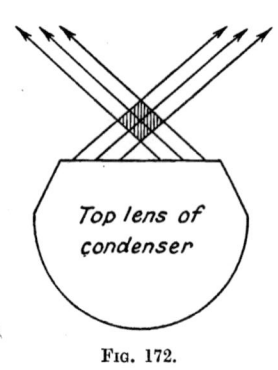

Another important feature is that
for normal work the position of the
condenser need not be altered for
different objectives or light sources,
or to allow for varying thicknesses
of object slides, and the microscope
is not in fact provided with a substage focussing rack. This simplifica-
tion is achieved as follows. Below
the condenser there is a ground glass
screen—the diffuser—which scatters
the light so that the emergent rays pass in all directions, and which
also removes objectionable features from the image of the light
source which might otherwise appear in the plane of the object.
The light can thus be regarded as made up of bundles of parallel
rays, with the result shown in Fig. 172. Since the condenser is
wide, there is a region of considerable depth within which the
illumination is essentially the same (shaded in the figure), and
corresponds to the full numerical aperture of the condenser. Provided, therefore, that the object is anywhere in this region, it will
be correctly illuminated.

If very strong lighting is required, the diffuser can be removed
and the image of the light source focussed on the object plane in
the usual way by adjusting the height of the condenser in its
mounting.

FIG. 172.

Top lens of condenser

ILLUMINATION

A full description of the " Research " model (Fig. 171) has been given by A. F. Hallimond and E. W. Taylor.*

Illumination. The best conditions of illumination so far as resolving power is concerned obtain when the mirror, condenser, and substage diaphragm are so arranged that the light is brought to a focus on the object, and therefrom diverges as a cone just wide enough to fill the objective. (A wider cone produces glare.) This is shown diagrammatically in Fig. 173. With microscopes fitted with Nicols and an adjustable condenser, it involves adjusting the condenser to give an image of the source of light in the plane of the object, and then closing the diaphragm until it is seen by removing the eyepiece that the objective is just filled with light. The light source needs to be somewhat extended to give even illumination, and so it may not be easy to tell when it has been brought to a focus. This difficulty may be overcome, if an artificial source such as an opal electric-light bulb is used, by focussing some small object, a pencil for example, held immediately in front of it, and if daylight from a window is used, by focussing one of the window-bars, subsequently removing the image of this from the field of view by a slight tilt of the mirror. With the Cooke microscopes, using the diffuser (see above), it is only necessary to adjust the substage diaphragm until, with the eyepiece removed, it is seen that the objective is just filled with light.

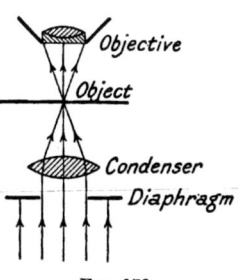

Fig. 173.

The condenser should always be used in conjunction with the plane side of the mirror, for if used with the concave side, the light is brought to a focus below the object. For low magnifications, however, the condenser may be removed and the light made to converge on the object by using the concave side of the mirror.

The foregoing conditions should be established as far as possible, when studying the outward form of crystals under the microscope in ordinary light, especially when using high powers, and when searching for minute structural details such as striations, and fine cleavage cracks. For the determination of many optical properties, however, it is desirable and sometimes necessary to depart from

* *Min. Mag.*, 1946, **27**, 175.

these conditions, because considerations other than resolving power are of superior importance. For example, the polarisation colours and extinction angles of crystal plates between crossed Nicols (see pp. 242-52) are properties relating to *parallel* light, and therefore when carrying out precision measurements on these properties the light should be rendered effectively parallel by closing the substage diaphragm almost completely. Effective illumination of the crystal may then require a specially strong source of light. (For ordinary determinations of polarisation colours and extinction angles, this precaution need not be rigidly observed, provided that the most oblique rays are cut out when using objectives of high numerical aperture.) Again, when making the Becke refractive index test (p. 225) a *weakly* convergent beam of light is required, necessitating the lowering of the condenser and the closing of the substage diaphragm, whilst in the van der Kolk refractive index test (p. 227), the crystals must be *obliquely* illuminated.

Illuminating Apparatus. Unless the source of illumination is adequately enclosed it will be necessary to erect in front of the microscope a screen to prevent light from falling on top of the stage, or dazzling the eyes. A suitable screen can easily be made from a sheet of three-ply wood, or cardboard, about 12" square with an opening 3" or 4" square at the bottom to permit the light to reach the microscope mirror. It is convenient to have wooden grooves fixed to the sides of this opening, for holding a sheet of ground glass or colour filters, when it is required to modify the light. The screen must be provided with supports to hold it vertically, and should be painted dead black.

For observations in daylight, the best source is a north window, and failing this, one facing east. If neither of these is available, windows facing in other directions may be used provided that direct sunlight is prevented from reaching the microscope by interposing a sheet of ground glass or thin white paper between the window and the instrument. Such a diffusing screen may also be used for the purpose of removing the image of window-frames from the field of view, in cases where the panes are so small that this cannot be done by tilting the mirror. Whenever a diffusing screen is used in this way, its position should be taken as that of the source of light for the purpose of adjusting the condenser to give the greatest resolving power, as described at the beginning of the preceding section.

ILLUMINATION

In general, however, it is preferable to use an artificial source of light, since the conditions of illumination are then under better control. The simplest source is an electric lamp with an opal or frosted globe. The light from this should be corrected for its yellow tint by passing it through a blue glass screen supplied for this purpose by makers of microscope accessories. Alternatively, a " Daylight " lamp may be used, in conjunction with a ground glass screen to diffuse the light. The lamp should be housed in a metal box, suitably ventilated, and having an opening directed towards the microscope mirror. There are many excellent forms of lamp housing on the market, in some of which the opening is provided with an iris diaphragm to alter the effective size of the light source, and grooves for the accommodation of ground glass, or coloured screens.

It is sometimes necessary to use a more intense and concentrated source of light than that afforded by an ordinary incandescent bulb, *e.g.* for photomicrography, or projection purposes, or to obtain critical illumination with objectives of very high numerical aperture. Lamps of the compact filament type and working on 6 or 8 volts at a current of 5 to 6 ampères are supplied for such purposes and are very convenient. In conjunction with a diffusing screen they may also be used for ordinary work at low powers. A still more powerful source may be required for certain purposes, *e.g.* for use with a monochromator (see below), and for this an electric arc is the best, preferably one with automatic control.

It must not be forgotten that powerful sources of light emit a considerable amount of heat radiation, and the possibility of the specimen heating up must not be overlooked. Regard to this point is particularly important in refractive index work, for many of the liquids used alter in index considerably with change of temperature. The heat rays may be absorbed by passing the light through a flat-sided glass cell filled with cold water ; or the cell can be made to combine the functions of cooler and light filter (to compensate for the yellow tint) by filling it with a dilute solution of cuprammonium sulphate prepared as follows : Ammonia solution is added to 25 c.c. of a 10% solution of copper sulphate pentahydrate until the precipitate of copper hydroxide is completely redissolved, and the solution diluted to half a litre. (More ammonia is added if the dilution reprecipitates the hydroxide.) The solution should be used in a layer 10–15 cm. thick, but somewhat thinner layers may be used if the solution is made proportionately stronger.

Filters made of special heat-absorbing glass (*e.g.* Chance ON19 made by Chance Bros., Ltd., Smethwick, nr. Birmingham) are also obtainable.

For advanced studies of crystal optics and for the double variation method of determining refractive indices (see p. 396), the most suitable source of (nearly) monochromatic light is a modified form of spectrometer known as a monochromator. By means of this instrument, a narrow band of wave-lengths from any part of the spectrum may be directed into the microscope. The intensity of such a band is only a small fraction of that of the light entering the monochromator, and in order that it shall be adequate to illuminate the crystal it is necessary to use an arc as the primary source. This consideration and the high cost of the instrument militate against its use for ordinary work.

For general purposes the most convenient source of monochromatic light is a type of electric discharge lamp which contains a volatile metal, and a small quantity of an inert gas, the purpose of which is to facilitate the initial discharge between the electrodes. As the lamp warms up, the metal vaporises and its characteristic frequencies are excited by the discharge. The light emitted also contains lines due to the inert gas, but their intensity is very low. A series of such lamps with sodium, cadmium, mercury, cadmium-mercury, and zinc fillings is supplied by the General Electric Company Ltd. (London), under the trade name " Osira ". These lamps give a number of strong lines well spaced throughout the whole of the visible spectrum, and most of the lines can be readily isolated, or nearly so, by means of single standard filters of the type consisting of dyed gelatine films mounted between glass plates.

The sodium lamp is the most generally useful, since it is customary to record optical data for sodium light (the doublet or D " line " at 5896 and 5890 Å). For most work with the microscope it is unnecessary to use a filter with this lamp, since the intensity of the lines due to the inert gas is only about 1% of that of the D line. If specially pure sodium light is required, however, the inert gas lines can be eliminated by means of a suitable yellow filter. For example, the Ilford No. 606 Spectrum Yellow filter (Ilford Ltd., Ilford, London *) cuts out all but a doublet in the yellowish green region at 5688, 5683 Å, and the intensity of this is much reduced.

* Full details of Ilford filters are given in a brochure, " Ilford Colour Filters ", published by the Company.

ILLUMINATION 187

Other parts of the spectrum are sufficiently represented for most purposes by the lines given by the cadmium and mercury lamps. The following table gives the strongest of these lines and suitable filters chosen from the Ilford series by means of which they may be isolated or partially isolated.

Lamp.	Wave-length (Å).	Colour.	Filter (Ilford Series).
Cadmium	6438	Red	No. 608 Spectrum Red isolates.
	5086	Green	No. 604 Spectrum Green isolates.
	4800	Sky-blue	No. 602 Spectrum Blue transmits both these lines and eliminates the rest except 5086, which is transmitted very weakly.
	4679	Blue	
Mercury	5791 5770	Yellow	No. 606 Spectrum Yellow isolates.
	5461	Green	No. 807 Mercury Green nearly isolates, but doublet 5791, 5770 is weakly transmitted.
	4357	Blue	No. 601 Spectrum Violet isolates.

The intensities of cadmium red 6438 and mercury blue 4357 isolated as above are rather low, and it may be found necessary, particularly when studying interference figures at these wavelengths, to concentrate the light on the microscope mirror by means of a condensing lens or lens system.

The lamp containing both cadmium and mercury gives all the above lines and thus covers a useful range, but the isolation of some of the lines is more difficult. Thus using the filters given above, while cadmium red 6438 and mercury yellow 5791, 5770 are still isolated, cadmium green 5086 is accompanied by mercury green 5461, and mercury blue by weak cadmium blue lines at 4800 and 4679. Better isolation can be achieved by using suitable combinations of filters, but the intensity of the transmitted light is then likely to be seriously reduced.

For many observations, light that is approximately monochromatic suffices, and this may conveniently be produced by passing white light through filters of the type mentioned above, or through coloured solutions. The Ilford range of " Spectrum " filters is satisfactory for this purpose. The limits of the transmission band and the wave-length corresponding to the peak intensity of the band are given for each filter below.

Filter.	Limits of Transmission Band (Å × 10).	Wave-length of Peak Intensity (Å × 10) (approximate).
No. 601 Spectrum Violet	385–475	430
No. 602 Spectrum Blue	445–495	470
No. 603 Spectrum Blue-Green	475–520	490
No. 604 Spectrum Green	500–545	520
No. 605 Spectrum Yellow-green	530–580	550
No. 606 Spectrum Yellow	560–610	575
No. 607 Spectrum Orange	575–720	600
No. 608 Spectrum Red	625 to infra-red	680 to limit of visible
No. 609 Spectrum Deep Red	640 to infra-red	710 to limit of visible

The figures given in the second column are the extreme limits of transmission, and with the normal illumination used for work with the microscope the light transmitted in the region of these limits will be of very low intensity. Thus the bands are effectively narrower than would at first appear.

Numerous recipes for coloured solutions have been suggested. The table on p. 189 gives a series proposed by Landolt.* For each colour, two or three solutions contained in separate flat-walled glass cells must be superposed. The distance between the cell walls (*i.e.* the thickness of solution through which the light passes) is given for each solution in the second column.

The crystal violet must first be dissolved in a little alcohol and then diluted with the necessary amount of water. The solution is not permanent in the light, and so when not actually in use should be kept in the dark. The $KMnO_4$ solution does not keep well and should therefore be freshly made up as required.

The authors have found that the following solution in a layer 8 to 10 cm. thick transmits light approximating to that of the D line, if illuminated with an electric lamp of 100 c.p. : 6 volumes

* *Berichte*, 1894, **27**, 2872.

of saturated potassium dichromate solution mixed with 1 volume of saturated copper sulphate solution.

Colour.	Thickness of Filter in mm.	Solution.	Gm. per 100 c.c. Water.	λ in $\text{Å} \times 10$.	Mean λ.
Red . . .	20	Crystal violet	0·005 }	639–718	665·9
	20	K_2CrO_4	10·0		
Yellow . .	20	$NiSO_4, 7H_2O$	30·0 }		
	15	K_2CrO_4	10·0 }	574–614	591·9
	15	$KMnO_4$	0·025		
Green . .	20	$CuCl_2, 2H_2O$	60·0 }	505–540	533·0
	20	K_2CrO_4	10·0		
Blue . . .	20	Crystal violet	0·005 }	410–479	448·2
	20	$CuSO_4, 5H_2O$	15·0		

The Use and Care of the Polarising Microscope. If both eyes are equally strong, fatigue may be avoided, when working for long periods, by using them alternately. The unoccupied eye should be kept open as less strain is thereby imposed, and with a little experience, it should be found possible to concentrate the attention upon what is seen through the microscope. It undoubtedly helps, however, if the vision of the unoccupied eye is blocked by a black shade fixed at the side of the upper end of the microscope tube. Such a shade can easily be made of cardboard, or thin wood, painted dead black, and suitably clipped to the tube, or one may be obtained cheaply from a microscope dealer.

It should be remembered that high-power objectives are very close to the object when they are in focus, and that therefore great care in focussing them must be exercised to prevent breaking the slide or ruining the objective itself. The safest method is to rack down the tube carefully until on looking sideways along the stage, only a thin streak of light can be seen between the bottom of the objective and the cover glass. This tube is then racked up slowly until the object is in focus.

When it is necessary to unscrew an objective from the microscope (as, for instance, with instruments provided with no other means of changing objectives), the tube should be racked up and the unoccupied hand held underneath so as to catch the objective if it should fall. As a further precaution against such accidents, it is a good plan to stand the microscope on a piece of thick cloth, *e.g.* baize, when it is in use.

The microscope must not be carried by any part above the fine adjustment mechanism, or the latter will be ruined. Most microscopes are best carried by holding the arm of the instrument with one hand and steadying the foot with the other.

When not in use, the microscope must be protected from dust either by replacing it in its case, or by covering it with a glass bell jar or a cardboard or cellophane hood. Such covers may be obtained from firms dealing in scientific apparatus. It goes without saying that the microscope must not be used in any room in which corrosive fumes are generated.

The microscope stand should be cleaned with a rag moistened with benzene or xylene. Alcohol should never be used for this purpose. To clean the coarse focussing mechanism, the tube should be removed and the slides cleaned with benzene or xylene, and then lubricated lightly with paraffin or clock oil, removing any surplus. The teeth of the rack and pinion should be cleaned with a small stiff brush, but not lubricated. A little paraffin or clock oil may be applied to the shanks of the knurled heads where these enter the arm. This attention should only be necessary at very infrequent intervals, if the instrument is kept free from dust. If anything goes wrong with the fine focussing mechanism, the instrument should be returned to the maker for attention.

Lenses and Nicols. The microscope must not be subjected to sudden changes of temperature, for such treatment is liable to crack the balsam films in the Nicols and lenses. It must also be remembered that these films will soften if the instrument is kept in too warm a place, *e.g.* near a radiator, or if powerful artificial light is passed through it for long periods without first passing the light through a cooling cell.

It is essential that all lenses and Nicols should be kept clean and free from dust. The upper ends of Nicols are usually provided with glass dust covers, but the lower ends sometimes need cleaning. This should be done only with a camel-hair brush, since more vigorous treatment is liable to scratch or otherwise damage them. It must constantly be borne in mind that the Nicols are an indispensable and expensive part of the polarising microscope, and must therefore be treated with the greatest care.

Lenses may be freed from dust with a camel-hair brush, and

USE AND CARE

cleaned by breathing on them and gently polishing them with a soft linen rag (not silk or cotton) kept specially in a closed box, or with the special lens paper sold for the purpose. If this is not effective, the rag may be *slightly* moistened with a little benzene or xylene, but not alcohol. Care must be taken that no liquid gets between the lenses. The eye and field lenses of the ocular should be unscrewed from time to time and their inner surfaces cleaned as above. This exposes the cross-wires and great care must be taken that they do not get broken during the operation.

Finally, ambitious adjustments and repairs such as resetting Nicols in their mountings, adjusting cross-hairs, and dismantling objectives should not be attempted. If attention to such points is necessary, the instrument should be returned to the makers. (Instructions for testing the setting of the Nicols and the cross-hairs are contained in Experiment 1, Chapter X.)

Recommended for Further Reading

Manual of Petrographic Methods, A. Johannsen (McGraw-Hill, 1918).
Handbook of Chemical Microscopy, Vol. 1, E. M. Chamot and C. W. Mason (Wiley, 1938).
Mikroskopische Physiographie der Mineralien und Gesteine, H. Rosenbusch and E. A. Wülfing (Schweizerbart, 1924).
Practical Microscopy, L. C. Martin and B. K. Johnson (Blackie, 1931).

CHAPTER VI

PREPARATION AND MOUNTING OF MATERIAL

It is difficult to make a satisfactory optical examination of dry crystals, unless they are in the form of plates orientated perpendicularly, or nearly so, to the path of the light, because, owing to the large difference between their refractive indices and that of air, much of the light is reflected from surface irregularities, inclined faces, etc., instead of passing through them. Crystals to be examined are therefore usually mounted between a slide and a cover slip in a liquid medium—naturally one in which they are insoluble. By using liquids of known refractive index this procedure also enables the refractive indices of the crystals to be determined, as will presently be described.

In general, when determining the optical properties of crystals or crystal fragments of irregular shape, or those with many inclined faces, the immersion liquid chosen should have an index as near as possible to the mean index of the crystals, so that the light will pass through them without appreciable deviation. For studying the outward form, however, it is necessary to use a liquid the index of which is not too close to the indices of the crystals, in order that their outlines shall appear in bold relief.

Slides and Cover Slips. The usual size of microscope slide, namely $3'' \times 1''$, is suitable for all ordinary work with crystals. The cover slips need not be larger than $\frac{1}{2}''$ in diameter and may be even smaller. As some immersion liquids are expensive, small cover slips make for economy in material and encourage neatness in practice. Their use also enables small crystals to be turned over more easily by shifting the cover slip with the point of a needle, because the layer of immersion liquid is deeper than that obtained when a large slip is used.

Large cover slips may readily be divided into smaller sizes by the following method. The large slip is placed on two or three thicknesses of filter paper laid on the working bench so as to form a slightly yielding surface. The ground edge of a 3" × 1" slide held in a slanting position is then pressed firmly and evenly on the slip. This results in a neat straight fracture along the line of pressure.

Slides and cover slips must be thoroughly cleaned and dried before use. This is most important, for much time may be wasted in studying specks of birefringent dust in mistake for part of the specimen, and in the determination of refractive indices, an inaccurate result will be obtained if the immersion liquid is contaminated with the remains of former liquids, or diluted with water. New slides may be prepared for use by washing them in hot soapy water, rinsing in hot distilled water, and finally drying and polishing them with a clean linen cloth, or a piece of well-washed cambric, holding them by the edges only. To prepare new cover slips it is usually sufficient to breathe on both sides and then dry and polish by gently rubbing them in a fold of well-washed cambric held between the finger and thumb, or alternatively by rubbing them between filter papers laid on a perfectly flat surface such as a piece of plate glass. Once cleaned they must be handled by the edges only, or more conveniently, especially if they are small, by means of forceps.

For cleaning slides after use, it is convenient to have two wide-mouthed glass-stoppered bottles on the working bench, one containing benzene and the other alcohol. Slides on which oily immersion liquids have been used are dropped into the benzene bottle, and after the lapse of a few hours are transferred by means of crucible tongs to the alcohol bottle. Slides contaminated only with aqueous immersion media (as when crystals have been examined in an aqueous mother liquor ; see below) may be placed at once in the alcohol bottle. The slides are removed as required from this bottle, rinsed in hot distilled water, and dried as described above. It is advisable to line the bottoms of the bottles with some soft material, such as a pad of cotton wool, to prevent occasional breakages when the slides are dropped in.

Usually it is hardly worth while to attempt to clean used cover slips, but if owing to a temporary shortage this should be necessary, they should be treated as for slides, that is with benzene (if they are oily), followed by alcohol, and hot distilled water, and then dried by rubbing between filter papers on a flat surface as described above for new cover slips.

194 PREPARATION AND MOUNTING OF MATERIAL

Apparatus. It will be assumed first that the material is in the form, or can be prepared in the form, of small separate crystals or crystal fragments as is usual in chemical practice; or that it is to be examined as a thin polycrystalline film prepared by melting it between a slide and cover slip. The apparatus required for pre-

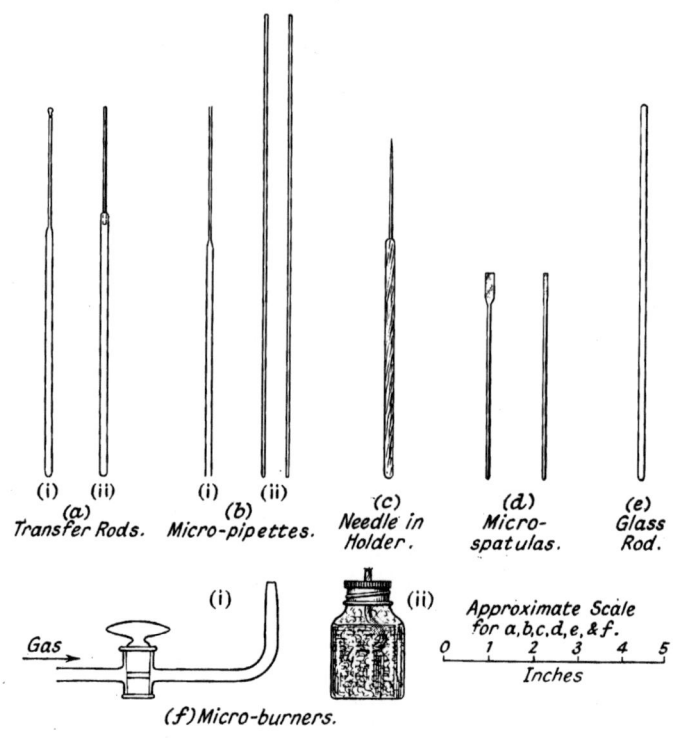

(i) (ii)   (i) (ii)
 (a)   (b)   (c)   (d)   (e)
Transfer Rods. Micro-pipettes. Needle in   Micro-   Glass
                                 Holder.   spatulas.   Rod.

(i)   (ii)   Approximate Scale
Gas   for a,b,c,d,e,&f.
                   0   1   2   3   4   5
                          Inches

(f) Micro-burners.

Fig. 174.—Apparatus for Preparation and Mounting of "Chemical" Material.

paring and mounting this type of material, which we will designate as "chemical", is shown in Fig. 174, and is described below. In this description the headings are lettered to correspond with the drawings in the figure.

(a) *Transfer rod.* This is used for transferring minute amounts of solid material, or small drops of solvent to the microscope slide;

APPARATUS

also as a stirring rod when crystallising material from a drop of solution on the slide. It may be made either by drawing out a piece of narrow glass rod, and rounding off the narrow end to a ball about the size of a pin's head (i); or by sealing a piece of stout platinum wire into a glass handle, and slightly roughening the end of the wire (ii).

(b) *Micro-pipette.* This is used for adding liquid to, or withdrawing it from, the slide. It is made as shown by drawing out one end of a piece of narrow glass tubing (i). Alternatively a number of glass capillaries about 8″ to 10″ long may be made (ii). These are sealed at each end to prevent the entry of dirt. When required for use, the sealed ends are broken off. After use the capillary is thrown away.

(c) *Needle in holder.* This has a variety of uses, e.g. for manipulating a single crystal in the operation of mounting it on a rotation apparatus (see later), for shifting the cover slip on a liquid mount in order to turn the crystals into other positions, and for teasing out fibrous material into its individual fibres preparatory to mounting it. A dissecting needle as used by biologists serves excellently, or a stout sewing needle may be fixed in a wooden holder drilled to receive it.

(d) *Micro-spatulas.* These are used for the transference of somewhat larger amounts of solid material than can be picked up by the transfer rod. It is convenient to have two sizes, one with a blade $\frac{1}{8}″$ to $\frac{3}{16}″$ wide, and a smaller one the blade width of which is about $\frac{1}{16}″$. They may be made by hammering out the end of a piece of copper, or better, nickel wire of suitable gauge, and trimming it to shape by filing or on a fine grinding wheel. It is important that the surfaces of the blade should be flat, and that its end be trimmed to a fine edge, otherwise difficulty will be experienced in picking up and retaining material upon it.

(e) *Glass rod.* This is used as a stirring rod when preparing solutions of material (otherwise than in a drop of solvent on the slide), and for transferring a drop of solution to the slide. The rod should not be thicker than about $\frac{3}{16}″$, otherwise too much liquid will be transferred. The ends of the rod must be evenly rounded.

(f) *Micro-burner.* This is used for heating a drop of solution on a slide in the technique of micro-recrystallisation from solution, and also for micro-sublimation, as will be described later. The burner should be capable of adjustment to give a flame not more than $\frac{1}{4}''$ high. The type shown in (*i*) is made from a pyrex glass tap with one limb bent upwards and drawn out to a jet, the best size of which may be found by trial. The jet should not be very fine, otherwise the flame will be too pointed and will crack the slide. The other limb of the tap is attached to the gas supply. A readily made burner using methylated spirits is shown at (*ii*). It consists of a small ink-bottle with a screwed metal cap which is drilled in the centre to fit a cigarette-lighter wick. The bottle is loosely filled with cotton wool. This burner gives a round " soft " flame, which has no tendency to crack the slide if the latter is kept in gentle motion.

(*g*) *Cooling block.* This is also used when crystallising material on the slide, its purpose being to cool the slide rapidly after it has been heated over the micro-burner, and thus to arrest evaporation of the solvent. The upper surface must be smooth and quite flat, and its dimensions somewhat greater than those of the slide, so that the whole area of the latter can, if necessary, be brought into contact with the metal. The block should be about $\frac{3}{16}''$ thick or more, and should preferably be made of copper or aluminium.

(*h*) *Hot plate.* This is used for preparing films of low melting substances by melting them between a slide and a cover slip; in making permanent mounts of crystals in Canada balsam; for micro-sublimations; and for other purposes as will appear later. The simple form illustrated consists of a metal slab about $8'' \times 3'' \times \frac{1}{2}''$ made of iron, copper, or brass, drilled to take a thermometer, and placed on a sheet of thin asbestos board supported on a tripod above a bunsen burner, the gas supply to which is controlled by the screw clip device shown. The purpose of the asbestos sheet is to shield from the flame the stem of the thermometer and also the projecting portions of slides which are not lying wholly on the plate. (Slides are sometimes laid partly on and partly off the plate, *e.g.* when it is desired to remelt a preparation in order to recrystallise it. The portion overhanging the edge of the plate remains unmelted and serves to inoculate the melt on cooling.) The upper surface of the plate must be smooth and flat, and, if the plate is of iron, should be coated with copper by first polishing with fine emery paper and then rubbing copper sulphate solution over it with a wad of cotton wool. This will prevent rusting, or at least retard it very considerably. A circular hole about $2''$ in diameter should be cut

APPARATUS 197

in the asbestos sheet immediately above the burner, so that there is no undue lag in raising the temperature of the plate.

Modifications of the above design of plate will no doubt suggest themselves to the reader, *e.g.* the use of electrical heating instead of gas. Whatever form is adopted it is very desirable to be able to maintain its temperature reasonably constant (say within 10°) without much trouble, and also to alter its temperature without undue waste of time. In these respects, gas heating

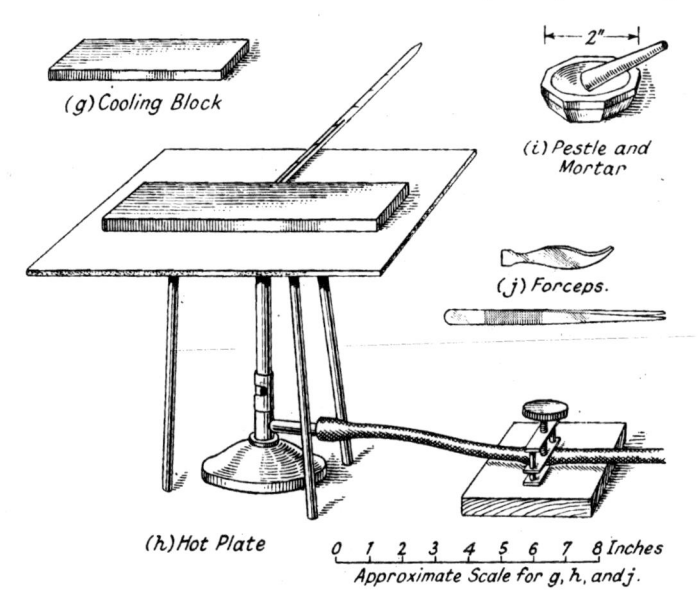

FIG. 174 (*cont.*).—Apparatus for Preparation and Mounting of " Chemical " Material.

probably has advantages over electrical heating, in the absence of a thermostatic control device.

(*i*) *Pestle and Mortar.* This is required for crushing coarse material to a suitable state of division for mounting. It should be quite small. One made of agate and about $1\frac{1}{2}''$ to $2''$ diameter serves admirably. Alternatively a glass block with a ground hemispherical cavity, used by botanists as a dissecting dish, may serve as the mortar, the pestle in this case being made of a piece of stout glass rod, the end of which has been rounded in the flame to as nearly as possible the same radius of curvature as that of the cavity.

(*j*) *Forceps.* These are convenient for handling cover slips. No special design is needed. The type supplied for manipulating analytical weights is very suitable.

Finally, it is advantageous to have a set of sieves, as used for fractionating mineral powders according to their particle size. The sieves should cover a range of about 120 to 325 meshes to the inch. With certain types of material which have to be crushed before examination, the sieves will be found very useful for separating particles of the optimum size for study.

Methods

Methods of preparing and mounting chemical material fall broadly into two groups according to whether they involve recrystallisation of the material or not. If recrystallisation is possible, then it usually pays to undertake it, for it is commonly found that the particles of the material as received are not sufficiently well developed to enable the optical characters to be related to the external crystalline form without ambiguity. Recrystallisation may, however, be forbidden by the nature of the material or of the problem, or both. Thus for example it may be necessary to examine a substance for the presence of a suspected impurity. In this case, recrystallisation will obviously defeat the object of the investigation. Many other problems arise, the essence of which is that the material must be examined, at least in the initial stages, " just as it is ", except in so far as it may be necessary to reduce the particle size by crushing in order to mount it.

I. Preparation and Mounting without Recrystallisation. The particles should not be larger than about 0·2 mm. in average diameter, for if they exceed this size they are difficult to cover with the immersion liquid, and are generally too thick to give characteristic polarisation colours between crossed Nicols (see later). On the other hand, their average diameter should not be appreciably less than 0·01 mm., for crystals much below this size cannot be usefully studied with the polarising microscope.

Crystals that are too large may be crushed by gently pounding them (not grinding) in a small mortar, and then, if sufficient material is available, sieving to obtain particles of about 0·1 to 0·05 mm. in size.* Particles of this size are easily rolled about on the slide by shifting the cover slip with a needle, unless they have a markedly platy habit. Many substances, however, tend to stick to the meshes of a sieve, and in such cases, or where sieves are not available, the crystals may be broken down until their diameter is about equal to that of a human hair (approximately 0·05 mm.), which, in order to make the comparison, may be mounted with a portion of the powder and viewed under the microscope.

A minute amount of the material is placed in the centre of a slide by means of a micro-spatula, and is then covered with a cover slip, which is gently pressed down with a slight rotary motion to spread out the crystal grains so that they do not overlie one another. This may be done with a light pressure of the finger, *if clean and dry*, but individuals who suffer from excessive perspiration are advised to use instead a rubber-tipped glass rod, for a clear impression of a finger-print on the cover slip is of no value in the subsequent examination of the crystals. A small drop of the immersion liquid is then brought to the edge of the cover slip by means of the dropping rod (with which bottles to hold immersion liquids are usually provided) when it will run in by capillary action. Further small drops are added if necessary until the space under the cover slip is just filled. Excess of liquid is to be avoided, as this results in a messy preparation, and the liquid may find its way to the top of the cover slip, where, besides interfering with observation, it is liable to contaminate objectives which have a small working distance.

The bulk of material taken should not in general exceed the size

* Standard sieve openings for this range are shown below.

Meshes per inch.	Size of Opening (mm.).	
	A.S.T.M. Scale (American Society for Testing Materials).	I.M.M. Scale (Institute of Mining and Metallurgy).
100	0·149	0·127
140	0·105	—
150	—	0·084
200	0·074	0·063
270	0·053	—
325	0·044	—

Note that the sieves will pass particles having their *smallest* diameters of the above values, so that the passage of prismatic or acicular crystals, for example, will depend on their cross-sectional diameters.

of a pin's head, and with a fine powder may be considerably less, for it is pointless, and may be confusing, to have an excessively large number of particles to view.

Immersion liquids are dealt with in the next chapter.

It is sometimes necessary to examine wet crystals, *e.g.* a crude reaction product, or a solid phase in a phase equilibrium study, suspended in its mother liquor. In such a case a drop of the material is simply placed on a slide, and covered with a cover slip. The liquid in the material acts as the immersion medium. The crystals may, of course, first be reduced in size by crushing, if they are too large to mount satisfactorily. Alternatively it may be possible, after putting the drop on the slide, to remove the liquid by means of filter paper and mount the crystals in a liquid of known refractive index, as described in II(*a*) below.

II. Methods of Preparation and Mounting which include Recrystallisation. (*a*) *Recrystallisation from Solution.* Substances may be recrystallised *on the slide,* if a suitable solvent is available. At the outset it should be said that it is only possible here, to give general directions, for substances vary so much as regards their solubility, temperature coefficient of solubility, and tendency to form supersaturated solutions in a given solvent, that the best conditions have to be found by trial in each case.

We will consider first the recrystallisation from water of a non-deliquescent substance, the solubility of which is moderate at ordinary temperatures and shows the usual increase with rise of temperature. The following apparatus will be required (ref. Fig. 174).

> Transfer rod
> Micro-pipette
> Glass rod
> Micro-burner
> Cooling block

By means of the glass rod or the micro-pipette, a drop of distilled water 5 to 7 mm. in diameter is placed in the centre of the slide. Minute quantities of the powdered solid are added by means of the transfer rod, and dissolved by stirring, taking care, however, not to spread the drop. Solid may readily be picked up on the tip of the rod if it has been breathed on first. Spreading of the drop is avoided by holding the rod in a nearly vertical position. If necessary the dissolving of the solid may be hastened by *gently* warming the slide over the micro-burner, keeping it in continual motion in a horizontal plane and about 1" above the flame to avoid any risk of cracking it. Addition of solid is continued until no more appears to dissolve. The slight excess remaining is taken up by further warming, or by the addition of a trace of water on the transfer rod. The object is to obtain a solution saturated at, or slightly above, room temperature.

If sufficient material is available, the above procedure can be avoided by making up a few cubic centimetres of the solution in a test tube, filtering from any suspended matter, and transferring a drop to the slide by means of the glass rod. It is worth emphasising, however, that the quantity of material required in the first method is extremely small, and this is likely to be an important consideration in many problems.

If the drop prepared in either of the above ways is warm, it may begin to deposit satisfactory crystals spontaneously on cooling. If examination under the microscope (using a low-power objective, e.g. 1") shows, however, that no crystals have formed, crystallisation may be induced as follows. The slide is warmed over the micro-burner to hasten the evaporation of solvent. Presently a crust will appear at the edges of the drop. This crust is of no value for study, for having grown inwards from the surface of the solution, it consists of poorly formed crystals matted together. The slide is immediately placed on the cooling block, and by means of the transfer rod the crust is broken up and gently stirred into the centre of the drop so as to inoculate it. It may happen that the crust dissolves when this is done, indicating that the original solution was not fully saturated. In this case, the drop must be evaporated further over the micro-burner and the operation repeated. Stirring is continued, again taking care not to spread the drop, and also not to rub the tip of the rod on the slide, for the latter action is likely to result in the formation of a multitude of very minute crystals along the tracks of contact. When crystallisation in the interior of the drop appears to be well under way, the slide is transferred to the microscope. If the crystals are well formed and are not so numerous that they overlie one another, their growth is watched until they have reached a suitable size for study. A cover slip is then placed over the drop to arrest further evaporation of the solvent, and the preparation is ready.

If, however, the crystals are unsatisfactory, they should be partially redissolved by adding a minute drop of water on the transfer rod and the solution stirred again, the undissolved solid acting as seed crystals. Alternatively a larger drop of water sufficient to cause the whole of the solid to redissolve may be added with the micro-pipette, and the solution concentrated by the use of the micro-burner as before.

If it is desired to mount the crystals in a liquid of known refractive index, it will be necessary to remove the mother liquor and this may be done as follows. As soon as the crystals have grown to the right size, the edge of the drop is touched with the end of the micro-pipette (dry) when most of the liquid will be drawn off by capillary action. The preparation is then gently dabbed with strips of filter paper applied edgeways, until microscopic examination shows it to be practically dry. (It is particularly important not to leave the crystals surrounded by small pools of liquid, for in the final drying these will deposit a confused mass of solid material which may interfere with observations of the crystal-liquid border after the immersion medium has been added.) The last traces of moisture are removed by gently warming the slide on the hot plate, or, if the material will not withstand this treatment, leaving it to dry off in the air. A cover slip is now placed over the crystals, and the immersion liquid run in from the edge as previously described.

If the substance is a hydrate, great care must be taken not to overheat it nor to prolong the drying unnecessarily, otherwise dissociation may occur. Moreover in the crystallisation stage, the temperature of the drop must not be allowed to rise above the transition point at which the hydrate ceases to be stable, otherwise crystals of the next lower hydrate or of the anhydrous compound may be deposited. If the transition point is low, it is better to use the hot plate instead of the micro-burner for preparing the solution, so that the temperature can be kept under control.

Sparingly soluble substances may be recrystallised by the above method, but only very small crystals are likely to be obtained. When preparing the solution, only the minimum amount of material should be added to the drop of water, for a large excess cannot be got rid of and is likely to hinder the examination of any crystals which do form from the solution. Crystallisation must be made to take place very slowly, either by allowing the drop to evaporate at room temperature, or cooling it very slowly by placing it on the hot plate, the temperature of which is gradually lowered.

Very soluble substances are apt to be troublesome because once the solution is saturated, a small further evaporation or fall of temperature may result in deposition of a relatively large mass of crystals overlying one another, which will be useless for study. This is particularly likely to happen if the solution supersaturates and then suddenly crystallises. The solution should therefore not be saturated at too high a temperature, otherwise the rate of cooling and of evaporation, and thus the rate of deposition of solid, will be rapid. Supersaturation must be prevented as far as possible, either by introducing a trace of the solid on the transfer rod, or in some cases by gently scratching the slide with the rod just at the edge of the drop. As another method of preventing supersaturation, Denigès * has suggested rubbing the slide with a crystal of the substance at the place to be occupied by the drop, whilst Howard and Stephenson † find that transfer rods previously used in the crystallisation of a substance may retain sufficient traces of it to prevent supersaturation in subsequent crystallisations of that substance.

Deliquescent substances cannot be recrystallised as described above because the vapour pressure of their saturated solutions is less than the normal partial pressure of water vapour in the atmosphere. Two methods are available for dealing with this difficulty. In the first, the slide bearing the drop of solution is placed in a desiccator, and as soon as crystallisation is well under way, it is removed and a cover slip is quickly placed over the drop. It is then examined as expeditiously as possible, though, as water vapour can only be absorbed at the narrow surface of solution around the edges of the cover slip, it is unlikely that the crystals will redissolve at an inconveniently rapid rate. As ordinarily carried out this method has the disadvantage that the drop cannot be stirred while it is crystallising, and so the crystals may be poorly developed. This, however, may be overcome by using a desiccator the lid of which has a tubulure (as in the ordinary vacuum desiccator) carrying a transfer rod in a flexible rubber connection, the slide being supported on a platform fitted with stops to prevent sideways movement. By means of this arrangement it is possible to stir the drop from outside. The second or " hanging drop " method is shown in Fig. 175. The slide carrying a small drop of the solution

* *Mikrochemie*, 1925, **3**, 33.
† *U.S. Dept. Agr. Bur. Chem. Bull.*, **137**, 189.

is inverted over a micro-desiccator, which in (a) consists of a thick glass slide with a flat-bottomed cavity ground in it, and in the alternative design (b) consists of a short glass cylinder with ground ends resting on an ordinary glass slide. The micro-desiccator contains a small quantity of a drying agent in powder form, *e.g.* phosphorus pentoxide, arranged around the periphery of the cell, and all junctions are made airtight by means of a thin film of vaseline, or in (b) the cylinder may be permanently cemented to the lower slide with Canada balsam, bakelite cement or other adhesive. The whole assembly is placed on the stage of the microscope, where the crystallisation of the drop may be watched, and the outward form and at least some of the optical properties of the crystals determined while this is in progress. The disadvantages of this method are that it does not permit of the observation of interference figures in convergent light (see Chap. VIII) since the

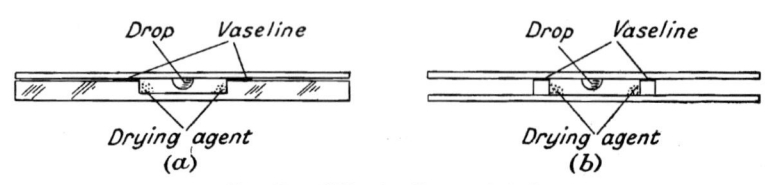

Fig. 175.—" Hanging Drop " Method.

condenser cannot be brought sufficiently close to the object, and if the slide carrying the drop is of the usual thickness, high-power objectives cannot be brought within their working distance of the crystals. This last objection can, however, be removed by using a large cover slip instead of a slide to support the drop. The curvature of the drop will also result in distortion and some loss of light by reflection in observations on crystals lying near its edge.

Recrystallisation on the slide from organic solvents may also be carried out by the methods described above for aqueous solutions. Solvents that are very volatile or mobile should be avoided, however, otherwise rapid evaporation or the uncontrollable spreading of the drop will occur, resulting in the deposition of imperfect crystals or continuous films of solid material, which will be of little value for study. Spreading of the drop may sometimes be prevented by previously surrounding the site with a film of vaseline. Alternatively a slide with a cavity ground in it, such as that shown in Fig. 175 (a) may be used.

Supercooled thymol (1-isopropyl : 2-hydroxy : 4-methyl benzene) is a useful solvent for many organic substances, for it does not spread on a slide. The freezing-point of this substance is $50°$, but it may be supercooled to room temperature and will remain indefinitely in that condition if the melt has been heated to about $90°$. Moreover the supercooling is not readily relieved when the liquid is stirred with a clean transfer rod. To recrystallise a substance from this solvent, the slide is placed on the hot plate and a few fragments of the solid thymol are placed in the centre of the slide. The thymol is melted and its temperature raised to $90°$ and then adjusted to the value at which it is desired to saturate the liquid, say about $60°$. (This temperature will depend on the temperature coefficient of solubility of the substance, and the optimum value will have to be found by trial.) By means of the micro-spatula the powdered substance is added, a little at a time, to the drop of thymol, and is stirred in after each addition with the transfer rod. When no more will dissolve, the excess is taken up by raising the temperature again, or alternatively it may be left to act as seed crystals, particularly if the substance is prone to form supersaturated solutions. The slide is then removed from the plate and the drop is stirred until crystals make their appearance, when the slide is transferred to the microscope stage for observation. If the crystals are satisfactory, a cover slip is placed over the drop.

In all the above methods of recrystallisation, crystallisation is brought about by the cooling or evaporation, or both, of a saturated solution of the substance in a single solvent. Other methods depending on *variation in the composition of the solvent medium* are also available in certain cases. For example, some substances may be brought into solution as complex ions by the addition of volatile acids or ammonia, and then crystallised out by allowing the volatile constituent to escape slowly. Thus silver chloride may be dissolved in ammonia, and if a drop of this solution is placed on a slide, the chloride will be deposited as the ammonia evaporates. (The slide should not be placed on the microscope stage until volatilisation is complete, or sufficiently so for the drop to be covered with a cover slip, otherwise the corrosive fumes will attack the instrument.) Substances soluble in alcohol but not in water may in some cases be caused to crystallise satisfactorily by allowing a drop of water to diffuse into a drop of the alcohol solution. Use may also be made of the principle of "salting out". Thus sodium benzene sulphonate may be crystallised from a drop of its aqueous solution by adding a little solid sodium chloride or a drop of a saturated solution of this salt.

When sufficient material is available, substances may be recrystallised, in a form suitable for microscopic examination, *otherwise than on a slide*, i.e. in a separate vessel. The usual methods of growing well-formed crystals, *e.g.* for measurement on the goniometer, which involve slow cooling or evaporation without disturbance of the solution, usually yield crystals which are too large for the present purpose, though these methods may have to be resorted to when single crystals for mounting on rotation apparatus (see pp. 213–16) are required. If, however, the solution is mechanically stirred during crystal growth, the result is usually a microcrystalline powder, the individual particles of which are well shaped and of a suitable size for mounting on a slide. The best conditions as regards rate of stirring, rate of cooling, etc., vary from substance to substance and must be found by trial. A suitable apparatus consists of a stout wide test tube fitted with a cork carrying a rotating stirrer driven mechanically. The tube may be immersed in a water bath, by means of which the rate of cooling can be controlled. Better results are sometimes obtained by evaporating the saturated solution at constant temperature instead of cooling it, and for this purpose the cork may be fitted with two tubes through which a current of air can be drawn over the surface of the solution while it is being stirred.

The microcrystalline powder obtained as above may either be examined in its mother liquor or be dried and mounted in liquids of known refractive index. For the former purpose a drop of the mixture may be transferred to the microscope slide by means of a micro-pipette, having a bore of sufficient size to allow the crystals to enter. If the crystals are to be dried, this may be done by removing the mother liquor from the drop on the slide, as already described, or alternatively the whole contents of the stirring apparatus may be filtered off and the powder dried by suction followed by gently pressing between filter papers.

(b) *Recrystallisation by Sublimation*. This method is particularly useful for organic substances and for volatile inorganic compounds, *e.g.* mercuric iodide, ammonium halides, and arsenious oxide. It also affords a means of separating a volatile constituent from a mixture. It must be borne in mind, however, that overheating, particularly in the case of organic material, may cause decomposition, and that with polymorphous substances a form which is unstable at room temperature may be deposited from the vapour. For example, the sublimation of mercuric iodide may yield crystals of the yellow form which is only stable above 127°, and these crystals are liable to undergo transformation into pseudomorphs of the stable red modification (see p. 5), particularly if they are touched or subjected to shock.

Microsublimation may be carried out in many ways. The simplest is to sublime the material from one slide to another held just above it, the former being heated with the micro-burner, or, with better control of the temperature, on the hot plate. It is advantageous first to fix the material to the slide by moistening it with a drop of water and then evaporating this off. Alternatively the substance may be dissolved in a suitable solvent and successive drops evaporated from the same place on the slide until a sufficient amount of solid has been accumulated. This brings the substance into more intimate contact with the glass, and prevents it being blown away by stray air currents, or scattered by accidental disturbance of the slide. If the micro-burner is used for heating, the slide carrying the material is held in one hand, and the receiver slide in the other, so that it lies across the former with one edge resting on it to keep the two steady in relation to one another. The vapour then diffuses across a wedge-shaped space between the two slides, and this space should be made as narrow as possible without allowing the receiver actually to touch the material. Heating should at first be cautious and as soon as the first sign of a sublimate appears, the slides should be raised well above the flame to reduce the rate of vaporisation, otherwise a very finely divided deposit is likely to be the only result. By moving the receiver into different positions a series of sublimates may be collected on it. If the hot plate is used, the receiver slide may be supported on strips of thin glass. Heating in this case should be even more cautious than when the micro-burner is used, at least until the optimum temperature for the substance has been ascertained, for it is not possible rapidly to reduce the temperature of the plate if sublimation should be proceeding too quickly.

With very volatile substances it may be necessary to cool the receiver. This may be done by putting drops of cold water on it from time to time by means of the micro-pipette ; or a chilled block of copper or aluminium may be used. On the other hand with substances which tend to form a very finely divided deposit, it may be advantageous to reduce the rate of crystallisation by warming the receiver, and this may be done by placing on it a warmed block of carbon, *e.g.* a piece cut from the ordinary charcoal block used in qualitative analysis, or from an arc electrode.

When only a very little of the material is available, losses from the vapour phase may be avoided by subliming in an enclosed space. For this purpose the substance may be heated in a shallow glass capsule with a carefully ground upper edge on which the receiver slide is laid. If these are held together with a suitable metal clip, the capsule can be heated over the micro-burner. Alternatively if the hot plate is used as described above, the two slides may be separated by a shallow glass ring cut from a piece of wide glass tubing and having its upper and lower surfaces ground perfectly flat.

Substances which sublime very slowly in air, may with advantage be sublimed *in vacuo*, or under reduced pressure. Various forms of apparatus appropriate for microscopic technique have been described for this purpose. One of these consists of a short glass tube fitted with a stopper and tube leading to the vacuum pump at its upper end, and with a constriction near the bottom supporting a circular cover slip which serves as the receiving surface. The substance is placed in the bottom of the tube and is heated by any convenient means such as a micro-burner.* In another apparatus which has been described a circular cover slip, the receiver, is affixed by a drop of liquid to the flattened under side of a hollow ground glass stopper, through which water for cooling may be circulated. The stopper fits into the upper end of the sublimation tube and is prolonged so that the cover slip is only just above the heated substance, which is placed on a small glass saucer resting on a layer of metal filings (to make the heating more uniform) covering the bottom of the sublimation tube. A side tube sealed to the sublimation tube leads to the pump.†

(*c*) *Recrystallisation from a Melt.* Substances having a low melting-point may with advantage be examined in the form of a thin film obtained by melting them between a slide and a cover slip.

* Eder, *Schweiz. Wochschr. Chem. Pharm.*, 1913, **51**, 228, 241, 253 ; Eder and Haas, *Mikrochemie, Emich Festschr.*, 1930, 43–82.
† Klein and Werner, *Z. physiolog. Chem.*, 1925, **143**, 141.

To make such a preparation the slide is placed on the hot plate, a small pile of the powdered material is put on the centre of the slide by means of a micro-spatula, and a cover slip laid on top of the pile. The temperature is then raised until the substance melts and fills the space between the slide and slip. The slide is then removed from the plate and the melt allowed to solidify, scratching the edge of the film with a needle, if necessary, to relieve supercooling. The crystals thus obtained, having grown rapidly, will probably be too small or narrow for convenient study, and in this event larger crystals may be obtained as follows. The bulk of the film is remelted by placing the slide partly on and partly off the hot plate, leaving an unmelted strip along one side. The source of heat is then turned off, and as the plate cools the melt will crystallise slowly, supercooling being entirely prevented by the presence of the unmelted strip.

The parallel-sided plates formed by this method admit of good microscopic examination being made without the use of an immersion liquid. A disadvantage is that the outward form of the crystals is largely obscured, because mutual interference along their boundaries during growth produces anhedral outlines, but the relation of the optical properties to prominent cleavage directions, when these exist or to a direction of elongation, is easily made out. The method does not permit the determination of the refractive indices, but such properties as pleochroism, extinction angles, birefringence, directions of fast and slow vibration, size and dispersion of the optic axial angle, and the optical sign, may be observed on suitably orientated plates. In general, however, the examination of crystals prepared in this way is only to be regarded as supplementary to their examination in a liquid medium.

Hard Compact Materials. For the examination of industrial materials such as refractories or slags, methods similar to those used by petrologists in the study of rocks and minerals have to be employed. These methods are described in detail in textbooks of practical petrology,* and it is only necessary to give an outline here.

---

* A. Johannsen, *Manual of Petrographic Methods* (McGraw-Hill, 1918), pp. 572–604. W. C. Krumbein and F. J. Pettijohn, *Manual of Sedimentary Petrography* (D. Appleton-Century Co., 1938), pp. 363–4. A Holmes, *Petrographic Methods* (Murby & Co., 1921). A. V. Weatherhead, *Petrographic Micro-technique* (Arthur Barrow, Ltd., 1947). G. R. Rigby, *The Thin-Section Mineralogy of Ceramic Materials* (British Refractories Research Association, 1948).

Massive material, *e.g.* firebrick, may be examined in two ways. First in the form of a thin transparent section, and second, by the crushing and separation of its constituent substances which may then be examined in immersion media in the usual manner. The first method reveals the texture, particle size, form, and relative abundance of the ingredients, and it is possible to determine a number of the optical properties of the various substances present. The method is particularly suitable in the study of thermal changes as for example, those undergone by silica brick when used as a furnace lining (see p. 438). The determination of the optical properties is circumscribed, however, by the fact that the optical section presented by any individual crystal is random, and cannot be altered except by the use of the universal stage, and also by the fact that the refractive indices cannot be determined except in a comparative way with adjacent substances. In many cases these limitations do not matter, because the possible ingredients are known in advance and differ so markedly in their optical properties, colour or cleavage, that they can be readily distinguished whatever the orientation of the grains. But with material about which little is known, or where ambiguities arise, the thin section method must be supplemented by crushing and the examination of the separate constituents. The following is a condensed account of the procedure to be followed for the examination of material by the two methods mentioned above.

A thin section may be prepared as follows. A thin chip about 1" square is broken from the edge of the specimen, or a parallel slice is cut from it by means of a power-driven soft iron disc armed with diamond dust, and a smooth surface is prepared on one side by grinding with successively finer carborundum powder on a metal plate or on a sheet of plate glass. Different plates must be used for the various grades of abrasive and the chip carefully washed when changing from one grade to another. Water is used as a lubricant unless any of the constituents is water soluble, when a thin oil may be used. A small quantity of natural Canada balsam is placed on a glass slide and heated slowly to a temperature of about 80° until it is "cooked", that is until a small portion removed on a needle-point or the pointed end of a matchstick and pressed against a hard surface such as the thumb-nail, has just lost its stickiness and hardens almost immediately. The prepared surface of the chip, which has been previously warmed, is now pressed down on the warm balsam so that all bubbles are excluded, and left to cool, when it will be found that the balsam has set quite hard. The other side of the specimen is now ground down in the same way until it is about 0·03 mm. thick. This can be judged by examining the slice from time to time under the microscope until some substance such as quartz (or any other substance of low double refraction), shows a grey or pale yellow of the first order of Newton's scale (see p. 245). Dirty balsam around the slice is cleaned off with a warm knife-blade and a little alcohol. The section is now covered with a cover glass. This is done by cooking a little balsam on a cover slip and inverting it upon the section, or by pouring a little warm balsam, already cooked, upon it. A warmed cover slip is now allowed to fall from one edge upon the section and gently pressed down. After cooling, the balsam which has been squeezed out around the cover slip is cleaned off with alcohol, and the preparation is ready for examination.

Friable material, which disintegrates under the grinding process, may be treated as follows.* A chip is placed in a dish or small beaker and covered with bakelite varnish diluted with methyl alcohol and a few drops of acetone. The dish is allowed to stand for several hours in an evacuated desiccator, after which the specimen is removed and heated in an oven at a temperature of 70° for about four hours to remove the solvents. The temperature is then raised and maintained at 100° for several days, or if the material is stable enough, at 200° for a few hours. The hardened specimen can now be ground in the usual way.

In the second method, the material is crushed in a cylindrical steel percussion mortar by tapping the pestle smartly with a hammer, and sieving to a convenient size. The material may then be mounted and examined in the manner already described for chemical substances, or, if desired a separation or partial separation of the ingredients by gravity methods may first be made. For this purpose, a small conical separating funnel is partly filled with a heavy liquid such as bromoform (sp. gr. 2·9), and a few grams of the sieved material are poured into the funnel. The lighter material will float and the heavier will sink, and can be drawn off after the separation is complete. Methylene iodide (sp. gr. 3·3) is also very useful for this purpose. To adjust the liquid to any desired specific gravity, bromoform may be mixed with methylene iodide, or either

* See also J. W. Fowler and J. Shirley, *Geol. Mag.*, 1947, **84**, 354.

may be mixed with benzene to give less dense solutions. Clerici solution * having a specific gravity of about 4·2 may also be used, and lower values may be obtained by dilution with water. While separation is in progress the material must be agitated with a glass rod from time to time to free heavy substances entangled amongst the lighter fraction. The heavier fraction is easily caught on a filter paper held in a conical funnel under the separating funnel, whilst the lighter fraction may be collected in the same way after removing the heavy material, and running off the major portion of the liquid. The use of the organic liquids mentioned above has the advantage that they evaporate without leaving any residue, and the grains do not need washing before mounting for the microscopic examination.

Permanent Mounts of Crystals. It is sometimes necessary to make permanent mounts of crystals for future reference or for instructional purposes. Inorganic substances which are not affected by moderate heating, *e.g.* anhydrous salts, may be mounted in Canada balsam as follows. The substance may first be recrystallised from a drop of solution on the slide and the mother liquor removed as previously described. The slide is then heated to about 80° on the hot plate, and the crystals are covered with a little natural Canada balsam which is added by means of a glass rod. Heating is continued until the balsam is cooked (see previous section), and any air bubbles may be drawn to one side and removed with a needle or a piece of thin wire. A cover slip which has been heated over a small bunsen flame is then placed on the preparation and gently pressed down, if necessary, to cause the balsam to fill the space between slide and slip, or to expel excess. When putting on the cover slip, it is advisable to allow it to fall on the balsam sideways so as not to entrap any air. The slide is then allowed to cool, and excess balsam cleaned off with alcohol.

If it is desired to mount material without recrystallising it, a small quantity of balsam may first be cooked on the slide, by heating it on the hot plate to about 80°, and the material then sifted into the balsam from a micro-spatula. A cover slip is added and the preparation completed as described above.

Inorganic substances which cannot be heated may be mounted in Canada balsam thinned with xylene, or in " Sira " mountant

* A concentrated aqueous solution of equal weights of thallium malonate and thallium formate.

(supplied by Stafford Allen & Co. Ltd., London). These mountants require some time to set.

For mounting organic substances (which are dissolved by the above mountants), the authors have found that " waterglass " (concentrated sodium silicate solution) is suitable. A little of the medium is added to the crystals on the slide, or alternatively it is placed on the slide first, the crystals sifted in, and the cover slip added. It is advisable to use just enough of the medium to fill the space between the slide and the cover slip. The waterglass does not set between the glass surfaces (though it is sufficiently viscous to prevent the crystals moving about), and it is therefore necessary to fix the cover slip in position. This may be done by means of a paper mask covering the edges of the cover slip and cemented to the slide with a suitable adhesive, e.g. " Durofix ". The mask should be cut from a piece of absorbent paper, e.g. duplicating paper, and gently pressed into place, when the waterglass around the edges of the cover slip will soak into the paper, and dry off in the course of a few days. Alternatively, the edges of the cover slip may be sealed with Canada balsam dissolved in a little xylene to a treacly consistency, or with " Sira " mountant.

Methods of Changing the Orientation of Crystals under Examination. Crystals mounted in a liquid medium may often be rolled from one position to another by gently moving the cover slip. This operation is facilitated by mixing the crystals with powdered glass, which helps to turn them over and supports them in slanting positions, and which being isotropic does not interfere with the examination. Alternatively the crystals may be mounted in a viscous medium, e.g. natural Canada balsam, or in " Araclor No. 1260 ".*

It is sometimes useful to mount single crystals on a rotation apparatus, when, for example, the extinction angles in a whole zone may be determined, or the position of the optic axial plane be found. Crystals to be examined in this way may be larger than those suitable for mounting on a slide, and, since the main object is

---

* The Araclors are a series of chlorinated diphenyls manufactured by the Monsanto Chemical Co., St. Louis, U.S.A. (and London). No. 1260 is a light yellow soft resin with a refractive index of about 1·64, which becomes fluid on gentle warming. The Araclors are non-oxidising, permanently thermoplastic, of low volatility, non-corrosive to metals, and are not hydrolysed by water or alkalis.

214 PREPARATION AND MOUNTING OF MATERIAL

to find the relationship of the optical vectors to the outward form, should be reasonably well developed.

Reference has already been made to the Miers stage goniometer (p. 176), which may be used for this purpose. A simpler form of stage goniometer, without tilting or traversing adjustments, is also

FIG. 176.

made by Messrs. J. Swift & Son and is shown in Fig. 176. The crystal can be mounted either in small forceps or by means of wax on a needle-point, and may be observed dry or in a liquid placed in the small cell shown. A divided semicircle gives readings of the amount of rotation by means of a vernier to 5' of arc. A modifica-

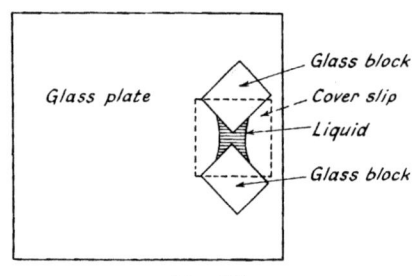

FIG. 177.

tion of this apparatus without the cell can be obtained, in which the crystal is immersed in a drop of liquid held by capillarity between two glass blocks as shown in Fig. 177. This has the advantage of economy in the use of immersion liquids, and facilitates changing from one liquid to another. Furthermore the needle can be removed from the apparatus to facilitate the mounting of the crystal (see below). If such apparatus is not available, simple

CHANGING THE ORIENTATION OF CRYSTALS 215

substitutes as shown in Fig. 178 (a), (b), and (c), may be made. The base in (a) and (b) is a thin wooden slip perforated in the centre, such as is supplied by microscope dealers for making dry cells for holding micro-preparations. Upon this a cork is glued (Fig. 178 (a)), and through its base a needle is thrust so that it may revolve with its point above the central hole. The crystal is mounted upon the end of the needle with wax.* The blunt end of the needle may be thickened with solder to make a handle. The revolving support may also be made of a fine glass rod held in position between two slips of cork with **V** notches (Fig. 178 (b)), the whole being clamped in position by an inverted **U**-shaped piece of brass upon which the spring clip of the microscope stage rests as shown. In the first

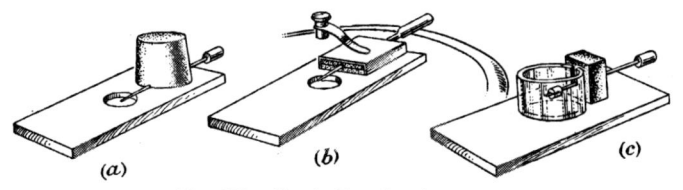

FIG. 178.—Simple Rotation Apparatus.

case where the needle cannot be unshipped in order to mount the crystal, the latter may be brought against the wax upon a piece of stout paper or card, the edge of which may be bent below the end of the needle. Types (a) and (b) are intended for the examination of crystals in the dry condition, while type (c) is meant for examination in an immersion liquid, and may be made as follows. A piece of stout glass tubing about 2 cm. in diameter, and 1·5 cm. in height, is ground flat on one side, the upper and lower surfaces also being made smooth by grinding upon a hone or upon a glass plate with carborundum or emery. With a cutting disc, a notch is cut half-way down the flattened side. The tube is cemented to a glass slide beside a piece of cork which presses closely against the flattened surface. The cement may be cooked Canada balsam.

* "About two parts to one part of beeswax and pitch, melted up together and thoroughly blended; when cold it is sufficiently adhesive and soft to permit of ready attachment of the crystal, while being adequately solid not to allow slipping after attachment."—A. E. H. Tutton, *Crystallography and Practical Crystal Measurement.* 1st Edit., Macmillan, 1911, p. 30.
  A "Soft red wax" supplied by Becker & Co., Hatton Garden, London, E.C., has been found very satisfactory for the purpose.

The balsam may be attacked by some fluids. If this occurs, a cement may be made by dissolving shellac in alcohol. The needle is mounted as before, and the crystals are fixed to it with wax, or by holding them in a small piece of cork with a slit at one end into which the crystal may be thrust, and held tight with a small wire clip, which grips the jaws of the cork vice together. The crystal should be first of all mounted on the cork, which may then be placed upon the needle with a pair of forceps, a small hole having already been made in the cork, into which the needle-point may easily enter. In place of the cell just described, the simple arrangement of glass plates shown in Fig. 177 may be used.

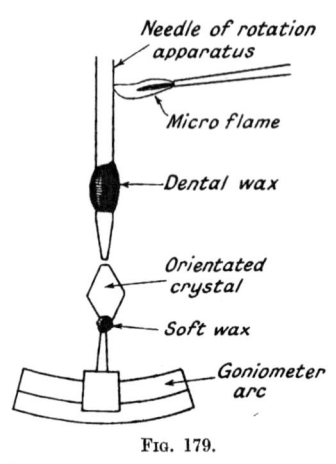

It is often difficult to mount a crystal in a desired orientation on the needle of a rotation apparatus, if it is not provided with tilting and traversing adjustments. This difficulty may be overcome by a method due to Wood and Ayliffe.* The crystal is first mounted with soft wax on the needle of an ordinary reflecting goniometer arranged with its axis vertical, and the crystal brought to the desired orientation by means of the adjustments on the goniometer head. A small quantity of white dental wax is melted on to the needle of the rotation apparatus about $\frac{1}{4}''$ from its point, and the needle is then fixed vertically above the crystal so that its point nearly touches it (see Fig. 179). This adjustment may be conveniently made by sighting on a plumb line. The upper part of the needle is now gently heated by means of a small flame, whereby the dental wax is made to run down to the tip of the needle, so gripping the crystal firmly on cooling. The goniometer head is now gently lowered, thus breaking the soft wax connection between the crystal and the goniometer needle.

FIG. 179.

Section Cutting (Single Crystals.) The cutting of accurately orientated sections of crystals entails the use of complicated and

* *J. Sci. Instr.*, 1935, **12**, 194.

SECTION CUTTING (SINGLE CRYSTALS) 217

expensive apparatus, the description and use of which lies outside the scope of this book. It is, however, sometimes necessary to make observations along the length direction of markedly prismatic or acicular crystals, and suitable sections may be obtained upon soft material in the following way. A stick of hazel pith, such as is used by botanists for embedding specimens to be sectioned with a microtome is taken, and with a razor, one side of it is cut flat. A longitudinal slit is then made for a short distance down the centre parallel to the flat side. One of the crystals is inserted in the slit so that its

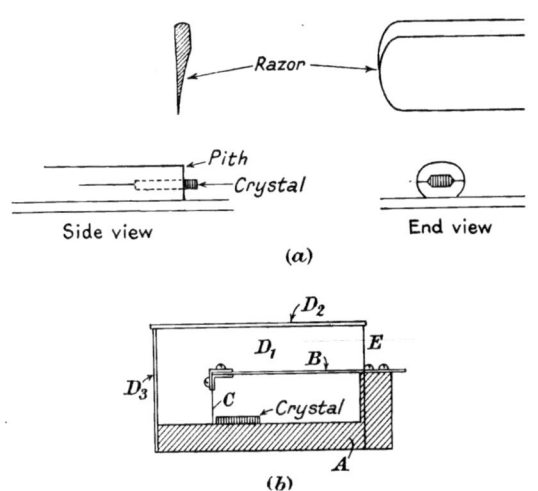

FIG. 180.—Apparatus for Section Cutting.

length is parallel to the length of the stick, and the whole is pressed down firmly on a glass slide, the flat side of the pith being in contact with the glass. A number of transverse sections through both pith and crystal may now be cut, making them as thin as possible. Fig. 180 (a) should make the method clear.

An alternative way of cutting transverse sections of small crystals has been described by J. Parry, A. F. Williams, and F. E. Wright,[*] which makes use of a small "guillotine" shown in Fig. 180 (b). A is a block of brass, to which is clamped a bronze spring B, which carries a piece of razor-blade C, fixed as shown. The apparatus, which is about 5 cm. long, is enclosed in glass slips D_1, D_2, D_3, the

[*] Min. Mag., 1932, 23, 158.

side slips D_1 being permanently cemented to the block A, and the others D_2 and D_3 being removable. The object of these slips is to prevent the sections flying off and getting lost. The instrument is worked by raising the spring by inserting, say, a screwdriver blade under the spring through the opening E, and suddenly releasing the blade on to the crystal, which has been arranged previously in the proper position.

Recommended for Further Reading

Handbook of Chemical Microscopy, Vol. I, E. M. Chamot and C. W. Mason (Wiley, 1938).

CHAPTER VII

THE MICROSCOPIC EXAMINATION OF CRYSTALS

(i) *PARALLEL LIGHT*

The Optical Examination. The complete optical examination of any substance comprises the following four main stages, which it is convenient to summarise here.

(i) *Ordinary Light (Nicols not inserted)*

Observations on colour; crystalline form, if developed; cleavage; fracture; measurement of edge angles; and the determination of the refractive index of isotropic crystals.

(ii) *Polarised Light (Lower Nicol inserted)*

The determination of the principal refractive indices of anisotropic crystals and observations on pleochroism and "twinkling".

(iii) *Crossed Nicols*

Distinction between isotropic and anisotropic substances. Observation of the extinction, whether straight, symmetrical or oblique, if the latter, whether dispersed or not; measurement of the extinction angle. Determination of the birefringence from interference colours and thickness; and of the fast and slow directions by means of compensating plates.

(iv) *Convergent Polarised Light between Crossed Nicols*

Observation of interference figures, and determination of uniaxial or biaxial character. If the latter, estimation of the size of the optic axial angle. Determination of the positive or negative character of the substance. Dispersion of the optic axes and bisectrices.

In practice, the properties listed under ordinary light are made above the polariser, the removal of which in some microscopes is a time-consuming operation. Very often it will not be necessary to go through the whole of the above process, the desired result of the problem in hand frequently emerging after a partial examination.

I. EXAMINATION IN ORDINARY LIGHT

Colour. Care ought to be exercised in describing the colour of a substance under the microscope, for, whereas the quality of the colour of a pure substance is a definite property, its appearance in transmitted light varies greatly with the thickness of the crystal, and some substances which are strongly coloured in crystals of a few millimetres in diameter may appear almost colourless in minute particles under the microscope. The colour displayed by any substance under the microscope should always be observed (due regard being paid to thickness), in order to compare it with any changes of colour due to pleochroism (see below).

Crystal Form. The shapes of crystals observed under the microscope will depend upon the system to which they belong, and the habits they affect, the latter determining the orientation of the crystals upon the slide. A crop of platy crystals will tend to orientate themselves similarly with their larger faces normal to the axis of the microscope, and will provide good material for optical study, because the orientation is fixed, and the observations will not be hampered by internal reflections from sloping prism and pyramid faces. On the other hand the examination in other directions may be difficult. Disproportionate development of similar faces of individual crystals often masks the true crystallographic form, but in a group of them grown on a slide, some approximating to ideal shape are almost certain to be seen. Crystals from manufacturers' stock may be badly developed or be too large to be mounted. These may be crushed, when, if cleavage is developed, characteristic shapes will be seen, and optical properties may be determined in relation to the cleavage directions. Alternatively, they may be recrystallised in the manner previously described. Fig. 181 shows some examples of characteristic shapes which occur in the various crystal systems, the orientation being given by naming the most important faces according to the Millerian notation. The crystals are shown in the positions they would be most likely to adopt upon the microscope slide. In studying the shapes of crystals under the microscope, the slope of faces and edges may be seen by focussing on different parts of them. No apparent shift will be observed on altering the

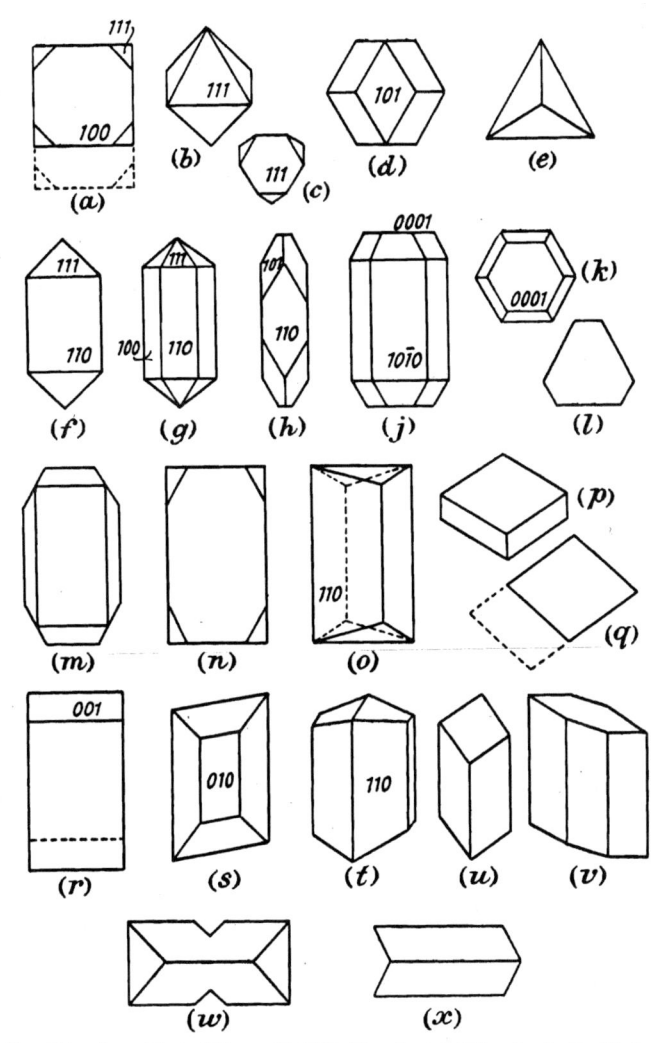

Fig. 181.—Some Typical Shapes Exhibited by Crystals Belonging to the Various Crystal Systems as seen under the Microscope.

Cubic: a, cube modified by the octahedron; b and c, octahedra; d, dodecahedron; e, tetrahedron.
Tetragonal: f, g, h, various combinations of prisms and pyramids.
Hexagonal: j, prism, pyramid and basal pinacoid; k, basal pinacoid and pyramid.
Trigonal: l, thin basal plate and two trigonal prisms; p and q, thick and thin rhombs.
Orthorhombic: m, n and o. The two former lying on pinacoids, and the latter on the prism (110).
Monoclinic: r, s, and t. The first a pinacoid form, lying on the (100); the second, pinacoid prism and dome, lying on (010); and the last, prism pinacoid and pyramid, lying on (110).
Triclinic: u and v, both crystals of copper sulphate, the first a pinacoidal form and the second, pinacoids and a prism.
The twinned forms are w, an orthorhombic twin of magnesium ammonium phosphate, and x, a twin of the monoclinic substance gypsum.

C.P.M.   H*

focus above the edge of a truly vertical face. Some distorted crystals may resemble at first sight normal forms of other systems, but a further examination of the optical properties of the crystal will reveal its true nature. For example, distorted cubes may resemble tetragonal or orthorhombic forms (Fig. 181 (a)) but between crossed Nicols show their isotropism (p. 241). Distorted rhombs are differentiated from monoclinic and triclinic forms by their decentralised uniaxial figure in convergent light (p. 283), and so on. Crystals of the tetragonal and hexagonal systems usually lie on their prism faces, but may display square tetragonal or hexagonal basal plates respectively. Trigonal crystals may develop prism forms with rhombohedral terminations or crystallise in stout or exceedingly thin rhombohedral plates (Fig. 181 (p), (q)). Twinned forms may be recognised by the presence of re-entrant angles as shown in Fig. 181 (w), (x), but must not be confused with parallel growths of optically similarly orientated crystals (p. 80).

Cleavage. Cleavage and type of fracture (p. 76) should be observed under the microscope wherever possible, as they are of diagnostic value. The former may be recognised by the regular shapes of broken fragments, or as fine lines which are traces of the cleavage planes upon the faces of entire crystals.

The Measurement of Edge (Profile) Angles. The measurement of edge angles may be undertaken on well-formed crystals, and is frequently of diagnostic value. It must be understood, however, that accurate values can only be obtained by confining attention to those edges lying in a horizontal plane, or to the intersection of planes parallel to the axis of the microscope. The method of measuring such angles is as follows : The junction of the two edges which contain the angle to be measured is brought near the intersection of the cross-wires, and one of the edges is adjusted parallel to, and nearly touching a cross-wire (Fig. 182). This enables a more accurate adjustment than if the edge was placed on the wire, because the thickness of the latter may obscure a small inclination. The reading on the revolving stage is now taken, and the specimen is rotated until the other edge forming the angle to be measured coincides with (i.e. becomes parallel to, and nearly touches) the same cross-wire. The reading on the stage is again taken and the angle obtained by difference. The stage may be turned either clockwise or anti-clockwise, in one case the supplement of the angle is measured and in the other, the angle itself.

Measurement of edge angles often reveals the true symmetry of a crystal which is unequally developed, but it must be borne in mind that these measurements are not more accurate than about $\pm 0.5°$, and therefore conclusions as to symmetry based on such measurements should always be confirmed by the optical properties of the section. For example, in crystals of carnallite, $KCl, MgCl_2, 6H_2O$ (see Expt. 4, p. 367), grown from solution on a slide, all the angles are very near to 120°, and as measured under the microscope might seem to be exactly so. The section is, however, birefringent, showing that it cannot be a basal section of a hexagonal crystal which would be isotropic in parallel light; actually the crystal is orthorhombic.

Donnay and O'Brien * have shown how it is possible, by the aid of stereographic projection, to correlate the results of normal goniometrical measurements and the profile angles as measured under the microscope. Two main problems are dealt with : (1) Given the morphological constant of a crystal (axial ratios, and interfacial angles), to determine the apparent angle between any two edges that will be seen when the crystal lies on any face ; (2) Given the apparent edge angles measured under the microscope (the number of angles depending on the crystal system), to determine the morphological constants. These problems are solved entirely by graphical methods.

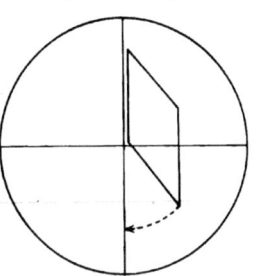

Fig. 182.—Measurement of Edge Angles.

Measurement of Size and Thickness. Accurate linear measurements of crystals under the microscope may be made by using a graduated mechanical stage ; or by means of a transparent scale at the focus of the eye lens in the ocular, calibrated as described overleaf ; or by using the more complicated screw-micrometer ocular, of which there are several types. In the first case, one side of the object to be measured is placed at the inter-section of the cross-wires and then moved until the other side occupies a similar position. The difference between the two readings on the vernier of the stage gives the required length. The eyepiece micrometer

* *J. Ind. Eng. Chem. (Anal.)*, 1945, **17**, 593.

scale is a piece of glass upon which has been engraved a scale consisting of fine lines one-tenth or one-twentieth of a millimetre apart. This is placed in the focal plane of the ocular. By placing upon the stage of the microscope a scale of known magnitude, the eyepiece scale may be calibrated for different combinations of objectives and oculars. The reader should familiarise himself with the size of the field seen with various combinations of oculars and objectives, as this will enable rapid approximate estimations of the sizes of crystals to be made. In drawings of crystals, the magnification in diameters of the object should always be stated. This is the ratio between the actual size of the object and the diameter of the drawing.

It is sometimes necessary to measure the thickness of crystals or crystal plates, as for example in the measurement of birefringence. This may be done by using the fine adjustment screw of the microscope where this has been calibrated by the makers. With a high-power objective, say $\frac{1}{4}''$ or $\frac{1}{8}''$, a fine scratch made upon a glass slide is carefully focussed by *raising the objective*. The crystal plate is then pushed into position above the scratch, and the upper surface of the plate is brought into focus by raising the objective again. The difference between the first and second positions of the micrometer screw gives the desired result. It is very important to note that the micrometer screw must be turned in one direction during a determination so as to get rid of " lost motion " which would introduce serious error into the result. It is obvious that such measurements must be done upon dry and uncovered crystals.

Refractive Index. The accurate determination under the microscope of refractive indices must be done in polarised light in the case of aniotropic substances, but the principles underlying the determination may conveniently be explained here, first of all with reference to isotropic crystals, which need only the use of ordinary (unpolarised) light.

If a colourless isotropic substance like glass is immersed in a colourless fluid, both having the same refractive index, the glass will become invisible in the fluid owing to the disappearance of refraction and reflection effects at the junction of the two substances. If the refractive indices are different, the *relief* or strength of the border between the two substances depends upon the magnitude of the difference. Practical use is made of this fact in the determination of refractive indices under the microscope, by immersing substances in different fluids of known refractive index, until the border phenomena disappear. When this happens, both substances have the same refractive index. In practice, when the border phenomena appear, it is necessary to know whether the refractive index of the fluid is higher or lower than that of the crystal. For, the relief of a substance with an index of 1·65 will be the same if it is placed in fluids with refractive indices of 1·70 or 1·60, the difference being the same in each case. Two methods are available, namely, (1) that of *central illumination*, commonly called the " Becke method ", and (2), that of *oblique illumination*, or the " Schroeder van der Kolk method ".

(*i*) *The Method of Central Illumination.* If the junction between

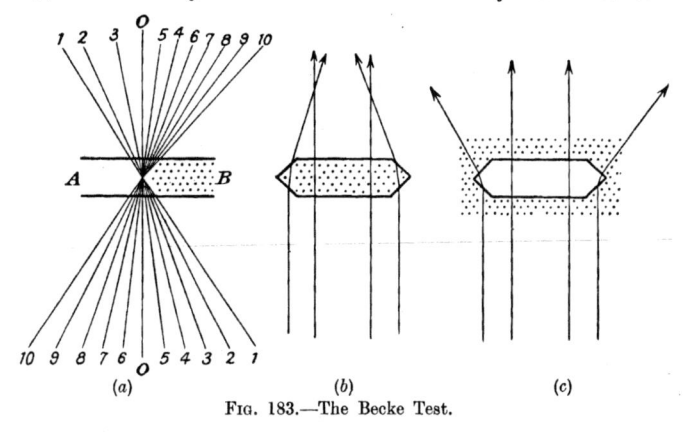

FIG. 183.—The Becke Test.

a crystal and an immersion fluid is sharply focussed with a high- or medium-power objective (say $\frac{1}{4}''$ or $\frac{1}{8}''$ focal length), the two substances having different refractive indices, and then the image is thrown slightly out of focus by raising the objective, a bright line will appear near the junction and will move laterally towards the substance having the higher refractive index, broadening the while and becoming fainter.

On depressing the tube, the phenomenon is reversed. The explanation for different types of junctions between crystals and liquids is as follows, the diagrams Fig. 183 (*a*), (*b*), (*c*), given in illustration of it, having the substance with the higher refractive index shaded.

At a vertical junction, the bright line (or " Becke line " as it is commonly called), is seen in slightly convergent light. In Fig. 183 (a), such a convergent beam of light is seen directed against the vertical junction between A and B, the latter having the greater refractive index. The ray O, which is parallel to the junction, will pass through without deviation. Those rays which pass from the medium with the smaller refractive index to the denser medium will be refracted as shown, being bent towards the normal to the junction within B. All rays which pass from B towards the junction will not, however, pass through it, but those which impinge within the critical angle are totally reflected and emerge on the same side of the junction. In these circumstances, the side of the junction towards the substance having the greater refractive index will be more brightly illuminated, and as the objective is raised the bright band will appear to move towards B, becoming more diffuse. Reference to the figure will show that the narrower the cone of light falling upon the junction the clearer will be the effect, as practically all the light emerging from A may be cut out. This may be done by means of a substage diaphragm or by lowering the condenser.

Vertical junctions such as have been described are not often met with in practice. More often, crystals or crystal fragments present a rude lens-like shape due to the edges being thinner than the centre. In cases like these, when the refractive index of the fragment is greater than the immersion fluid, rays of light converge above it, the effect being similar to that of a double convex lens (Fig. 183 (b)). In the opposite case, where the refractive index of the substance is less than the fluid, the fragment acts as a diverging or double concave lens (Fig. 183 (c)). If such crystals or grains are observed with a fairly high-power objective it will be seen that in the first case, on raising the tube of the microscope from the position of sharpest focus, that the interior of the grain will be illuminated more strongly than the surrounding field, the bright line at the margin moving inwards. In the second case the bright line moves outwards and the crystal appears darker than the field. The effect may be observed upon grains of this kind in either parallel or weakly convergent light, the latter always being used in practice.

It sometimes happens that two bright lines are seen, and it is difficult to determine which is the "true" Becke line, since they move in opposite directions on raising the objective. The effect usually occurs with crystal fragments, the edges of which slope irregularly in a number of directions. For this reason it tends to be more prominent with large fragments. The "false" line can usually be eliminated by closing the substage diaphragm or by using the method of oblique illumination described in the next section.

(ii) *The Method of Oblique Illumination.* The effect of the lens action of the fragments may be observed in another way, using a low-power objective, say 1", and inclined illumination. The latter may be obtained either by tilting the mirror to one side; by inserting a card between the condenser and the stage so that the aperture is half covered; or by placing a finger or a card over one side of the mirror. It will be seen that in the first case, when the substance has a greater refractive index than the immersion fluid, on viewing the appearance in the microscope, a shadow appears on the same side *of the grain* as that upon which the screen was inserted, the other border being more brightly illuminated than the field (Fig. 184 (a)), and *vice versa* Fig. 184 (b). Care must be taken to see that if the condenser is not removed during the determination, it always occupies the same position, for the phenomenon as seen in the microscope can be reversed by placing the object above or below the focus of the condenser.

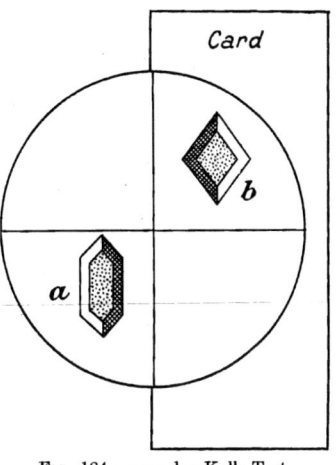

FIG. 184.—van der Kolk Test.

A method giving more critical oblique illumination has been described by Saylor.* In this two light stops are used. One is in the focal plane of the objective and covers half its aperture. The other, which consists of a straight-edged card or thin metal sheet, is located above the low-power component of the condenser and on the same side as the upper stop. By means of the Bertrand lens the focal plane of the objective is viewed, and the lower stop adjusted vertically until its image is brought into this plane. This stop is then pushed in until only a narrow beam of light appears

* *J. Res. Nat. Bur. Stand.*, 1935, **15**, 277.

between the parallel edges of the two stops (see Fig. 185). When the Bertrand lens is removed, the field is seen to be faintly but uniformly illuminated. Under these conditions of illumination it is claimed that much greater accuracy in comparing the refractive indices of crystal and liquid is obtained. In some microscopes the construction of the condensing system may make it difficult or impossible to achieve the optimum conditions for this method.

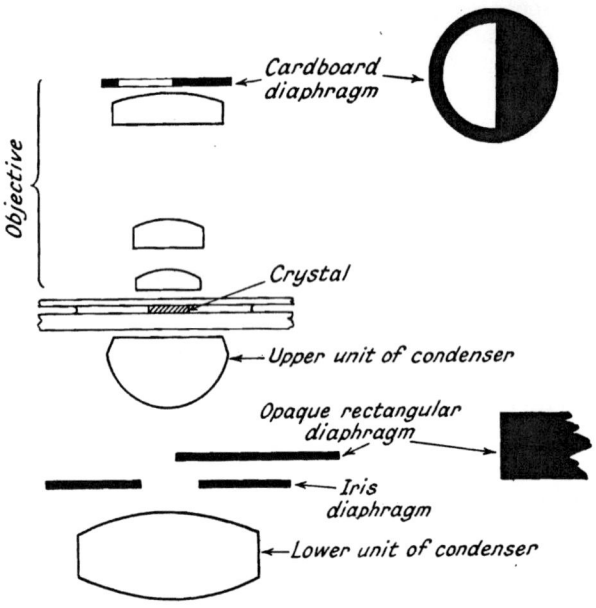

Fig. 185.—Simplified Diagram Illustrating Saylor's Double-Diaphragm Method.

Most of the liquids used in refractive index work have a greater dispersion than crystals. When both crystal and liquid have the same refractive index for a wave-length near the middle of the spectrum, say for sodium light, the substance will have a higher index than the liquid for the red end of the spectrum and a lower refractive index than the liquid for the blue. If, now, the observation is made in white light, instead of the usual dark and bright borders appearing, coloured fringes will be seen, one side of the crystal fragment being bluish, and the other reddish in colour.

The blue fringe occupies the position which in blue light would be the bright border, and the red fringe the position of the bright border in red light.

The limit of accuracy of the immersion method is ordinarily about ± 0.002 in monochromatic light, but this may be improved by careful temperature control (see p. 398).

Immersion Media. The liquids used in refractive index determinations should be colourless, or nearly so, and chemically stable. They must neither dissolve nor react with the substance undergoing test. It is convenient to make up a permanent set of liquids covering a range of refractive indices between about 1·35 and 1·78, with intervals of not more than 0·01. Media with indices higher than 1·78 should be made up as required. The liquids used should be miscible with those above and below them in the set, so that mixtures having intermediate indices can be prepared from them when it is desired to make determinations to the third decimal place. It is inadvisable to mix liquids with boiling-points which differ greatly, as differential evaporation will cause the refractive index to change rapidly. The bottles in which the liquids are kept should have well-fitting ground glass stoppers with dropping rods attached. The set may be arranged in a series of holes drilled in a thick board, provided with a cover to protect the liquids from the action of light when they are not in use—an important point with some liquids, *e.g.* methylene iodide, which darkens rapidly if left uncovered for any length of time. Bottles enclosed in wooden cases with screw tops can be bought. With these it is unnecessary to provide a cover for the set, except as a protection against dust.

Many different refractive index liquids have been suggested by different authors according to the special needs of their work, and the reader will not, as a rule, have much difficulty in making a selection within the limits mentioned above. A list of suitable liquids is given below. (This does not include many of the special liquids used in index-variation methods which are described on p. 394). The indices for nearly all of the single chemical compounds in the list are given to the third decimal place and refer to the D line at the temperatures stated. The data for the other liquids (mainly natural oils and solutions) are only approximate values for room temperature, since the indices in these cases vary with the source, concentration, etc. Some data for the decrease of refractive index with rise of temperature (dn/dt), and for the dispersion are also

EXAMINATION IN ORDINARY LIGHT

Liquid.	n.	° C.	dn/dt.	Dispersion.
Water	1·333	20	slight	—
Acetone	1·359	19·4	—	slight
Methyl acetate	1·359	20	—	—
Ethyl alcohol	1·362	18·35	−0·0004	slight
Methyl butyrate	1·386	20	—	slight
Ethyl butyrate	1·393	18	—	slight
Ethyl valerate	1·393	20	—	slight
Paraldehyde	1·405	20	—	—
Amyl alcohol	1·410	14·6	−0·0004	slight
Ethyl bromide	1·424	20	—	—
1,2-Ethylene dichloride	1·444	20	—	—
Chloroform	1·446	18	—	—
Petroleum *	1·45	—	—	—
Lavender oil	1·46	—	—	—
Carbon tetrachloride	1·466	12·3	—	—
Olive oil	1·47–1·48	—	—	—
Almond oil	1·47–1·48	—	—	—
Turpentine	1·47	—	—	—
Glycerine †	1·47	—	—	—
Medicinal paraffin ‡	1·47–1·49	—	−0·0004	slight
Castor oil	1·48	—	—	—
Toluene	1·498	16·35	—	—
Benzene	1·501	20	—	—
p-Xylene	1·497	16·2	—	—
m-Xylene	1·500	14·85	—	—
o-Xylene	1·508	15·5	—	—
Lubricating oil §	1·50	—	−0·0004	slight
Sandalwood oil	1·51	—	—	—
Cedarwood oil	1·51–1·52	—	—	—
Ethyl iodide	1·522	7	—	—
Ethyl salicylate	1·525	14·4	—	—
Monochlorobenzene	1·525	20	—	—
Clove oil	1·53	—	−0·0005	moderate
1,2-Ethylene dibromide	1·538	20	−0·0006	—
Methyl salicylate	1·532	21	—	—
o-Nitrotoluene	1·547	20·4	—	—
Aniseed oil	1·55	—	—	—
Mononitrobenzene	1·553	20	−0·0005	—
Dimethylaniline	1·559	20	—	—

* Ordinary lamp oil.
† Takes up water from the air rather rapidly.
‡ Any medicinal oil, such as Nujol, may be used.
§ Any clean oil, as used for motor-car engines, is suitable.

IMMERSION MEDIA

Liquid.	n.	° C.	dn/dt.	Dispersion.
Monobromobenzene	1·560	20	—	—
Benzyl benzoate	1·57	—	—	—
o-Toluidine	1·573	20	−0·0005	—
Cinnamon oil	1·58–1·60	—	−0·0003	strong
Aniline	1·586	20	−0·0005	—
Bromoform	1·598	19·0	−0·0006	—
Cassia oil	1·59–1·60	—	—	—
Cinnamic aldehyde	1·619	20	−0·0003	strong
Monoiodobenzene	1·621	18·5	—	—
Carbon bisulphide	1·6295	18	—	strong
α-Monochloronaphthalene	1·632	21·6	—	moderate
α-Monobromonaphthalene	1·659	19·4	−0·0005	moderate
Cadmium borotungstate solutions *	up to 1·70	—	—	—
Potassium mercuric iodide solutions †	up to 1·71	—	—	—
Methylene iodide ‡	1·74	—	−0·0007	rather strong
Methylene iodide and sulphur	1·74–1·79	—	—	—
Methylene iodide, sulphur and iodides §	1·74–1·87	—	—	—
Phenyldiiodoarsine	1·84	—	—	—

* Klein solution. The concentrated solution is supplied commercially as a heavy liquid for separating minerals (its sp. gr. is 3·28). It may be diluted with water to give liquids of lower index, and these solutions may be concentrated again by evaporation if required. The solution is decomposed by lead, zinc, and iron.

† Sonstadt (or Thoulet) solution may be obtained commercially or prepared as follows: 6·75 gm. of HgI_2 and 5·75 gm. of KI are dissolved in 20 c.c. of water with stirring. The solution is evaporated on a water bath until a crystalline film forms on the surface. After cooling, the clear liquid is decanted from any solid which has crystallised out. The solution may be diluted with water to give liquids of lower index, and these may be concentrated by evaporation. It is very poisonous, acts upon metals, and attacks the skin.

A modification of this solution proposed by Wherry is as follows (*U.S. Dept. Agr. Bull.* 679 (1918)): Mix equal volumes of glycerine and water, and dissolve in the liquid 3·38 gm. of HgI_2 and 2·88 gm. of KI per c.c. Evaporate on a water bath to the point of incipient crystallisation, and decant from any solid. Dilute with glycerine to make liquids of lower index.

‡ Liberates iodine on exposure to light. The iodine may be removed by adding a piece of clean copper or tin, or a little mercury.

§ To 50 gm. of methylene iodide add 17·5 gm. of iodoform, 5 gm. of sulphur, 15·5 gm. of stannic iodide, 8 gm. of arsenious iodide, and 4 gm. of antimonious iodide. Warm to hasten solution, allow to stand, and filter off undissolved solids (Merwin, *J. Wash. Acad. Sci.*, 1913, **3**, 35).

included. It cannot be too strongly emphasised that the indices given in the list are only to be used as a general guide : the index of every medium selected must be measured on a refractometer or otherwise, because material from stock may not have exactly the correct values, owing to the presence of impurities. The liquids in a permanent set should also be tested periodically—say, every three months.

It will be noted that the average value of dn/dt is about -0.0005. Therefore when aiming to determine the refractive index of crystals to the third decimal place, it is essential that the index of the liquid should be determined at a temperature as near as possible to that prevailing on the microscope stage.

In actual practice it is not necessary to use a large number of single substances. It is more usual to make up a set covering the required values by using various mixtures of a small number of liquids, bearing in mind that it is not advisable to mix liquids of widely differing boiling point, as mentioned above. For example, F. E. Wright [*] proposed the following combinations, which have proved very serviceable for inorganic compounds and for a number of organic compounds as well.

Mixtures of—	n_D
Petroleum and turpentine	1·450–1·475
Turpentine and 1,2-ethylene dibromide or clove oil	1·480–1·535
Clove oil and α-monobromonaphthalene	1·540–1·635
α-monobromonaphthalene and α-monochloronaphthalene	1·640–1·655
α-monochloronaphthalene and methylene iodide	1·660–1·740
Methylene iodide and sulphur	1·740–1·790
Methylene iodide, iodoform, antimony iodide, arsenic sulphide, antimony sulphide, and sulphur	1·790–1·960

Many organic compounds would of course be rapidly soluble in these liquids. For such compounds the solutions of cadmium borotungstate or potassium mercuric iodide given in the previous list will usually be suitable.

The following mixtures covering a wider range were used by Bryant [†] in a study of the optical properties of derivatives of the lower alcohols and aldehydes, and would no doubt be applicable to a wide range of organic compounds.

[*] *The Methods of Petrographic Microscopic Research*, Carnegie Inst. Publ., No. 158 (1911), p. 96.
[†] *J. Amer. Chem. Soc.*, 1932, **64**, 3758.

Mixtures of— n_D

Glycerol–glycol–water	1·36–1·40
Potassium mercuric iodide–glycerol–water	1·40–1·72
Methylene iodide–sulphur	1·74–1·78
Methylene iodide–arsenic trisulphide	1·85–2·10

Some organic substances are appreciably soluble in both aqueous and organic media, and in such cases it may be necessary to prepare a special set of liquids saturated with the substance.

Most immersion liquids of low refractive index are very volatile, and difficulty may be experienced in their use, owing to their rapid evaporation during a determination. This may be avoided by using slides in the upper surface of which is ground a small concave depression in which the crystals and liquid are placed. The cover slip, laid over the cavity, seals the liquid much more effectively than when used with an ordinary slide, since it is not lifted away from the glass by the crystals. Such cavity slides can be obtained from dealers in microscope equipment.

For substances with refractive indices above about 1·87, amorphous mixtures of sulphur and selenium, and of piperine with arsenic and antimony triiodides have been used. The crystals are heated on a slide with the immersion medium until the latter melts, a cover slip being then placed over the liquid. On cooling, the medium sets to a glass, in which the crystal particles are embedded. This method is only applicable to substances which are unaffected by the heating required to melt the medium, and is not therefore of general application. A concise summary of the preparation and properties of such media is given by Larsen.*

In mixing two liquids of known refractive index to get an intermediate one, the following formula may be used:

$$v_1 n_1 + v_2 n_2 = (v_1 + v_2) n$$

where v_1 and v_2 are their respective volumes, and n, the required refractive index.

Liquids having refractive indices below, say, 1·74 are easily standardised on a refractometer, one of the Abbé type being most suitable as only a drop of liquid is required. For liquids of higher refractive index it is necessary to use other methods, since the upper limit of the ordinary refractometer is determined by the refractive index of its prism and this is rarely greater than 1·74. The most

* *Microscopic Determination of the Non-opaque Minerals*, Bull. Geol. Survey, U.S.A., No. 848 (1934), p. 18.

accurate of these methods to use is a hollow prism and determine the index by the method of minimum deviation as described in any textbook of physics. The usual type of hollow prism used for this purpose requires a relatively large amount of liquid and so it is better to use a small one of special construction for which only a drop or two of liquid is necessary. Such a prism described by Larsen (*op. cit.*) is shown in Fig. 186 (*a*), and a modification with temperature control, described by Butler,* in Fig. 186 (*b*). The general construction of these is easily seen from the diagrams. In both cases the glass plates of the hollow prism are fixed with Bakelite varnish, or similar cement, to shapes of brass or copper, and the edges forming the angle of the prism are ground to fit

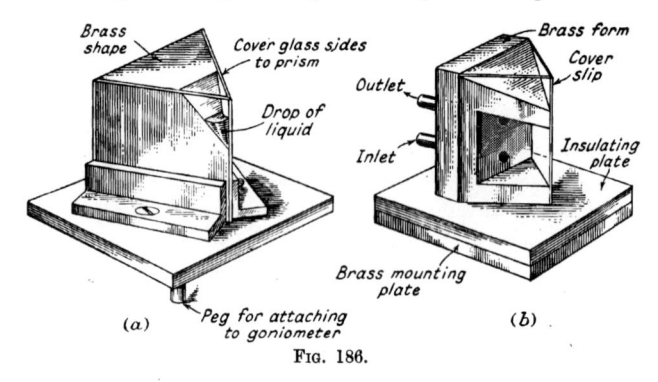

Fig. 186.

closely. For maximum accuracy the glass plates should be optically true, but for most purposes it is sufficient to use slides or thick cover slips, which have been carefully selected as follows. The reflection of a vertical cord, such as that of a window blind, seen when the slide is laid on the bench, is observed across the whole surface of the glass, by moving the eye slowly from one side to the other. A satisfactory slide will give only one undistorted reflection. It should be noted that for high index liquids the prism angle should be much smaller than the usual value of 60°, not only because of the large deviation, but also because high index immersion liquids containing, for example, iodides are so deeply coloured that sufficient light can only be made to pass through the thinnest parts of even a narrow angle prism. It has been found

* *Amer. Min.*, 1933, **18**, 394.

necessary to use prisms with an angle of between 10° and 30° with some liquids. It is therefore convenient to have two or more prisms of different angles.

A hollow prism the angle of which can be varied has been described by Cogswell.*

A simple method of determining the refractive index of a liquid is shown in Fig. 187. A source of monochromatic light A directs a beam through a slit B in a screen C, upon which is a vertical scale. The slit and scale are viewed through an aperture D in a stage plate carrying a glass plate E to which is cemented a small prism F. A drop of liquid is placed in the angle of the prism as shown. Light

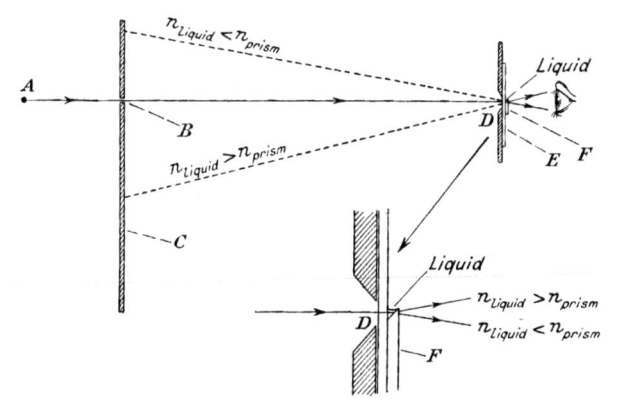

FIG. 187.—Principle of Leitz-Jelley Refractometer.

passing through the prism is refracted upwards or downwards according as its index is greater or less than that of the prism. In the former case the image of the slit appears on the screen, below its actual position; in the latter case it appears above this position. The scale may be calibrated with substances of known index. By using white light the image of the slit is drawn out into a coloured band, thus affording a means of making a rough estimate of the dispersion of the liquid (strictly, relative to that of the prism). This is the principle on which the Leitz-Jelley micro-refractometer is based (E. Leitz, Mortimer Street, London). In this instrument the scale is calibrated for the direct reading of indices for sodium light from 1·33 to 1·92, the third decimal place being obtained by

* *Amer. Min.*, 1945, **30**, 541.

estimation. Another refractometer using the same principle is made by the Fisher Scientific Company (Pittsburgh, U.S.A.). Another and more approximate method useful for the higher

Substance.	n_D.	Remarks.
Ammonium iodide, NH_4I	1·703	Cubic habit and cleavage
Ammonium bromide, NH_4Br	1·710	Tetrahedral. Low temperature form, stable below 109° C.
Spinel, $MgAl_2O_4$ . . .	1·72	Cleavage poor, octahedral. Made from fusion
Magnesium oxide, MgO	1·734– 1·737	Cubes and octahedra, perfect cubic cleavage
Arsenious oxide, As_4O_6	1·755	Isotropic form made by sublimation. Octahedral cleavage
Sodium iodide, NaI .	1·774	
Lead nitrate, $Pb(NO_3)_2$	1·781	Tetrahedral form made from aqueous solution
Lithium bromide, LiBr	1·784	
Cæsium iodide, CsI . .	1·787	Phase stable at room temperature
Zinc spinel, $ZnAl_2O_4$.	1·805	Octahedra
Potassium platinichloride, K_2PtCl_6	1·825±	Octahedra and cubes
Calcium oxide, CaO. .	1·837	As made by fusion in round colourless grains. Hydrates easily
Strontium oxide, SrO .	1·870	Hydrates easily. As made by fusion
Lithium iodide, LiI .	1·955	
Cuprous chloride, CuCl	1·973	Tetrahedral
Barium oxide, BaO .	1·98	Hydrates easily. As made by fusion
Silver chloride, AgCl .	2·071	As made by fusion
Stannic iodide, SnI_4 .	2·106	Colour red
Silver bromide, AgBr .	2·252	Cubes
Cuprous iodide, CuI .	2·346	Tetrahedra
Thallous bromide, TlBr	2·418	
Thallous iodide, TlI . .	2·78	
Selenium glass . . .	2·92	

Reference may be made to A. N. Winchell's *Microscopic Characters of Artificial Inorganic Solid Substances* (Wiley, 1931), p. 346 *et seq.* for other substances which may be more accessible than some of those given above.

indices is to immerse isotropic substances of known refractive index in the liquid until a satisfactory match is obtained. A series of artificial glasses may be used, but the cubic substances in the table on p. 236 will be more easily obtained.

Various methods have been devised for determining the refractive indices of liquids under the microscope. One of the most convenient is that of Wright.* The apparatus (Fig. 188) consists of two pieces of highly refracting lead glass (n_D 1·92), with an edge of each bevelled at 60°. One of these edges is polished and the other finished with a matt surface. The two pieces of glass are held in a metal slide above a circular opening, with the inclined edges in contact, the polished one uppermost. A drop of liquid is placed between the inclined surfaces, being held there by capillarity and its index of refraction is found by noting the position of the limiting refracted ray between the light and dark portions of the field. The observa-

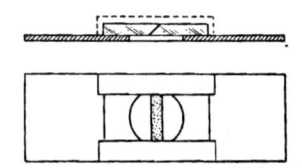

FIG. 188.—Wright Stage Refractometer.

tion should be made with the converging lens, a 16-mm. or $\frac{2}{3}''$ objective, the Bertrand lens, and a micrometer scale in the eyepiece. The instrument may be calibrated by using three or four liquids the refractive indices of which have been accurately found upon a refractometer. The graph which results when the refractive indices are plotted against the divisions of the scale is a simple sine curve.

It should be understood that a fixed combination of objective, ocular, and tube length must be preserved when using this method. The accuracy is 0·001 units.

Index Variation Immersion Methods. Improved immersion methods which give greater accuracy than the usual ± 0·002 of the method already described have been proposed by certain authors. The principles upon which these methods are based are (1) that the index of refraction of liquids increases with a decrease in the wavelength of light, and this dispersion is generally greater in liquids than in solids ; (2) the index of refraction of liquids decreases as

* F. E. Wright, *J. Wash. Acad. Sci.*, 1914, **4**, No. 11.

the temperature rises to a much greater extent than it does in solids for a similar rise, change in the latter being negligible over the range of temperature employed.

The methods are, briefly : the *dispersion method*, proposed by Merwin and Larsen,* by which the index of refraction of the substance is matched with the immersion fluid by altering the wavelength of the light used by means of a monochromator ; the *single variation* method of Emmons † by which the indices of fluid and substance are matched by changing the temperature of the mount upon a warm stage ; and the *double variation* method of the same author ‡ in which both the wave-length of light and the temperature are changed, thus making it possible to determine the refractive index of the substance for any kind of light from the dispersion curves, which can be drawn after the refractive index of liquid and crystal has been matched at three or four different temperatures. Index variation methods are dealt with in greater detail in Chapter XI.

II. EXAMINATION IN POLARISED LIGHT (LOWER NICOL INSERTED)

When the polariser has been inserted in its proper position beneath the microscope stage, all the light which reaches the latter is plane polarised. If an anisotropic crystal section or plate is rotated above the polariser in the plane of the microscope stage, it will bring its two directions of vibration for light passing through it, successively parallel to the plane of vibration of the polariser. In these positions all the light from the polariser will pass through the crystal plate without any change in the direction of vibration. In other positions, the vibrations from the Nicol will be distributed between the two vibration directions of the crystal plate according to the angle these make with the polariser (see Fig. 138, p. 101).

Variation in the Selective Absorption of Light. Pleochroism. The variation in the selective absorption of light with its vibration direction in coloured crystals has already been referred to in Chapter IV. It may be studied by rotating the crystal above the polariser, when changes in both intensity and tint will be observed in most cases. The phenomenon is called in uniaxial crystals *dichroism* (since such crystals have two principal absorption

* *Amer. J. Sci.*, 1912 (4), **34**, 42–7.   † *Amer. Min.*, 1928, **13**, 504.
‡ *loc. cit.*, and A. N. Winchell, *Elements of Optical Mineralogy*, Part 1 (Wiley), 1931, 216–34.

axes), and, in general, *pleochroism*. In cubic crystals, absorption is the same for all directions of vibration, and these crystals therefore show no change when rotated above the polariser. Striking examples of pleochroism are afforded by magnesium platinocyanide, $Mg[Pt(CN)_4],7H_2O$, tetragonal, which is bluish-red for light vibrating along its length, and carmine red for vibrations at right angles to it ; and the orthorhombic ferro-magnesian silicate, hypersthene, $(Fe,Mg)SiO_3$, which is brownish red parallel to X,* and for the Z axis, green. The effect is more intense the thicker the crystal plate.

In the case of coloured uniaxial crystals, as for example in tourmaline, the maximum difference in absorption may be observed at right angles to the optic axis. When the c axis is parallel to the vibration plane of the lower Nicol, the crystal is pale yellow (if a brown tourmaline is used), and in a position at right angles to this it becomes dark brown or almost black. The pale yellow is that colour due to the extraordinary ray which vibrates parallel to the vertical axis of the crystal, and the dark brown is that due to the ordinary ray, which vibrates at right angles to the vertical axis, and is largely absorbed. In intermediate positions the colour observed is due to a combination of the two rays.

The absorption of the ordinary ray is, in this case, greater than that of the extraordinary ray. This fact is expressed by the *absorption formula* $O > E$, or, using the symbols for the corresponding refractive indices, $\omega > \varepsilon$. The change in colour due to the absorption of different parts of the spectrum is expressed by means of the *pleochroism formula* thus : O, dark brown ; E, pale yellow.

If a basal section of a uniaxial substance is rotated above the polariser (that is, with the optic axis of the crystal parallel to the axis of the microscope), no change in colour is seen. This is because light passing through such a plate normally has the same ease of vibration in all directions and the section therefore displays uniform absorption properties. In uniaxial crystals the maximum and minimum absorption effects are always seen parallel to the directions of vibration of O and E, but either O or E may have the greater absorption. In other words, if the directions in which the maximum and minimum changes are observed are called the *principal absorption axes*, these are at right angles and one of them will coincide with the vertical axis c, and the optic axis.

* The three axes of the indicatrix corresponding to α, β, and γ are often referred to as X, Y, and Z respectively.

The pseudo-uniaxial mineral *biotite* (see Expt. 1, p. 362) may be taken as another good example for the study of pleochroism. Cleavage plates which are nearly normal to the pseudo-uniaxial optic axis show very little change in colour above the polariser. Vertical sections such as may be seen in rock sections of some granites show well-marked cleavage and strong absorption. The pleochroism is similar to that of tourmaline, the greatest absorption being in a direction at right angles to the optic axis, but the elongation of the biotite sections is in the direction of maximum absorption, whereas in tourmaline the direction of greatest absorption is across the prism.

In biaxial crystals there are three principal absorption axes at right angles, light vibrating parallel to which may be absorbed in a different way for both intensity and colour. These axes are usually parallel to the axes of the indicatrix ellipsoid. In the orthorhombic system they coincide with both the ellipsoid axes and the crystallographic ones. In the monoclinic system, one of them, together with one of the ellipsoid axes, coincides with the crystallographic axis b. The other two, usually, but not necessarily,* coincide with those of the ellipsoid. In the triclinic system the absorption axes may not coincide with any of the ellipsoid axes.

The absorption formula for biaxial crystals is expressed X (or Y, or Z) $\gtrless$ Y (or Z or X) $\gtrless$ Z (or X or Y). Or similarly, the refractive indices may be used to indicate the absorption, e.g. $\alpha > \beta > \gamma$. The pleochroism formula is expressed, to take a hypothetical case, X red, Y orange, Z light yellow.

In describing the pleochroism of crystals the changes in absorption should be correlated with the fast and slow directions of vibration in a crystal plate, as determined with the quartz wedge or otherwise (p. 260). For example, it may not have been possible to determine the directions of the axes of the ellipsoid, but the directions of fast and slow vibration having been determined upon a face which is described, the pleochroism could be written, for example, " length direction of the face, fast, pale green ; across the face, slow, yellow " (see Expt. 13, p. 378).

The study of coloured crystals by X-ray methods has revealed how their pleochroism is related to their structure. In organic crystals, the absorption is greatest when the light is vibrating along the direction of the bonds in the chromophoric groups,

* H. Laspeyres, *Z. Krist.*, 1879–80, **4**, 444.

such as $-\mathrm{C}\!=\!\mathrm{C}-$, $-\mathrm{N}\!=\!\mathrm{N}-$, and $=\mathrm{C}\!\begin{smallmatrix}\diagup\mathrm{C}=\mathrm{C}\diagdown\\\diagdown\mathrm{C}=\mathrm{C}\diagup\end{smallmatrix}\!\mathrm{C}=$. Thus in

p-azoxyanisole, $\mathrm{CH_3O}\!\!\left\langle\quad\right\rangle\!\!\overset{\overset{\mathrm{O}}{\uparrow}}{\mathrm{N}}\!=\!\mathrm{N}\!\!\left\langle\quad\right\rangle\!\!\mathrm{OCH_3}$, when the vibrations are parallel to the length of the molecule, *i.e.* to the $\mathrm{N}\!=\!\mathrm{N}$ double bond, the light is yellow, and when the vibrations are transverse, it is colourless.

In inorganic compounds, absorption is greatest for light vibrating along directions in which the ions are distorted.

Variation in Relief. It has been stated that the relief of a crystal in an immersion medium depends upon the difference in refractive index between the substance and the liquid. Also in anisotropic crystals, the refractive index varies according to the direction of vibration of light. When such a substance is rotated above the polariser its relief will change according to the direction that the plane of the polariser makes with the ellipsoid axes. Such changes in relief can only be detected by the eye when the birefringence of the section is great, and the immersion fluid has a refractive index near to one of the refractive indices of the crystal plate.

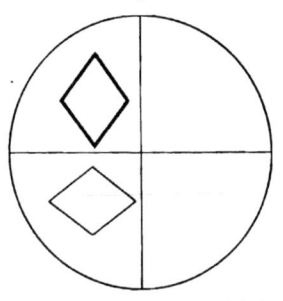

FIG. 189.—Variation in Relief.

This phenomenon, known as " twinkling ", is well exhibited by the common substance calcite, $\mathrm{CaCO_3}$, a cleavage fragment of which, when immersed in a liquid with a refractive index of 1·56, appears successively on rotation above the polariser as shown in Fig. 189.

III. Examination between Crossed Nicols

Isotropism and Anisotropism. By their behaviour between crossed Nicols, all non-opaque substances can be divided into two groups, namely, isotropic and anisotropic substances (Chap. IV, p. 89). The former, when examined thus, remain dark, like the rest of the field of the microscope no matter how they may be orientated upon the stage. On the other hand, anisotropic substances will appear coloured in most positions and only in certain definite orientations will become dark like the crossed Nicol field.

The reason for this difference of behaviour is readily understood when it is remembered that in isotropic substances light vibrates with equal ease in any direction, the wave surface being a sphere, and so, when such a substance is placed between the Nicols, it does not interfere in any way with the direction of light vibration, and the field remains dark, exactly as if the microscope stage were empty. In anisotropic substances the plane polarised light emerging from the polariser is doubly refracted on entering the crystal plate, the two rays vibrating in planes at right angles to one another, and travelling with different velocities. If such a substance is turned by rotating the microscope stage, it will become dark, or *extinguish* in four positions at intervals of 90°, and between these positions of darkness the crystal is illuminated, being brightest at 45° from the extinction position. The section extinguishes when the traces of the vibration directions in it are parallel to the planes of vibration of the Nicols, for in such positions, light from the polariser is not resolved in the plate, but passes on to the analyser as if no crystal were on the stage, and darkness results.

The colours shown between the positions of extinction are known as *interference*, or *polarisation colours*. The formation of these will now be explained.

Interference Colours. Consider first the passage of monochromatic light from polariser to analyser through an anisotropic crystal not in the extinction position, placed on the microscope stage. The light emerging from the polariser is plane polarised, and this beam, E, Fig. 190, upon entering the crystal above, is resolved into two plane polarised beams O' and E', which travel with different velocities, and vibrate at right angles to one another. At any one point on the upper surface of the crystal plate, two rays may emerge belonging to separate incident rays. Of these two rays, one has suffered a certain amount of retardation behind the other, but this, even if it has a value of half a wave-length ($\frac{1}{2}\lambda$), cannot give a resultant of zero since the waves are not vibrating in the same plane. The effect is, in general, to produce elliptically polarised light, since the waves will usually have different amplitudes and emerge in different phases (see p. 108). This elliptically polarised beam will be further resolved into four components in the analyser, two, O'' and E'' from O', and two O''' and E''' from E'.

INTERFERENCE COLOURS 243

The rays E″ and E‴ are totally reflected and do not emerge from the analyser, but the rays O″ and O‴, now vibrating in the same plane combine or interfere, producing effects which depend upon their respective amplitudes and phases. Ordinarily, a retardation of $\frac{1}{2}\lambda$ in one of two similar waves which follow the same path and vibrate in the same plane will produce darkness. In the polarising microscope, however, the analyser increases the phase difference by $\frac{1}{2}\lambda$ so that a retardation of one whole wave-length, or any whole number of wave-lengths is necessary to produce the same effect. This will be explained by means of Fig. 191.

Let AA′ be the direction of vibration of the analyser and PP′ be

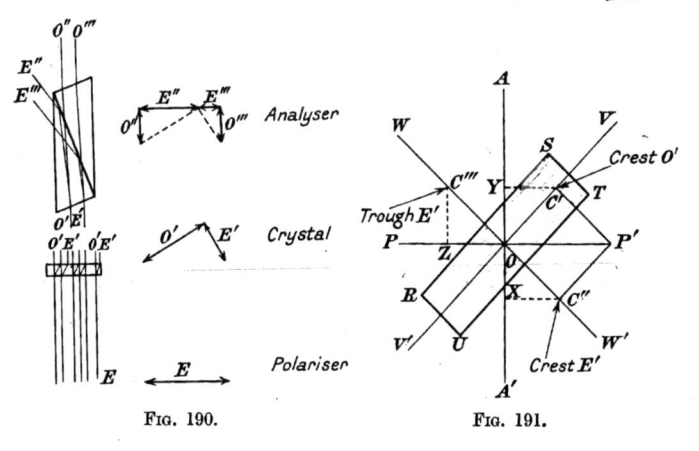

Fig. 190.   Fig. 191.

the direction of vibration of light emerging from the polariser, OP′ being its amplitude. In the crystal section RSTU, the light vibrates along VV′ and WW′ with amplitudes OC′ and OC″ respectively. When both waves have no phasal difference, they will pass through O together and reach their crest C′ and C″ simultaneously. When such is the case the vertical components which may interfere in the analyser, OY and OX, are seen to be equal and opposite, and darkness results. On the other hand, if a difference of phase of $\frac{1}{2}\lambda$ or any odd multiple occurs, it will be seen that when one wave reaches its crest at C′, the other will be in its trough at C‴. In these circumstances the vertical components OY, ZC‴ will be equal and of the same sign, the resultant wave having twice the amplitude and therefore four times the intensity. It is important to notice that in the first case darkness will result no matter what the value of the angle between the directions of vibration in the crystal and those of the Nicols, for in any position the components passing through the analyser are equal and opposite. In all cases where the phase difference is not a whole wave-length, the section is illuminated except in the extinction positions, being most brightly lit up in the 45° position when the vertical or analyser components are greatest.

The amount of retardation of one wave behind the other in a crystal plate depends upon two factors: (1) the difference in wave velocity of the beams in a direction normal to the surface of the section and (2) the thickness of the crystal plate. The wave velocities are related inversely to their respective refractive indices in the section—α' and γ'. If d be the thickness of the crystal plate, t_1 and t_2 the times taken by the two beams to pass through the section, and V_0 the velocity of light *in vacuo*, then

$$\text{velocity of the slow beam}, \frac{d}{t_1} = \frac{V_0}{\gamma'}$$

$$\text{velocity of the fast beam}, \frac{d}{t_2} = \frac{V_0}{\alpha'}$$

$$\therefore t_1 = \frac{\gamma' d}{V_0} \text{ and } t_2 = \frac{\alpha' d}{V_0}$$

and
$$t_1 - t_2 = (\gamma' - \alpha')\frac{d}{V_0}$$

which expression gives the retardation in time of one beam behind the other. In terms of the thickness of the section the *relative retardation* or *optical path difference*, R, is derived by multiplying both sides of the equation by V_0 thus:

$$V_0(t_1 - t_2) = (\gamma' - \alpha')d$$

The relative retardation of a section is therefore its birefringence (the difference between its greatest and least refractive indices), multiplied by the thickness. Thickness is generally expressed in microns ($\frac{1}{1000}$ mm.), symbol μ, and relative retardation in micromillimetres ($\mu\mu$), so that

$$R\ \mu\mu = 1000(\gamma' - \alpha')\mu d.$$

The differences in the appearance of a crystal section due to variations of thickness are well illustrated by viewing a wedge of quartz or other anisotropic substance between crossed Nicols in

INTERFERENCE COLOURS 245

monochromatic light, when a series of alternating dark and light bands will be seen in the 45° position. The dark bands correspond in position to those parts of the wedge which give a relative retardation of none or any whole number of wave-lengths of the light used, and the bright portions half-way between the bands have a relative retardation equal to $\frac{1}{2}\lambda$ or any odd multiple. This is illustrated in Fig. 192.

When white light is used with which to view a wedge under the same conditions as those above, the dark bands for each colour of the spectrum are superposed as in Fig. 193 and a series of interference colours is seen, similar to that produced by the interference between reflections from upper and lower surfaces of a thin film of air or liquid in the Newton's ring experiment as described in most textbooks on light.

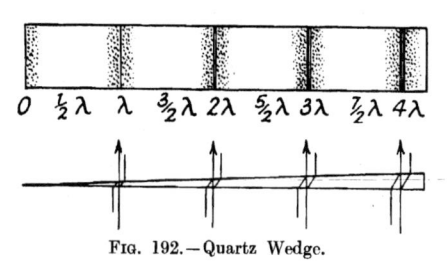

$0 \quad \frac{1}{2}\lambda \quad \lambda \quad \frac{3}{2}\lambda \quad 2\lambda \quad \frac{5}{2}\lambda \quad 3\lambda \quad \frac{7}{2}\lambda \quad 4\lambda$

FIG. 192.—Quartz Wedge.

At the thin end of a perfectly made wedge (never achieved in practice), the phase difference is zero, and darkness results. As the increasing thickness introduces a phasal difference between the fast and slow rays, greatest at first for the blues, a grey-blue tint is seen which merges into white at about $R = 230$ to $260 \ \mu\mu$, where all colours have some phasal difference but none are as yet extinguished. As the point $R = 400 \ \mu\mu$ is approached, and the middle of the violet extinction band is reached, yellow becomes strong, merging through orange into red, which is prominent about 500 to $560 \ \mu\mu$, because blue, green, and yellow are weak, and the red is still about $200 \ \mu\mu$ away from the centre of its own extinction band. The colours mentioned comprise the *first* and *lowest order* of interference colours, the sequence of colours in the second, third, and fourth orders is given in the table included in Fig. 193, each order terminating with and including a red. As the thickness of the wedge increases, the colours become fainter and more complex,

R in $\mu\mu$	Wedge at 45°. Crossed Nicols.	R in λ. Colours extinguished.	Resulting interference colours. Crossed Nicols.	Order.	Colours with Parallel Nicols.
0			Black		Bright White
			Iron Grey		White
100			Lavender Grey		Yellowish White
			Greyish Blue		Brownish White
200			Grey		Brownish Yellow
			White		Light Red
300			Light Yellow	1st. Order	Indigo
			Yellow		Blue
400		1λ Violet			Blue Green
		1λ Blue	Orange		
500			Red		Pale Green
			Violet		Greenish Yellow
600		1λ Yellow	Indigo		Yellow
			Blue		Orange
700		1λ Red	Green		Light Carmine
					Purplish Red
800		2λ Violet	Yellow Green		Violet Purple
900		2λ Blue	Yellow	2nd. Order	Indigo
			Orange		Dark Blue
1000			Orange Red		Greenish Blue
1100			Dark Violet Red		Green
		2λ Yellow	Indigo		Pale Yellow
1200		3λ Violet	Greenish Blue		Flesh Colour
1300		3λ Blue	Green		Violet
1400		2λ Red	Greenish Yellow	3rd. Order	Greyish Blue
1500			Carmine		Green
1600		4λ Violet	Dull Purple		Dull Sea Green
			Grey Blue		Greenish Yellow
1700		3λ Yellow	Bluish Green		Lilac
1800		4λ Blue	Light Green		Carmine
1900			Greenish Grey		Greyish Red
2000		5λ Violet	Whitish Grey		Bluish Grey
			Flesh Red	4th. Order	Green
2100		3λ Red			
2200		4λ Yellow			
2300		5λ Blue			

Increasing thickness of wedge →

Violet
Blue } Shading indicates the limits
Yellow } between which light of the
Red   ) wave-lengths shown, is extinguished.

FIG. 193.

due to the overlapping of the extinction bands for different parts of the spectrum, the fifth and sixth orders consisting mainly of pale pinks and greens. In still higher orders, these colours merge into " whites of the higher orders ".

If parallel Nicols are used, *i.e.*, the polariser or analyser is rotated 90° from the position of extinction, the dark and light bands in monochromatic light change places, and in white light the colours listed in the right-hand column of Fig. 193 will be seen. The colours in parallel polarised light are useful in confirming the order of polarisation colour seen in any section between crossed Nicols, and especially in distinguishing between whites of high orders and that of the first order. In the high orders, the character of the colours seen between crossed and parallel Nicols is the same, but the pinks and greens change places, and gradually, with increasing phasal difference, in both positions of the Nicols, high whites are seen. It is important to note that true complementary colours will only be seen when the wedge is truly in the 45° position.

Before going on to the next section, a brief summary of the facts already dealt with in connection with interference colours may be given :

(1) Polarisation colours are produced in white light when the analyser-passing components of the slow and fast beams from an anisotropic crystal interfere in the analyser because of a difference of phase. In monochromatic light a phasal difference of $n\lambda$ (where n is an integer) results in darkness.

(2) The actual colours seen are due to the *relative retardation* of the section, and this in turn depends upon the difference in velocity of the beams (hence upon the birefringence), and the thickness of the section.

The reader is advised to procure a copy of Michel-Lévy's chart of birefringences published in colours by Ch. Beranger, 15, Rue de Saints Pères, Paris 6e. On it are shown the relation between the interference colours, the birefringence of a section, and its thickness. An outline of the chart is given in Fig. 194. From top to bottom of the chart are shown the colours listed in Fig. 193 due to the relative retardation, the value of which is marked in $\mu\mu$ along the left margin. Thickness of sections is shown in millimetres from left to right and the value of the birefringence down the right-hand side. The colour which a section of known birefringence and thickness will show between crossed Nicols may be found at the inter-

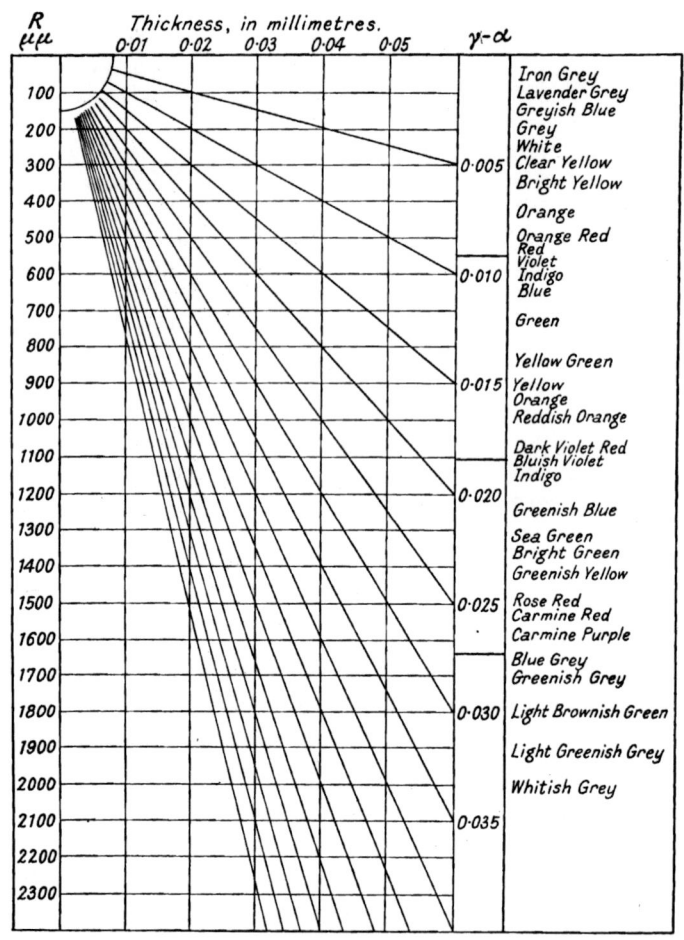

Fig. 194.—Outline Diagram of Michel-Lévy's Chart of Birefringences.

section of the line connecting the given birefringence value to the origin at the top left-hand side of the diagram, and the ordinate indicating the thickness of the section. With such a chart, if the thickness of the section and order of the colour is known, the birefringence may then be found and *vice versa*.* The method of

* Crystal grains to be examined will frequently be thicker than 0·06 mm., but the diagram may be used for these by pinning it upon a bench on which

determining the exact order of the colour seen in any section will be dealt with in a later paragraph.

In the above discussion, the term *biréfringence of the section* was used. This should not be confused with the maximum birefringence for any one substance, or *birefringence of the substance*, i.e. $(\gamma - \alpha)$ or $(\varepsilon \sim \omega)$. For the determination of this property, it is necessary to examine vertical sections of uniaxial crystals (*i.e.* sections parallel to the optic axis), and in biaxial crystals sections parallel to the optic axial plane. These sections may be recognised by their characteristic interference figures in convergent light (see Chap. VIII). A qualitative estimate of the birefringence of any substance may be made if the particles are orientated at random by observing the highest colours shown for least thickness. The quantitative determination of birefringence is described on p. 263.

Anomalous Interference Colours. The interference colours shown by some substances cannot be matched in Newton's scale of colours, and are therefore *anomalous*, or abnormal. These abnormal interference tints are due to different causes in different substances. In some crystals, light of a certain narrow band of wave-lengths may be largely or entirely absorbed, and since the band is narrow the substance may not be markedly coloured, but the resulting anomalous interference colour cannot be perfectly compensated by means of the quartz wedge (p. 262). Strongly coloured substances modify the interference tints in a marked manner, masking them in certain cases so strongly that the colour between crossed Nicols appears very little different from that of the substance in ordinary light (for example, *o*-nitroaniline). Another cause of abnormal interference colours is marked dispersion of the birefringence. For example, calcite, which shows the phenomenon, has a birefringence of 0·185 for violet light, and 0·167 for extreme red. The effect is also commonly seen in organic crystals. For example, *p*-nitrobenzaldehyde crystallised as a thin film formed by fusion between a slide and cover slip shows abnormal yellows, blues, and purples.

Extinction and Extinction Angles. We have seen that a

the scale of thicknesses may be extended. The radial lines may be extended by fixing to the origin a piece of thin string or thread by means of a pin, the birefringence being read as before along the intersection of the radial lines with the ordinate marking a thickness of 0·06 mm.

crystal section will extinguish when placed between crossed Nicols in parallel light (1) if it is parallel to a circular section of the indicatrix, and (2) if it is parallel to an elliptic section of the indicatrix and its vibration directions are parallel to those of the Nicols. The former case is fulfilled in the case of cubic crystals, basal sections of uniaxial substances, and optic axial sections of biaxial substances. The second case is fulfilled in all other sections of anisotropic substances, the vibration directions in these being the axes of the indicatrix section, parallel to the crystal plate.

The distinction between circular indicatrix sections of anisotropic substances and cubic crystals is made in convergent light, when the former show characteristic interference figures (see next chapter). Basal sections of uniaxial substances extinguish completely and remain dark on rotation unless they are enantiomorphous and rotate the plane of polarisation (p. 126). Biaxial optic axial sections do not extinguish so completely as do uniaxial basal ones, but are faintly and uniformly illuminated on rotation. This phenomenon is not properly understood, but it may be connected with the fact that the optic axes are not directions of single ray velocity. In certain cases the dispersion of the optic axes for different wave-lengths of light is a contributing factor when white light is used. The appearance of such sections between crossed Nicols in parallel light is very characteristic, and even before the investigation in convergent light begins they can be recognised and the conclusion reached that the crystal is biaxial.

The *directions of extinction* in any crystal section are determined by the orientation of the indicatrix within the crystal (p. 120). If a direction of extinction is parallel to the length of the section, or to a single cleavage or prominent edge, it is said to be *straight* or *parallel*; if it bisects the angle between two cleavages or prominent edges it is called *symmetrical*; and in the case where it is neither symmetrical nor straight, it is *oblique* or *inclined*.

In all non-basal uniaxial sections straight and symmetrical extinctions obtain, since the axes of the indicatrix are parallel to crystal axes. The same applies in the orthorhombic system to all sections for which at least one vibration direction corresponds to a principal refractive index, *i.e.* all faces belonging to zones the axes of which are crystallographic axes. Faces not belonging to these zones which have symbols of the general type (hkl) will show oblique extinction. This is illustrated in Fig. 195, in which HKL represents a (*hkl*) face, and AA' and BB' the optic axes, the optic axial plane being parallel to (010). The vibration directions for light normal to the section HKL can be derived by the application of a relationship discovered by Biot and Fresnel. This states that the vibration, or extinction, directions in any section for light of perpendicular incidence are given by the trace of the plane which bisects the angle between the two planes normal to the section and

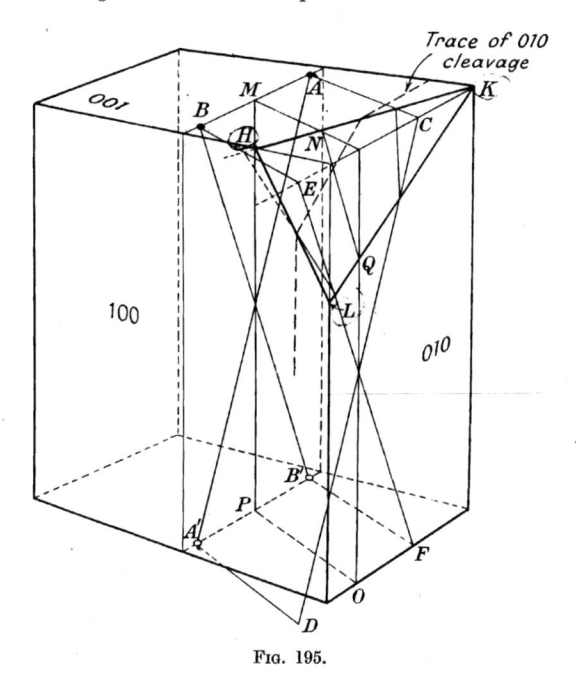

FIG. 195.

which contain the two optic axes respectively. Thus in the figure the plane ACDA' is normal to HKL and contains the optic axis AA', and the plane BEFB' is also normal to HKL and contains the other optic axis BB'. The plane MNOP bisects the angle between these two planes and its trace NQ on HKL is one vibration direction, the other being at right angles to it. It will be seen that the extinction directions are not parallel to any of the edges of HKL, nor to the traces on the section of any common cleavage, such as that shown parallel to (010).

In the monoclinic system, one ellipsoid axis only is parallel to the crystal axis b, and the other two lie in the plane parallel to $b(010)$, with the result that in the zone $a(100):c(001)$ straight extinction is given. On faces belonging to any other zone, oblique extinction obtains, the maximum value for the substance being given on the face $b(010)$. Fig. 196 illustrates this by showing the orientation of the ellipsoid in two monoclinic crystals. In one the plane containing the optic axes, or *optic axial plane*, is parallel to the one possible symmetry plane of the system, *i.e.* the (010), and in the other case it is normal to the (010). The extinction angles in the prismatic

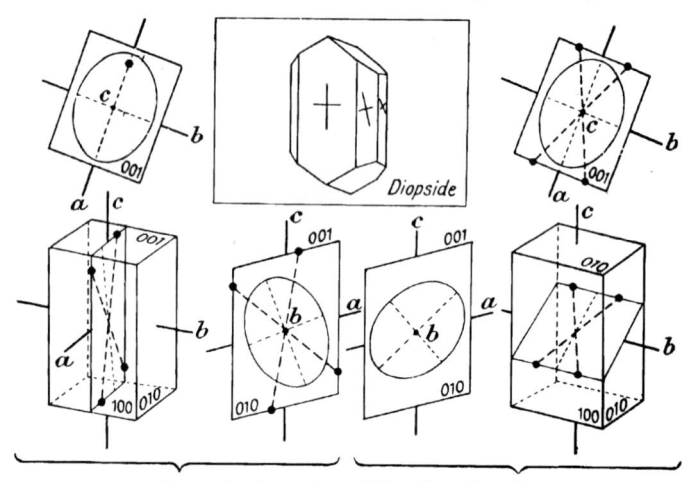

FIG. 196.—Extinction of Monoclinic Crystals.
The optic axes are represented by broken lines terminated by large dots.

zone for the mineral diopside, a monoclinic pyroxene, $CaMg(SiO_3)_2$, are also shown.

In the triclinic system, in general, no axis of the ellipsoid coincides with a crystal axis and oblique extinction results, although occasionally straight extinction may be found on one face ; for example in a series of solid solutions formed from two isomorphous substances, the swing of the ellipsoid in sympathy with changing chemical composition may bring an ellipsoid axis parallel to one of the crystal axes.

In any section, complete extinction is only obtained in white light when the directions of extinction are the same for all colours,

DETERMINATION OF EXTINCTION ANGLES

i.e. when the section is normal to a plane of optical symmetry for all colours. These conditions hold for all birefringent sections of uniaxial and orthorhombic crystals. In the monoclinic system, only one axis of the indicatrix coincides with the crystal axis b, the other two lying in the one possible crystal symmetry plane of the system. Only this plane, therefore, acts as an optical symmetry plane, and only one axis, that parallel to b, is the same for all colours. Complete extinction will be given in the $a(100):c(001)$ zone, and in other zones, *i.e.* on all faces which give oblique extinction, dispersion may be noticed. This is made clear by Fig. 153, Chapter IV.

In the triclinic system, there is no coincidence between optic and crystallographic planes of symmetry, for the latter do not occur. In this case the extinction directions for different colours will not be the same.

When such dispersion occurs, it is said that the extinction is dispersed, or alternatively, that the bisectrices are dispersed. It should be mentioned that although this kind of dispersion exists in all monoclinic and triclinic crystals, yet in certain cases it is so slight as to be difficult of detection. Nevertheless, if many birefringent grains of a substance, orientated differently, give sharp and complete extinction in white light, it is extremely probable that it is uniaxial or orthorhombic.

The Graphical Determination of Extinction Angles. It has already been pointed out (p. 251) that the extinction angle on any crystal face is determined by the orientation of the optical ellipsoid within the crystal and the directions of the optic axes. It is possible to calculate the extinction angle therefore if these data are known. The formula for the general case is, however, complicated and time consuming to use. The student who is interested is referred to the works in which the subject is treated more fully.* In actual practice the value of an extinction angle is rarely required to be more accurate than to the nearest degree or half a degree, and it is easier and quicker to solve such problems by a graphical method based upon Biot and Fresnel's Law, using the stereographic projection. The simpler case of the extinction angles in the (100) : (010) zone of the monoclinic system will be taken first. In Fig. 197 the

---

* Johannsen, A., *Manual of Petrographic Methods* (1918), pp. 399–405.
Rosenbusch and Wülfing, *Mikroscopische Physiographie* (1921/24), Band I, Erste Hälfte, pp. 487–97.

254 EXAMINATION BETWEEN CROSSED NICOLS

construction is given for the determination of the extinction angle on a face—say the (110). A and B are the optic axes, C the emergence of the vertical axis (e) of the crystal and D the acute bisectrix. The angle CD is therefore the maximum extinction angle on (010), with reference to c, the plane ADCB being both the optic axial plane and the cyclographic projection of (010). SS′ is the trace in the projection of the (110) face and NN′ the normal to it. The two planes which each contain an optic axis and the normal to the section are represented by the great circles NAN′ and NBN′ and the plane bisecting the angle between these two planes is NEN′. E marks the emergence of the extinction direction on the section SS′ (which is (110)) and the required extinction angle is EC. By taking a number of sections

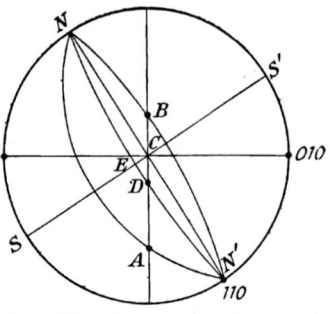

FIG. 197.—Stereographic Construction for Extinction Angle on (110). Optic Axial Plane parallel to 010. *Note:* The great circle NEN′ does not actually pass through the point D though very close to it in the case figured.

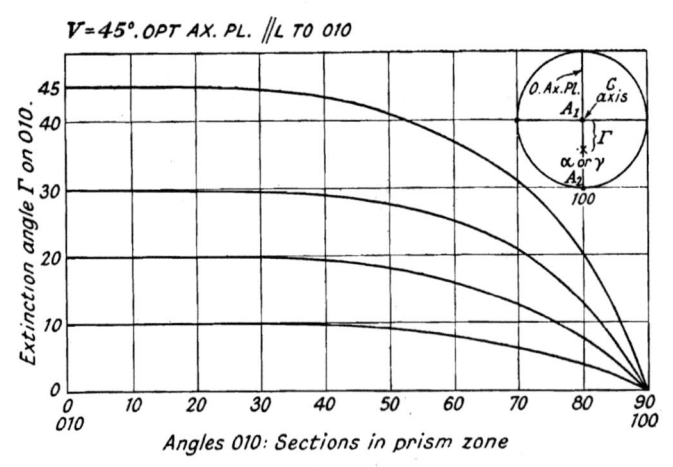

FIG. 198.

DETERMINATION OF EXTINCTION ANGLES 255

around the vertical zone the variation in the extinction angle may be tabulated or plotted as in Fig. 198. The same construction applies when the plane of the optic axes is normal to (010) as shown in Fig. 199, DC being the maximum extinction angle and EC the extinction angle on the section (110). A typical series of extinction angles for this case is shown in Fig. 200.

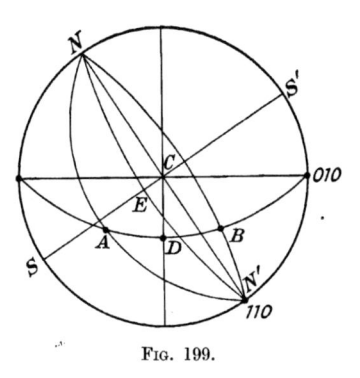

The examples shown in Figs. 198 and 200 have been selected to demonstrate the extremes of change in extinction angles around the (100) : (010) zone. In the former, V is large, the optic axial plane is parallel to (010), and it will be seen that this combination of properties results in an initial slow decline in extinction angle as the section changes from (010) towards (100). In the second case, V is small, with the acute bisectrix, and therefore the optic axial plane, normal to (010), and it will be seen that the maximum extinction angle falls rapidly as the section changes away from

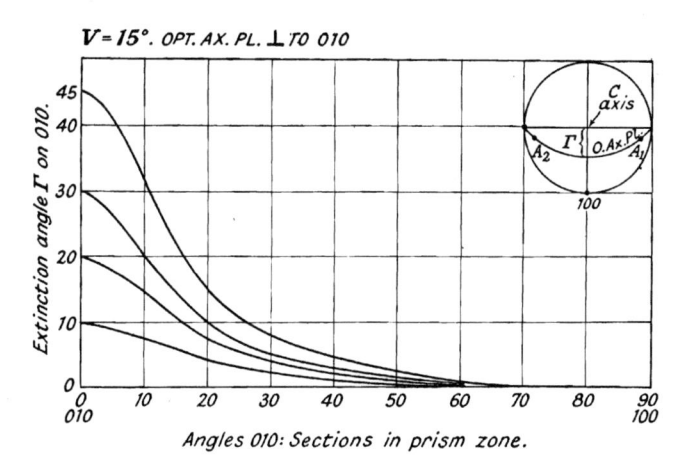

FIG. 199.

FIG. 200.—(*Note*: Acute bisectrix is normal to (010)).

(010). Other combinations of optical properties will give types of curves that are intermediate in form between those shown.

The general case is not quite so simple because the cyclographic projection of the section and its normal are not now straight lines but great circles. In Fig. 201 AB is the optic axial plane, and the angle AB = 2V. The plane of the section is the great circle SS' and its normal emerges at N, 90° away on the diameter KN. Two great circles FANG and HBNJ are drawn, each containing an optic axis and the normal to the section, N. The angle between these circles is bisected (using the construction of the pole of a zone) by the plane LENM, and E is the emergence of one extinction direction on the section SS', and E' the emergence of the other extinction direction, 90° away on the great circle bisecting the obtuse angle between the planes which contain the normal and the two optic axes respectively. If the plane PORQ represents the (010) cleavage, then ER, the angle between the trace of that cleavage R on the section and E, is one extinction angle, and E'R' (its complement) the angle between the other extinction direction and the trace of the same cleavage.

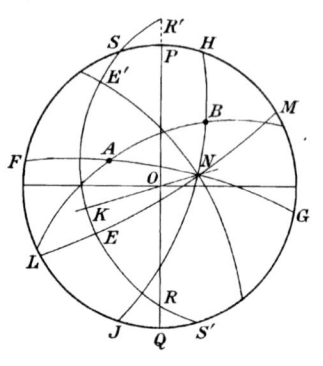

Fig. 201.

The student should use this method to demonstrate that the extinction angle on (hkl) faces measured from a pinacoidal cleavage in crystals of the orthorhombic system is really oblique (see p. 251).

Only one or two practical applications of these methods will be given here, but with experience the student will be able to adapt them to his own problems. It might happen that a substance is being studied and is suspected to be one of which the crystallographic constants (morphological and optical) are cited in the literature. For example, a monoclinic substance might have a maximum extinction angle of $\gamma\char`\^c$* = 35° and an optic axial angle of 60°, (010) being the optic axial plane. If, however, the crystals under examination have a prismatic habit with a good development of (110), and (010) is absent or poorly developed, it will not be possible

* Angles between crystallographic vectors can also be expressed thus—$\gamma : c$.

to measure the maximum extinction angle. It would be possible, however, to construct a stereogram of the crystal and its optical data, and to determine whether the extinction angle measured is that to be expected on the (110) face of the suspected substance.

The beginner is apt to take short cuts in such operations as the determination of principal refractive indices. In monoclinic crystals the (010) sections show the maximum extinction angle, and also two principal refractive indices. Suppose a study is being made of a monoclinic substance, the maximum extinction angle of which is say, 36°, which has been determined by statistical methods on a group of crystals lying at random around the prism zone. A student may assume that any crystal which gives an extinction angle of nearly 36° should be very nearly a (010) section, and therefore a proper one on which to determine a principal refractive index. Fig. 198 shows the variation of extinction angles in the prism zone of crystals with large optic axial angle ($V = 45°$) and various values for the maximum extinction angle on (010) which is the optic axial plane. It will be seen that sections up to 40° from (010) will show extinction angles very little removed from the maximum, but may be very far from presenting principal refractive indices. It is therefore better to be sure, and look for the appropriate interference figure as described in Chapter VIII. On the other hand, with crystals showing properties similar to those in Fig. 200, the method of locating the (010) section by measuring extinction angles would be legitimate.

The Measurement of Extinction Angles. Extinction angles should be determined in parallel light, and so the sub-stage diaphragm should be almost completely closed (see p. 184). When the extinction directions are not dispersed, white light gives an accurate result, but if there is any dispersion of these directions, *i.e.* when the extinction angle varies with wave-length, a precise result can only be obtained by using monochromatic light. For many purposes, however, the mean value obtained in white light is sufficiently characteristic of the section for determinative purposes. To determine an extinction position, the analyser should first of all be removed, the better to see the section, and the crystal edge or cleavage brought parallel to a cross-wire. The stage vernier is now read, and the analyser having been inserted, the crystal is turned to the nearest position of extinction, after which, the second reading on the stage vernier is taken, and the angle found by difference.

This should be done a few times, and the average taken; then the stage should be rotated through 180°, the angle found in the same way, and a final average taken. If one crystal only is being examined, it must be turned if possible on to different faces and the extinction angle on each face determined, or if a group of crystals is available, many should be examined, and the maximum angle taken as the extinction *angle of the substance*, for this differs as has been shown from the extinction angle of a face. The simple rotation apparatus described on p. 215 is very useful for examining the extinction angles in one zone, especially the prism zones of monoclinic and orthorhombic crystals which are immediately differentiated by this means.

In cases where the polarisation colours are very low, or are masked by the deep colour of the substance, difficulty may be experienced in locating the exact position of extinction. Methods of overcoming this difficulty are as follows. The crystal may be brought into the extinction position from opposite directions a few times, taking careful readings and averaging the result. Another method[*] is to place the crystal as nearly as possible in the extinction position, and then rotate the polariser through a small but definite angle, say 2° in either direction, and note the change in illumination which takes place. Next, rotate the polariser the same amount in the opposite direction, and if the illumination effect is the same in both cases, the correct position of extinction has been found. It is useful to have the polariser marked on each side of the arrow indicating its zero position, for this purpose. If the resulting illumination is not the same in both cases, the stage is rotated slightly towards the direction in which the crystal was least illuminated on rotation of the Nicol, and the experiment is repeated. It goes without saying that the Nicols should be properly crossed between each part of the test.

Other methods using sensitive plates have been proposed. Two of these only will be described here. In the first, a *unit retardation plate*, described on p. 261, is placed in the diagonal slot below the analyser, and the crystal is rotated to the extinction position, in which it will show the same sensitive violet tint as the plate.

If now a slight rotation be given to the crystal, a striking difference in the colour of the latter will be seen, it being blue when moved in one direction and yellow for a slight turn in the other direction.

[*] F. E. Wright, *Amer. J. Sci.*, 1908 (iv), **26**, 349.

The exact position of extinction may thus be determined. This method is only really satisfactory on colourless material of low birefringence and upon isolated fragments, which allow the matching of the colour of the field and crystal to be made without trouble.

In the second method a twinned plate of mica, known as the *Bravais plate*, is used in the slot of the microscope. It is made as follows. A small square is cut from a cleavage plate of muscovite mica of such a thickness as to show a sensitive violet tint (approximately 0·1 mm.), so that its edges are parallel to the vibration directions. It is then cut across a diagonal, and one half is inverted. The two portions are cemented to a glass slip with the diagonals in contact, and the edges of the square parallel to those of the slip. The fast and slow vibration directions in the two halves are now opposed, as shown in Fig. 202. Between crossed Nicols both halves of the plate show the same colour, which remains unaltered when a crystal in the extinction position lies on the stage. A slight rotation of the crystal from this position results in the two halves of the Bravais plate showing different tints. It is important that the plate be accurately oriented in the slot, and for this reason it should be mounted in a well-fitting metal slider. Other methods will be found described in Johannsen's *Manual of Petrographic Methods* (McGraw-Hill, 1918) pp. 392 *et seq.*

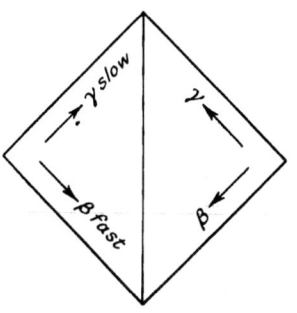

Fig. 202.—The Bravais Mica Plate.

The exact description of extinction angles necessitates discrimination between the two vibration directions of the crystal plate, and a statement not only of the angle, but the vibration direction from which it was measured. This involves the differentiation between the fast and slow directions in a section, and the angle is expressed in some such form as " slow direction : crystal axis $c = 20°$ ". If the orientation of the ellipsoid in the crystal has been elucidated, the vibration direction can be named according to the refractive index to which it corresponds, and the extinction angle described as, for example, " $\alpha : c = 20°$ ".

Distinction between the "Fast" and "Slow" Vibration Directions in Crystal Sections. To differentiate between the directions of vibration of a crystal section, the principle is adopted of placing between the Nicols a birefringent plate or wedge in which the vibration directions are known, above or below the section, and noting the effect upon the interference colours given by the latter. If the test plate or wedge is so placed that its vibration directions are similarly orientated to those of the crystal being examined, *i.e.* fast parallel to fast and slow parallel to slow, then the phasal difference between the two beams emerging from the plate on the stage is increased, and the interference colours will rise in Newton's scale, but if the fast direction of one plate is superposed on the slow of the other, the phasal difference is reduced and the interference colours as viewed under the microscope will fall in the scale. Should the phasal difference be reduced to zero, *compensation* results and darkness, or a dark band, will be seen across the wedge at the position occupied by the colour which has been compensated. Superposing fast on fast, or fast on slow, is therefore equivalent to thickening or thinning the crystal section respectively.

The accessories used are of two kinds, namely, simple parallel plates or wedges. The following three form the minimum equipment desirable.

1. The quarter undulation mica plate.
2. The unit retardation plate.
3. The simple quartz wedge.

The quarter undulation plate, or $\frac{1}{4}\lambda$ plate, is made of a cleavage flake of muscovite mica of such thickness that one vibration will be retarded $\frac{1}{4}\lambda$ (for yellow light) behind the other, and the transmitted beam will be circularly polarised. Between crossed Nicols in parallel light, the plate gives a pale grey interference colour of the first order (R = about 150 $\mu\mu$). Muscovite is negative in sign, the acute bisectrix emerging nearly normal to the cleavage face, and so the vibration directions will correspond respectively to β and γ. The plate of mica is mounted between two thin slips of glass, with one vibration direction parallel to the long edge of the slide. The slow direction of vibration is marked by inscribing an arrow on the glass slide as shown in Fig. 203.

To use the plate for the above purpose, the crystal to be examined is first placed in the extinction position and then turned through 45°, thus bringing one of the vibration directions parallel to the

"FAST" AND "SLOW" VIBRATION DIRECTIONS 261

slot in the microscope tube.* The mica plate is then inserted in the slot and the change in the interference colours is noted. A rise in the scale of colours shows that there is coincidence in the corresponding directions of vibration.

The unit retardation plate is usually made of a cleavage fragment of gypsum (also called selenite) although mica or quartz may also be used. The plate is cut of such a thickness as to have a retardation of 575 $\mu\mu$ (equal to one wave-length of yellow light), thus giving the "sensitive violet" or first order red between crossed Nicols in white light. The orientation of the plate in relation to the cleavage piece from which it is cut is shown in Fig. 204. It is mounted similarly to the $\frac{1}{4}\lambda$ plate and used in the same way. When the

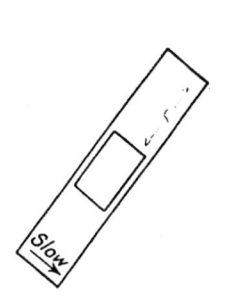

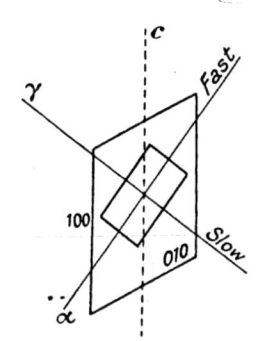

FIG. 203.—The $\frac{1}{4}\lambda$ Mica Plate.   FIG. 204.—Orientation of the Gypsum Plate.

birefringence of the crystal being examined is low, say grey of the first order, the difference in the resulting colours when vibration directions are (a) opposed and (b) coincident is very striking, being an orange of the first order in the first case, and a bright blue of the second order in the second case.

The fact that the mica plate has a relative retardation of only one quarter that of the gypsum plate makes the former more useful in examining sections of relatively high retardation (second to fourth orders), because the polarisation colours are moved by it only a short distance in the scale. Confusion is therefore less likely to occur than if the gypsum plate were used because this alters the

* In some microscopes of continental make, the slots are placed E-W. In this case, the vibration directions of the $\frac{1}{4}\lambda$ and other accessories must be at 45° to their length.

262 EXAMINATION BETWEEN CROSSED NICOLS

polarisation colour by a whole order of the scale. Thus, for example, it might be difficult to decide whether a green of the third order had been raised to green of the fourth order, or lowered to that of the second order.

The simple quartz wedge is usually cut parallel to the c axis of the crystal so that the slow vibration given by the extraordinary ray is along the length of the wedge, quartz being uniaxial and positive. On inserting such a wedge into the slot at 45° the various colours of Newton's scale will be seen usually up to about the third or fourth order. On superposing this wedge above a crystal plate orientated properly at 45° the colours will fall or rise according to whether the directions in the wedge and crystal are opposed or not.

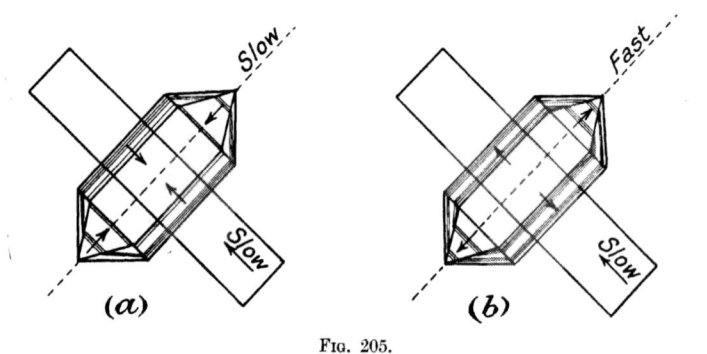

FIG. 205.

If compensation occurs (p. 260) the two are opposed, and not only are the directions of vibration determined but also the exact position in Newton's scale of the interference colour shown by the crystal plate as described more fully in the next section. If the wedge is pushed in beyond the point of compensation, colours appear but in ascending order. If compensation is not obtained in the first position adopted, the section should be turned through 90°.

When crystals with pyramidal ends or with wedge-shaped edges are examined, the interference colours will be seen to be arranged in contour-like form as shown in Fig. 205 (a) and (b), the higher colours being inside, where the relative retardation is greater. On inserting the quartz wedge these bands will move outwards if the colours are raised, and inwards if the colours are lowered. The figures show the two cases, the smaller arrows indicating the

DETERMINATION OF BIREFRINGENCE OF CRYSTAL PLATE 263

direction of movement of the interference bands, a "slow" wedge being used in each case.

In elongated sections, the character of the length is regarded as positive or negative according to whether it is slow or fast respectively, and this is termed the *sign of elongation* of the section. In the case of uniaxial crystals of prismatic habit, *i.e.* elongated parallel to the vertical axis *c*, the sign of elongation determines the sign of the crystal, for positive crystals have a slow length (E vibrating parallel to *c*), and negative crystals a fast length.

Determination of the Birefringence of a Crystal Plate. The value of the birefringence of any section may be found if the relative retardation of the section is known and also its thickness (p. 224). The basis of the method is first of all to determine the relative retardation of the section by means of the quartz wedge. This is done by compensating the polarisation colour of the crystal as described above. The crystal is then removed from the stage,* and the order of the compensating colour is determined by noting the sequence of colours shown as the wedge is slowly withdrawn. The relative retardation of the section may now be found by using the Michel-Lévy chart of birefringences (p. 247). Sometimes it is possible with wedge-shaped crystals to determine the order of the polarisation colour by counting the number of red bands from the margin of the crystal, *i.e.* without using the quartz wedge at all. More refined methods of determining the relative retardation depend upon the use of graduated quartz wedges of different kinds. In one type the relative retardation in $\mu\mu$ is shown on a scale engraved on the upper glass surface of the wedge. In order to be able to read the scale, the wedge must be inserted in a slot which is either in the focal plane of the objective (in which case the scale may be focussed by using the Bertrand lens), or in the focal plane of the ocular. In a modification of this compensator, the wedge is superposed on a plate made for example of gypsum, arranged with its vibration directions in opposition to those of the wedge.† When this combination is viewed between crossed Nicols in the 45° position, a black compensation band is seen where the relative retardations of plate and wedge neutralise one another, this point being the zero of the scale, and the colours rise in opposite directions on either side. When viewed in conjunction with a crystal plate

* Not necessary if part of the field is unoccupied by crystals.
† F. E. Wright, *Amer. J. Sci.*, 1910, **29**, 517.

also in the 45° position, the black band is displaced either up or down the wedge by an amount depending on the relative retardation of the crystal, and according as its vibration directions are opposed or not to those of the compensator.

The relative retardation may be determined by the above methods with a considerable degree of accuracy. But the other quantity necessary to complete the determination of the birefringence, namely the thickness of the crystal, cannot in general be measured so precisely. The measurement of the thickness of a dry crystal has already been described on p. 224. Where the preparation has been made by melting the substance between the slide and the cover slip (see p. 208) the thickness may be estimated as follows. The thicknesses of the slide and cover slip are previously measured by means of a micrometer screw gauge, and are subtracted from the total thickness of the preparation, measured in the same way. Alternatively, the microscope may be focussed first on a scratch or ink mark made on the upper surface of the slide to one side of the cover slip, and then upon the upper surface of the cover slip, which has been marked in the same way. The difference in the readings of the fine focussing micrometer for these two positions, less the thickness of the cover slip which must be known, gives the thickness of the preparation. A number of measurements should be made in different parts of both slide and cover slip in both these methods, and the average taken, to allow for variations of thickness of the glass and the crystal film.

In the case of crystals mounted in an immersion liquid no really satisfactory method of measuring the thickness is available. The *apparent* thickness can be measured by focussing on the upper and lower surfaces of the crystal if the habit permits this to be done, but the true thickness can only be obtained by taking into account the refractive indices of the cover slip, the immersion liquid, and, if observations are made through the crystal, of this also. Johannsen has discussed these corrections and for further details reference should be made to his *Manual of Petrographic Methods*, pp. 240–1. It should, however, be pointed out that the birefringence of crystal sections is usually determined by measuring the two refractive indices (see p. 316) so that the limitations mentioned above in the measurement of thickness are not so important as would at first appear. In some cases, however, where one index is inaccessible by the immersion method owing to its high value, a direct measure-

CRYSTAL AGGREGATES AND TWINNED CRYSTALS 265

ment of the birefringence from polarisation colour and thickness is useful because from the other index and the birefringence the inaccessible index may be calculated.

Crystal Aggregates and Twinned Crystals. The optical properties of columnar or fibrous aggregates of crystals need no special mention, but the special case in which fibres or needles are arranged radially to form spherulites is interesting. When these are formed of anisotropic material, a black cross is seen between crossed Nicols in parallel light, similar in appearance to the isogyres of a uniaxial interference figure in convergent light (p. 278). The cross remains stationary on rotation of the stage. The sign of elongation of the fibres composing the spherulite may be determined in the usual way, even when they are too fine to be separately distinguished, for on the insertion of a plate or wedge, the interference colours in the two diagonal quadrants parallel to the accessory used will be raised or lowered with respect to the colours in the two opposite quadrants.

Twinned crystals of anisotropic substances are easily recognised between crossed Nicols because adjacent units of such crystals exhibit, in general, differences of illumination due to their different orientation to the planes of the Nicols.

Contact twins lying with their composition plane normal to the stage present no difficulty, but if the plane of junction is oblique to the microscope axis, a certain amount of overlap between the crystals will take place. In such a case, extinction will be impossible unless the vibration directions in each unit coincide. In interpenetration twins where no regular plane of junction exists, a similar effect will be produced.

The appearance of different types of twins is illustrated in Figs. 206 to 213, the directions of vibration of the Nicols being taken to be parallel to the edges of the page.

In crystals which are truly cubic, twins can only be detected by their outward form, but in the case of pseudo-symmetric cubic crystals, as leucite, $KAl(SiO_3)_2$, the twinning is seen on (001) as a series of fine bands alternately light and dark, lying parallel to the traces of the twin planes, which are those of the dodecahedral form (110) (Fig. 206).

Uniaxial crystals twin commonly in " elbow " or geniculate shapes due to the twinning axis being normal to a pyramid face. Each unit shows straight extinction (Fig. 207).

In the orthorhombic system common types are interpenetration twins of cruciform and stellate shapes (Figs. 208, 209).

In the monoclinic and triclinic systems twinning becomes of greater importance, both simple contact and multiple kinds being common. Fig. 210 shows a twin of gypsum lying on $b(010)$, each

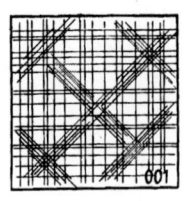

Fig. 206.—Twinning in Leucite.

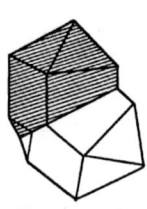

Fig. 207.—Twin of Cassiterite.

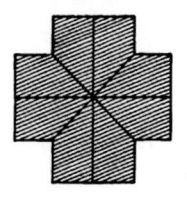

Fig. 208.—Cruciform Twin of Staurolite.

Twinned Crystals between Crossed Nicols. 1.

unit having an extinction angle of 52° 30', symmetrical about the composition plane. The twin of o-nitroaniline shown in Fig. 211 has no re-entrant angles, and might be taken for a hemimorphic form, because of its different terminations, but examination between

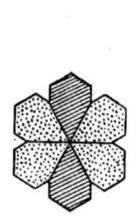

Fig. 209.—Stellate Twin of Aragonite.

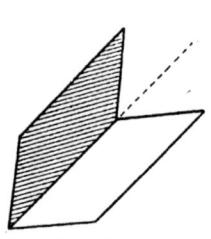

Fig. 210.—Twin of Gypsum.

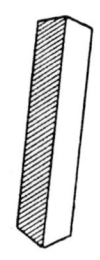

Fig. 211.—Twin of o-Nitroaniline (γ form).

Twinned Crystals between Crossed Nicols. 2.

crossed Nicols shows at once that it is a simple contact monoclinic twin on (100), each part of which has an extinction angle of 7°. In a crop of these crystals, however, some showing re-entrant angles are always seen, the case selected above being an isolated one.

In both the monoclinic and triclinic systems, in addition to contact twins on the pinacoid faces, multiple twins of various types

ZONING

are common. The latter may occur in conjunction with a simple contact twin, or two sets of multiple lamellæ may occur together. Fig. 212 shows lamellar twinning parallel to the basal plane in the monoclinic pyroxene diopside, $CaMg,(SiO_3)_2$, and Fig. 213 that of the substance albite, parallel to (010). Fig. 214 shows the

FIG. 212.—Multiple Twinning in Diopside.

FIG. 213.—Multiple Twinning in Albite.

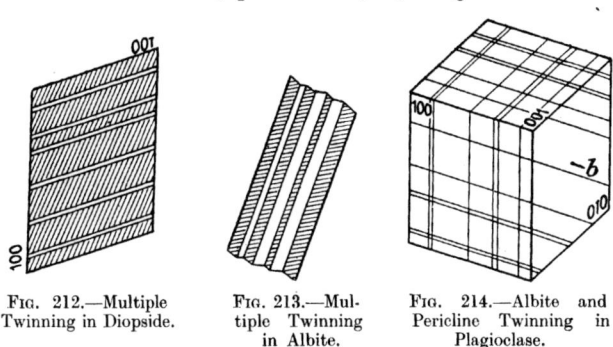

FIG. 214.—Albite and Pericline Twinning in Plagioclase.

Twinned Crystals between Crossed Nicols. 3.

combination of the albite type of twinning with another set which has as its twinning axis, b, and is known as *pericline* twinning. The latter combination is characteristic of the potash felspar microcline, $KAlSi_3O_8$, and the plagioclase felspars, of which albite is an end member, display it commonly. Between crossed Nicols, sections in the (100) : (001) zone will show cross hatching, due to the twin planes being nearly at right angles.

Zoning: During the crystallisation of substances which form solid solutions, imperfect reaction between the solid and liquid phases may result in zoning of crystals. This may be continuous

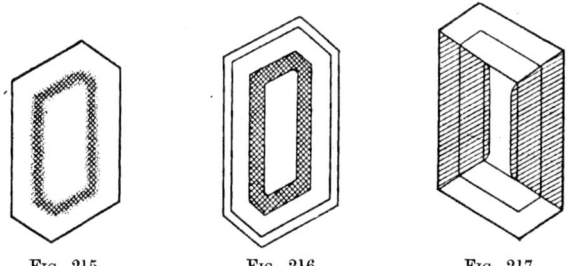

FIG. 215.   FIG. 216.   FIG. 217.

Sections through Zoned Crystals of Different Types between Crossed Nicols.

or discontinuous, that is, the crystal may vary in chemical composition gradually from centre to outer margin, or be composed of a series of distinct shells of different composition. In the first case, the crystal will give a continuous expanding wave of extinction from centre to margin when revolved between crossed Nicols, and in the second case, each shell will have its own distinct extinction angle (Figs. 215, 216). In other cases, certain ions of the mixed crystal may show a preference for the faces of a certain zone, thereby producing an " hour-glass " structure (Fig. 217).

Optical Anomalies. Some crystals may show optical properties which are not in keeping with their outward symmetry. For example, some crystals which are outwardly cubic show varying amounts of double refraction between crossed Nicols, and certain uniaxial crystals may show biaxial interference figures in convergent light. Such anomalies may be due to various causes. Should a crystal be under strain its properties will be anomalous. The effects of mimetic twinning have already been referred to (p. 84), and the appearance of such crystals under the microscope between crossed Nicols reveals their true nature. Crystals of polymorphic substances sometimes undergo transformation without a change of external form, the result being an aggregate of very small crystals of the stable modification confined within the boundaries of the original unstable form. Such *pseudomorphs* as they are called exhibit confused optical properties which cannot be correlated with the external symmetry. For example, crystals of yellow mercuric iodide grown from the vapour are orthorhombic tablets parallel to (001). On being disturbed, transformation begins by the sudden appearance of the red modification in the form of bands parallel to the edges of the tablets. These bands spread sideways until the whole crystal is transformed, there being no change in the outward form of the plates. Another example is afforded by the unstable form of potassium nitrate which is deposited as rhombohedra from a warm solution of the salt (see Expt. 9, p. 373). When these come in contact with a crystal of the stable modification, transformation sweeps rapidly through the rhombohedra but they retain their shape. In both cases the crystals lose much of their transparency and acquire a finely granular appearance. Between crossed Nicols they give no definite positions of extinction, owing to the random orientation of the crystals and to the fact that they overlie one another, the effect being known as *aggregate polarisation*.

When optical anomalies are suspected in a crystal, various changes in the illumination should be made, such as inserting the converging lens, and moving the position of the condenser, the stage being well shaded from top light. It may then be found that the suspected anomalies have been nothing more than internal or external reflections from inclined faces. Irregular grains of cubic substances are sometimes difficult to make truly isotropic in appearance, until the above expedients have been tried, or the crystals have been immersed in a liquid with the same refractive index.

CHAPTER VIII

THE MICROSCOPIC EXAMINATION OF CRYSTALS

(ii) *CONVERGENT LIGHT*

Introduction. The examination of a crystal in parallel light between crossed Nicols reveals its optical character in one direction only, that parallel to the axis of the microscope. Very important additional information may be obtained by passing a strongly convergent beam of light through the crystal, when it is possible by various means, shortly to be described, to examine its optical character in many directions at one and the same time, by viewing between crossed Nicols not the image of the crystal, or *object image*, but another optical image formed in the principal focus of the objective by the strongly convergent beam of light. This image is called variously, the *directions image* (as opposed to the object image), the *image in convergent light*, or the *interference figure*. The appearance of some of these figures is illustrated on pp. 278, 285.

The Microscope as a Conoscope. Special instruments, essentially wide angle telescopes, have been devised for the study of interference figures, and are called *conoscopes* or *hodoscopes*. They differ in use from an ordinary microscope in that they are focussed on infinity and not on an object, and parallel rays are collected at the same point in the directions image.

The polarising microscope may be simply converted into such an instrument by inserting a lens of short focal length (the *converger*) above the condenser (see Chap. V, p. 162), employing at the same time a medium- or high-power objective, say $\frac{1}{4}''$ or $\frac{1}{8}''$. No other auxiliary is absolutely necessary with which to view the figure, but various methods of doing so are described below (p. 271 *et seq.*). The angular aperture of both converger and objective should be large, for this enables complete interference figures to be brought within the field of view even when (in biaxial crystals) there is a large angle between the optic axes (Fig. 218).

In microscopes fitted with wide aperture " Polaroid " plates in place of Nicol prisms, such as those made by Messrs. Cooke, Troughton, & Simms, the fixed condensing system provides a cone of light of sufficiently wide angle for viewing interference figures without the use of a special converging lens, as already explained in Chapter V.

Fig. 219 shows the principle of the use of the polarising microscope as a conoscope. Light is reflected from the mirror and passing through the polariser P is made by the condenser C and the con-

"Brushes" or Isogyres

Interference bands

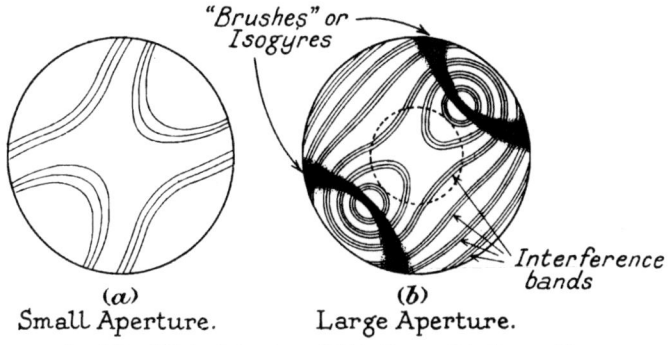

(a) Small Aperture.   (b) Large Aperture.

FIG. 218.—Effect of Aperture of Objective on Interference Figure.
In (a) only the part of the figure within the dotted circle in (b) appears in the field of view.

verger N to focus upon the object X. The divergent rays pass through the objective and converge again in the curved focal surface F (the principal focus of the objective) lying immediately above it, to form a small real image, *each point in which is the focal point of light which has passed in a definite direction through the crystal*, hence the name *directions image*. On viewing this image through the analyser, certain optical properties of these directions are revealed and a characteristic interference figure is seen. This cannot be seen in the ordinary way through the eyepiece because it lies far removed from the focal plane of the latter. Other methods are therefore used for examining the figures and will now be described.

Methods of Viewing the Interference Figure. The simplest method of all (due to Lasaulx) is to remove the ocular and view

272   EXAMINATION IN CONVERGENT LIGHT

the small image in the focal plane F (Fig. 219) through the analyser direct. The interference figure is sharply defined but small, and as the ocular with its cross-wires or scale has been removed, accurate measurement on the figure such as the apparent angle between the optic axes (see later) cannot be made. *It should be noted that the interference figure as seen by this method is inverted with respect to the object image seen when the ocular is in place.* This makes no difference when the crystal presents a centred figure, but must be taken into account when studying crystals giving inclined figures.

The most usual method employed is to insert in the body tube an auxiliary lens — the Bertrand lens (B, Fig. 219)—most often situated above the analyser, and by means of it the interference figure may be brought into the focal plane of the ocular. The Bertrand lens is sometimes fitted into a tube, the vertical movement of which enables interference figures to be focussed when different combinations of objective and eyepiece are used. The figures seen when the Bertrand method is used are not so sharply defined as they are by the Lasaulx method but are larger. The outstanding advantage of this method is that the ocular with its cross-wires or micrometer scale is retained, and these can be viewed simultaneously with the figure.

A third method due to Klein,* is to view the figure by focussing the Ramsden disc of the ocular with a small hand lens. A more convenient method is to mount a small Ramsden eyepiece (usually called a Becke lens when used for this purpose) in a collar and fit it above the ocular so that the Ramsden disc of the latter

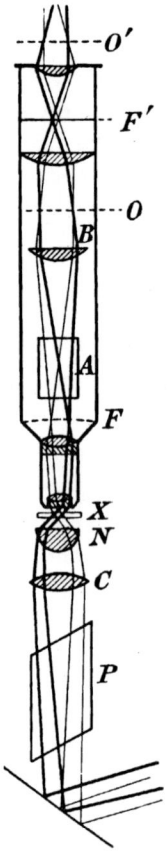

FIG. 219.—The Microscope as a Conoscope (diagrammatic).

Rays which pass through the crystal X in the same direction are emphasised equally in the drawing. Where such rays are brought to a focus, viz. at F and F', an interference figure is formed. Where rays which emanate from the same point on the crystal are focussed, viz. at O and O', an object image is formed.

* A. Johannsen, *Manual of Petrographic Methods* (McGraw-Hill, 1918), p. 450, which see for a full account of these methods.

falls in the focal plane of the magnifying lens system. This may be set once for all to focus the figure for a fixed combination of objective and ocular or may be provided with a focussing arrangement. The disadvantage of this method, like that of Lasaulx, is that although the interference figure is sharp if the magnification is not too great, the cross-wires or scale in the ocular cannot be seen as they are not in the focus of the Becke lens. This difficulty may be overcome by having a special Becke lens made with an engraved scale on glass in its focal plane, or by using another ordinary Huygens' ocular with cross-wires. The latter is placed above the eyepiece of the microscope, when the Ramsden disc in some instruments may be made to coincide with the focal plane of the upper ocular.

The interference figure as seen by the Bertrand and Klein methods is *not* inverted with respect to the object image of the crystal.

Methods of Isolating Interference Figures of Small Crystals. It sometimes happens that in a mounted sample the particles are so small that more than one appears in the field, even when a fairly high-power objective is being used. In these circumstances difficulty may be experienced in isolating the interference figure of a selected grain. Various methods have been adopted to achieve this result, the principle underlying most of them being to use diaphragms either beneath the stage and so illuminate only the desired crystal, or in various positions above the stage with the object of cutting out all light which has not passed through the crystal. In order to do this satisfactorily, it is essential, as pointed out by J. W. Evans,* that the diaphragm shall coincide with a real image of the object, or with a focus conjugate to it. Only when this condition is satisfied is it possible to separate critically a portion of the object image, that is, the selected crystal.

The various methods will now be described. In all of them, if the instrument is not provided with simultaneously rotating Nicols, it is imperative that the condenser, object, objective, diaphragm, and Bertrand lens be correctly centred.

1. In many microscopes an iris diaphragm is fitted below the condenser, and this may often be used to illuminate a small central portion of the field. After reducing the size of the aperture the diaphragm is lowered with the condenser until the image of the aperture is focussed in the plane of the object and appears simultaneously in the field with it. In this position the diaphragm is

* *Min. Mag.*, 1916, **18**, 48.

in the conjugate focus of the object (see Fig. 220). The size of the opening is then adjusted so that only the selected crystal is illuminated, after which the interference figure may be viewed by any of the methods previously described. In some microscopes this method of isolation cannot be used, because the position of the diaphragm relative to the condenser may be such as to make the focussing of the aperture image on the object impracticable. This method was used by J. W. Evans.*

2. Diaphragms are commonly fitted immediately above or below the Bertrand lens. In such a position it is very difficult to obtain

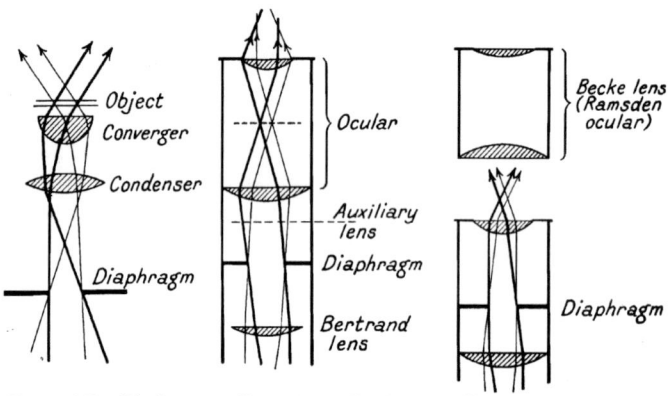

FIG. 220.—Diaphragm at Conjugate Focus of Object below Condenser (diagrammatic).

FIG. 221.—Diaphragm at Object Image above Bertrand Lens (diagrammatic).

FIG. 222.—Diaphragm at Focal Plane of Eye Lens of Ocular (diagrammatic).

Methods of Isolating Interference Figures of Small Crystals. 1.

coincidence of the diaphragm and object image, and where this can be done by raising the tube, the effective angular aperture of the objective is largely reduced. Such a position is not satisfactory even if complete isolation is effected.

3. A diaphragm may be placed above the Bertrand lens coincident with the object image formed there (O in Fig. 219). A small auxiliary lens may be used to focus this image for the purpose of adjusting the size of the diaphragm. This having been done, the lens is removed and the interference figure is viewed through the ocular in the ordinary way (Fig. 221).

* loc. cit.

ISOLATING INTERFERENCE FIGURES OF SMALL CRYSTALS

4. The central portion of the field may be isolated by using a diaphragm in the focal plane of the eye lens of the ocular of the microscope (the Bertrand lens not being used), and the interference figure which is formed in the Ramsden disc of the ocular may be viewed by the Becke method (Fig. 222). In this case the diaphragm occupies the position of the cross-wires or scale, which may be fixed instead in the focal plane of the Becke lens.

5. Czapski * devised a method (which like No. 4 does not involve the use of the Bertrand lens) whereby the isolation of the central

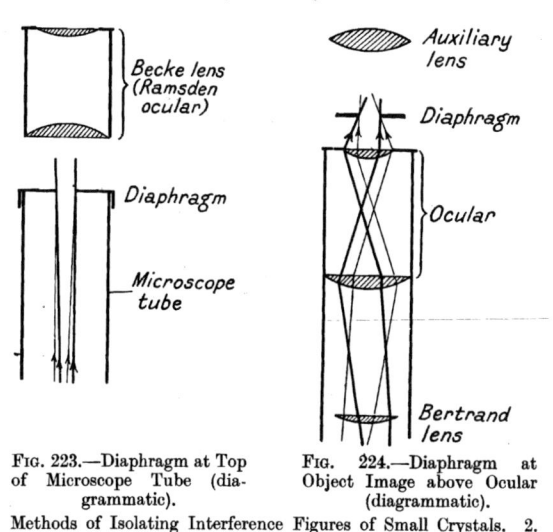

Fig. 223.—Diaphragm at Top of Microscope Tube (diagrammatic).

Fig. 224.—Diaphragm at Object Image above Ocular (diagrammatic).

Methods of Isolating Interference Figures of Small Crystals. 2.

portion of the field is effected by a diaphragm placed at the upper end of the microscope tube, where an object image may be focussed by adjusting the position of the objective, the ocular having been removed, and observing the aperture with a Ramsden ocular or positive lens (Fig. 223). On removing this lens the interference figure may be viewed in the focus of the objective. Instead of an iris diaphragm, caps with small circular apertures and made to fit the top of the tube, or a revolving disc perforated with holes of various sizes, may be conveniently used.

6. When the Bertrand lens is in position a small object image is

* *Zeit. fur Krist.*, 1898, **22**, 158.

formed a short distance above the ocular of the microscope (O', Fig. 219) and a diaphragm may be placed in a cap to coincide with it. An auxiliary lens above the diaphragm is necessary with which to view the aperture while focussing and isolating the selected crystal (Fig. 224).

In all the methods described above diaphragms with circular apertures are used, and no advantage is to be gained from other types when the stage rotates. But in microscopes with simultaneously rotating Nicols (such as the Swift " Dick ", and Leitz " SY " models; see p. 171), diaphragms which can be adjusted to the shape of the object may be used, and thus complete isolation coupled with more effective illumination may be obtained.

Following the discovery by Bertrand that interference figures could be seen in bubbles lying in the mount above small crystals, Schroeder van der Kolk and Johannsen elaborated the method.* The small bubbles act as lenses of short focal length and the objective and ocular of the microscope take the place of the Bertrand lens and ocular in the usual method of viewing interference figures. Instead of the bubbles imprisoned within the mount Johannsen substituted minute spherical lenses of glass less than 0·1 mm. in size, made by rotating for an instant in a bunsen flame the end of a short rod of glass of hair-like fineness. When such a lens is held closely above a crystal and viewed between crossed Nicols with a low- or medium-power objective, a small interference figure is seen. The use of the converger is not necessary but is useful in reducing the dark border around the lens. The figure is best seen on slightly raising the objective after focussing on the lens. The advantages of the method are that the optical system of the microscope need not be altered; the whole field may be kept in view; the optical orientation of many crystals may rapidly be studied by simply placing the lens over each grain; and the isolation and study of the interference figures of crystals so small as to be difficult of study by other methods is easily effected undisturbed by surrounding particles. For convenience, a glass rod with its lens may be fixed with wax to the microscope stage so that it lies in the centre of the field very slightly above the cover slip of the slide, which may then be moved about to bring selected crystals beneath the lens.

General Discussion of Interference Figures. Since all the rays of light passing through the crystal plate are not parallel,

* A. Johannsen, *op. cit.*, pp. 451–5.

the paths followed by different rays will not only be longer the greater their inclination from the normal to the section but they will be normal to different sections of the indicatrix. It therefore follows that the interference figure from a parallel-sided plate will not show a uniform interference colour as the section does when its image is viewed in parallel light between crossed Nicols, but a series of curved interference bands is seen, coloured in white light and dark in monochromatic light, identical with those shown by the quartz or other wedges described in Chapter VII, p. 245. These interference bands are symmetrically arranged around the optic axis (or axes), and the central portion of the field shows the same interference colour as the crystal would do in parallel light, because the central rays are quite or nearly parallel to the axis of the microscope. In addition to these bands there are dark " brushes " or *isogyres* (Fig. 218 (b)), the shape of which is determined by the positions of those points on the interference figure, where rays vibrating in directions parallel to the vibration directions of the Nicols are brought to a focus.

By the study of sections giving suitable interference figures the following optical characters may be determined:

In general

1. Isotropic substances may be differentiated from sections of crystals normal to an optic axis.

2. Anisotropic substances may be classified optically as belonging to uniaxial or biaxial crystal systems.

3. The positive or negative sign of the substance may be determined.

4. If the thickness of the section is known, an estimate of the birefringence may be made in sections normal to an optic axis.

In biaxial crystals

5. The direction of the optic axial plane may be found and the vibration directions of α, β, and γ determined.

6. The size of the optic axial angle may be measured.

7. The dispersion of the optic axes may be studied and the crystal system determined.

The Study of Interference Figures. Truly isotropic substances when viewed in the conoscope will not show interference figures because all directions are optically identical. Those sections of anisotropic crystals which remain dark on rotation between

278   UNIAXIAL CRYSTALS

crossed Nicols, *i.e.* basal sections of uniaxial crystals and the faintly illuminated biaxial optic axial sections (p. 250) reveal their true nature at once in convergent light giving respectively the figures shown in Fig. 225 and Fig. 240.

UNIAXIAL CRYSTALS

Basal or Optic Axial Sections. If a basal plate or section of a uniaxial substance is viewed in monochromatic light in one of the ways described above (p. 271), an interference figure will be seen consisting of concentric dark rings which are more closely spaced towards the margin of the field, and two isogyres which lie parallel to the cross-wires and intersect in the centre of the field (Fig. 225). If the stage is rotated the figure will remain unchanged, but if the Nicols are rotated simultaneously as in the Dick pattern

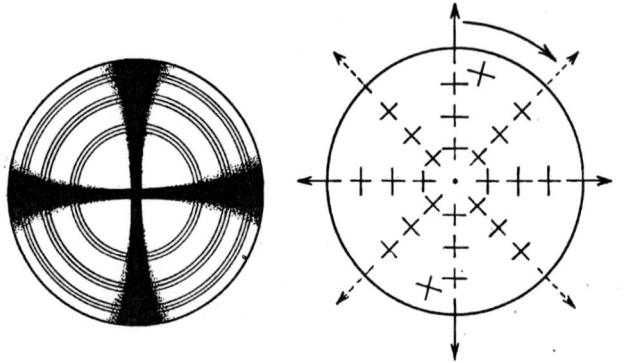

FIG. 225.—Basal Uniaxial Interference Figure. FIG. 226.—Vibration Direction in Basal Uniaxial Interference Figure.

The arrows represent two positions of cross-wires when the Nicols revolve.

microscope for example (p. 171), the isogyres will rotate with them, preserving at all times their right-angle relation. This will be understood from Fig. 226 in which the traces of the directions of vibration for different parts of the figure are shown.

In white light the isogyres remain dark but the rings are coloured, the colours seen being the ordinary interference colours of Newton's scale, the lowest order being nearest to the centre of the figure.

The explanation of such a figure is as follows:

In Fig. 227 let ABCD be a section cut normal to the optic axis which is imagined to be in the plane of the paper, and let EE be parallel rays of a monochromatic convergent polarised beam of light,* entering the section. It will be seen that as the rays (except those travelling perpendicularly to the section) do not travel in the unique direction of single velocity they are doubly refracted. At the upper surface there must emerge at all points rays O' and E' (as at X in Fig. 227), derived from a given pair of incident parallel rays EE, which from there onwards travel along the same path, vibrating of course in planes at right angles to one another. One of these rays (which one, depending upon the optical sign of the substance) will have been retarded behind the other by an amount

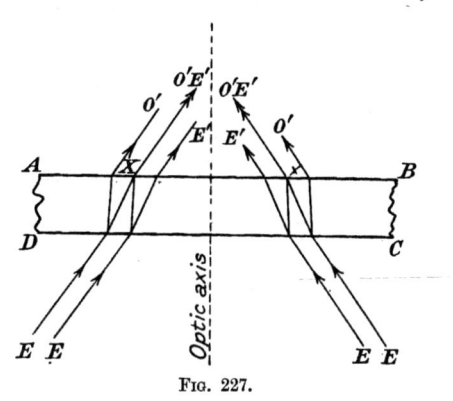

Fig. 227.

which depends upon the direction of their paths through the crystal. When the retardation of one ray behind the other is exactly one wave-length of the light used or any whole multiple of one wave-length, darkness results as described in Chapter VII, p. 245. Since the extraordinary wave surface is an ellipsoid of revolution about the optic axis, all emergent rays having a given inclination to this axis consist of ordinary and extraordinary components differing in phase by the same amount. Emergent rays so allied to one another lie on the surfaces of an infinite number of geometrically similar cones coaxial with the optic axis, and the locus of their focal points in the interference figure is a circle. This explains the series of concentric dark rings in the figure when monochromatic light is

* A convergent beam of light may be regarded as being composed of convergent *bundles of parallel rays*.

used, each ring corresponding to a group of cones for which the retardation is a whole number of wave-lengths. It is obvious that the retardation becomes greater the greater the inclination of the emergent rays to the optic axis, for with increase in the inclination, the extraordinary rays differ in velocity more and more from the ordinary rays, and also the thickness of crystal through which the light has to travel gets greater and greater. Fig. 228 shows diagrammatically cones of emergent light corresponding to different retardations, each group of cones of a given inclination being represented for simplicity by a single cone.

In Fig. 229 are shown the circular traces formed on the upper

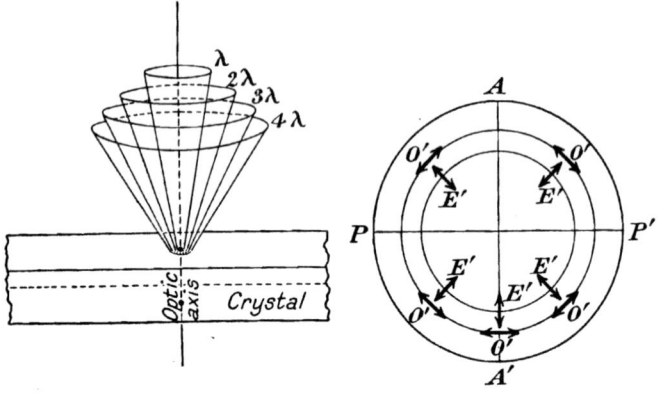

Fig. 228.—Cones of Equal Retardation around the Optic Axis of a Universal Crystal.

Fig 229.

surface of the crystal by two cones of O' and E' rays derived from the same cone of incident light. We have seen that an ordinary ray vibrates in the plane containing the ray and a line normal both to the ray and to the optic axis, whilst an extraordinary ray vibrates in the plane containing the ray and the optic axis. The traces of these planes of vibration are denoted in Fig. 229 by small double-headed arrows. The ordinary rays vibrate tangentially, and the extraordinary rays radially. It is obvious that along the directions PP' and AA' which represent the vibration planes in the polariser and analyser respectively, extinction will result, and at 45° to these directions, between the dark rings, the interference figure will be most brightly illuminated (see p. 244). In the central

SECTIONS PARALLEL TO THE OPTIC AXIS

portion of the field the rays are normal to the section and travel parallel to the optic axis. The field here will therefore remain dark. As far as the appearance of the figure goes, it does not matter whether the O' or E' rays are most refracted. This depends upon the relative velocities of O' and E' in the crystal and determines the sign of the substance.

The number of rings of equal retardation seen around the central cross in a uniaxial figure, for a given combination of lenses, depends upon the double refraction of the substance and the thickness of the section. The greater these two are, the more crowded around the cross will the rings be for a definite wave-length of light; and for

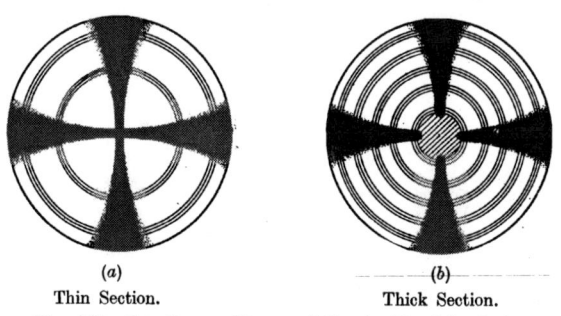

(a) Thin Section.   (b) Thick Section.
FIG. 230.—Interference Figures of Quartz (Basal Section).

a given thickness those for blue light will be more closely spaced than those for red.

Basal sections of uniaxial crystals which exhibit rotatory polarisation along the optic axis (Chap. IV, p. 126), give a characteristic figure in convergent light as has already been explained (p. 128). Figs. 230 (a) and (b) show two basal sections of quartz, the thinner one a appearing as a normal uniaxial figure because the section is not thick enough to rotate the plane of polarisation sufficiently for the eye to detect any change in white light. The other section (b) is thick enough for the plane of polarisation of light passing along the optic axis to be rotated sufficiently to produce an illuminated spot in the centre of the field.

Sections Parallel to the Optic Axis (Optic Normal Sections). Optic normal sections are presented very commonly by uniaxial crystals under the microscope, for prismatic habits are common, and such crystals (tetragonal, hexagonal, and trigonal) tend to lie with their optic axes normal to the axis of the microscope. The figure given is very like that of the biaxial optic normal interference figure (p. 289), and consists of four series of hyperbolic isochromatic bands (or dark bands in monochromatic light) which are disposed symmetrically in quadrants as shown in Fig. 231, and which rotate with the stage. The bands are frequently very diffuse in white light, and always appear more clearly and can be distinguished in greater number in monochromatic light. Faint hyperbolic isogyres (similar in shape to those presented by biaxial crystals) enter the field as the stage is rotated, form a broad diffuse cross in the centre when the optic axis is parallel to one or other of the cross-wires, *i.e.* when the crystal is in the extinction position for parallel light, and

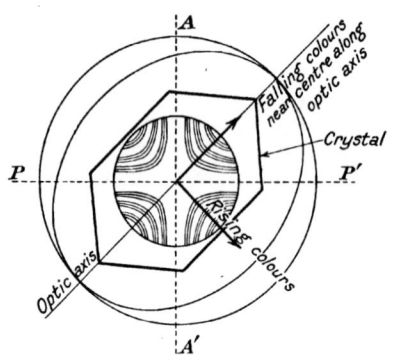

Fig. 231.—A Vertical Section of a Uniaxial Crystal viewed in Convergent Light (small central circle) with a Section of the Appropriate Positive Wave-surface Figure shown.
AA' and PP' are the directions of vibration of the analyser and polariser respectively.

swing out again in the direction of the optic axis. These isogyres, which are indistinct, move very rapidly and only occupy the centre of the field during the rotation of the stage through a few degrees, giving the impression of a momentary darkening of the whole field. Their direction of movement is often extremely difficult to detect.

In white light, if the section be turned into the "45° position", as it is commonly called (*i.e.* the position in which it is midway between, or at 45° to, two adjacent extinction positions), the direction of the optic axis is revealed by the fact that, starting from the centre of the figure, the polarisation colours fall in Newton's scale in the two quadrants through which the optic axis passes, and rise in the other two quadrants (see Fig. 231). With thick

SECTIONS OBLIQUE TO THE OPTIC AXIS

sections, or sections of high birefringence, the fall in the two former quadrants may give place to a rise as the margin of the field is approached. For this reason, the sequence of colours must always be traced from the centre. The explanation of this fall and subsequent rise is that the birefringence diminishes as the optic axial direction is approached, and this produces first a fall in the colours. As the rays become more oblique, however, the thickness of the crystal through which they must pass increases rapidly, and with thick or highly birefringent sections this may more than compensate for the fall in birefringence, and lead to a rise in the colours in the outer parts of the quadrants.

Sections lying obliquely to the Optic Axis. Certain uniaxial crystals, especially rhombohedral forms of the trigonal system, or pyramidal types of the tetragonal or hexagonal systems, may lie in such a way upon the slide that an *oblique, uncentred,* or *inclined*

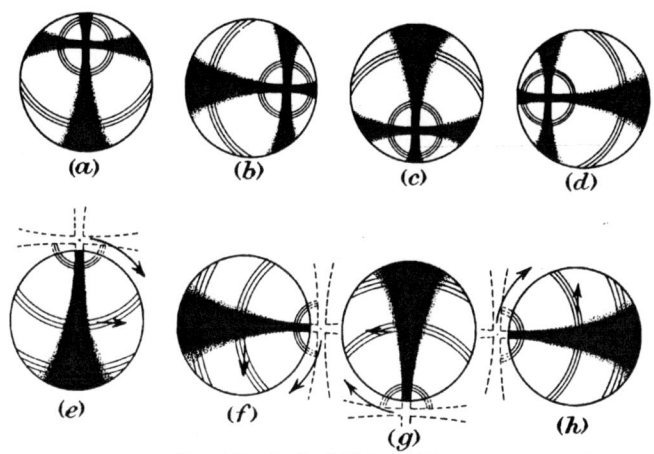

FIG. 232.—Inclined Uniaxial Figures.

In the lower diagrams (e), (f), (g), and (h), the larger arrows show the direction of movement of the centre of the cross and the smaller arrows the direction of movement of the isogyres.

figure is given in convergent light, so that the centre of the cross is displaced from the intersection of the cross-wires in greater or less degree, according to the angle which the optic axis makes with the axis of the microscope. On rotation of the stage the centre of the figure describes a circular path, either within or outside the boundary of the field, the isogyres (if the centre is not displaced much beyond the margin of the field) preserving their parallelism to the cross-wires while sweeping successively across the field as shown in Figs. 232 (a)–(h).

False Interference Figures. When using objectives of high power and crossed Nicols, a faint image appears resembling the dark cross of a basal uniaxial figure, even when there is no crystal in the field. This is due to the fact that the lower component of the objective has a strongly curved upper surface, and, as may be shown from optical theory, the plane of polarisation of the light from the polariser will be slightly rotated on refraction at such a surface, except where this plane is parallel or at right angles to the plane of incidence. These latter conditions only exist for rays travelling within those vertical principal sections of the lens which are parallel to the vibration planes of the Nicols, and so only these rays will be completely extinguished by the analyser; hence the dark cross. Elsewhere, owing to the rotation, the light will have a weak component which can pass the analyser. On rotating the objective in its mount the cross remains stationary, provided that the glass of the lens is entirely free from strain. If, however, the lens is under strain, irregular shadows or a much distorted cross will be seen, and these will change their position as the objective is rotated. Such an objective is unsuitable for critical work.

The cross seen with an unstrained objective does not interfere with the observation of interference figures in general because it is so much weaker than these, but when dealing with weakly birefringent material, care must be taken not to confuse it with a genuine figure due to the substance.

BIAXIAL CRYSTALS

Sections Normal to the Acute Bisectrix. When the section normal to the acute bisectrix is examined in convergent light, an interference figure similar to that shown in Fig. 233 is seen. This consists of two " eyes " or *melatopes*,* which mark the points of emergence of the optic axes, surrounded by bands of equal retardation, dark in monochromatic light and coloured in white light. The inner bands surround each " eye " separately and are oval in shape, or nearly circular if very close to the " eye "; the outer bands surround both " eyes ", and their form changes from " figure of eight " lemniscates to approximate ellipses as the margin of the

* The term is due to Johannsen, *op. cit.*, p. 420.

SECTIONS NORMAL TO THE ACUTE BISECTRIX

field is approached. Through the points of emergence of the optic axes pass two isogyres, narrow and well defined at their centres, broader and more diffuse towards the margin of the field. On rotation, the bands of equal retardation move with the stage. The isogyres form a cross when the trace of the optic axial plane (*i.e.* the line passing through both melatopes) lies parallel to either cross-wire (Fig. 233 (a)), that arm of the cross passing through the melatopes being narrower than the other. In this position the crystal would be extinguished in parallel light. Upon rotation from this position the cross breaks up into two hyperbolic brushes which are centred on the melatopes and lie as shown in Fig. 233 (b),

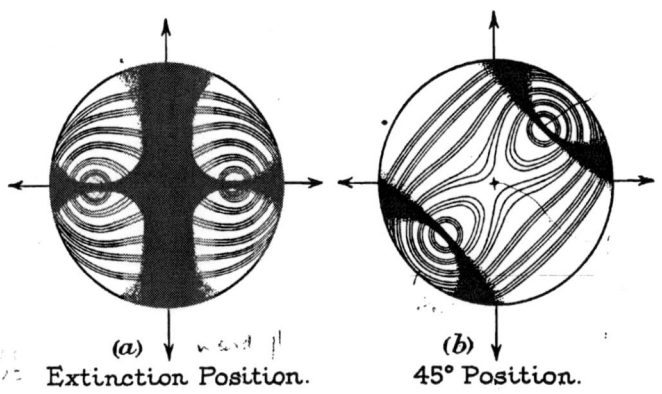

(a) Extinction Position.   (b) 45° Position.

FIG. 233.—Biaxial Interference Figure given by a Section Normal to the Acute Bisectrix.

when the trace of the optic axial plane is at 45° to the cross-wires; the convex sides of the isogyres being towards the acute bisectrix, which emerges in the centre of the field. The isogyres revolve in a direction opposite to the movement of the stage.

The explanation of such an interference figure is as follows. Rays of convergent light enter the crystal section, and, except along the optic axes, double refraction takes place. Those rays which travel along the optic axes are brought to a focus in the interference figure at two points—the melatopes—which, being extinguished by the analyser, appear dark. All other rays emerging from the crystal are made up of two components, differing in phase and vibrating in directions at right angles to one another (just as in the uniaxial case described above), and therefore in a position to interfere in the analyser. Emergent rays for which the retardation is the same lie on conical surfaces surrounding each optic axis, the sections of the cones being nearly circular when the inclination to the optic axes is small, and becoming more and more oval as this inclination increases; at still greater inclinations, the surfaces merge so as to surround both optic axes. The relative arrangement of representative surfaces corresponding to retardations of λ, 2λ, 3λ, etc., for a given wave-length of light is shown in Fig. 234. Each surface (together with its allied parallel surfaces) produces a ring of focal points in the interference figure similar in shape to its trace upon a horizontal plane, and, as a comparison of Figs. 233 and 234 should show, this accounts for the shape of the bands of equal

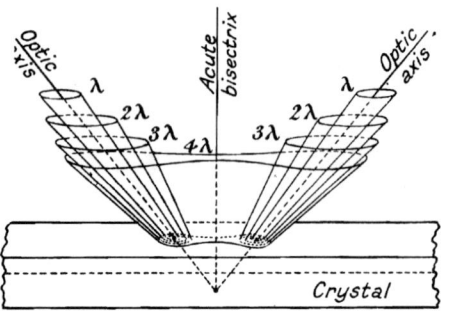

FIG. 234.—Surfaces of Equal Retardation around the Optic Axes of a Biaxial Crystal.

retardation in the interference figure. In white light the bands will be coloured.

The positions of the isogyres in the figure are determined by the directions of vibration at different parts of it. Fig. 235 shows the construction whereby these may be determined. MM' are the points of emergence of the optic axes. At any point on the figure, the directions of vibration may be found by joining the point to each melatope and bisecting the angles included between the lines thus drawn. This simple construction follows from the rule of Biot and Fresnel. It will be seen that when the trace of the optic axial plane is at 45° from the cross-wires as shown, the directions of vibration parallel to those of the Nicols fall on curves which coincide with the shape of the isogyres of Fig. 233 (b), whilst when it is parallel to a cross-wire, the isogyres form a cross as shown in Fig. 233 (a).

SECTIONS NORMAL TO THE ACUTE BISECTRIX 287

The apparent distance between the melatopes is determined by the angle between the optic axes in the crystal, and the optical system of the microscope. As in uniaxial crystals the number of

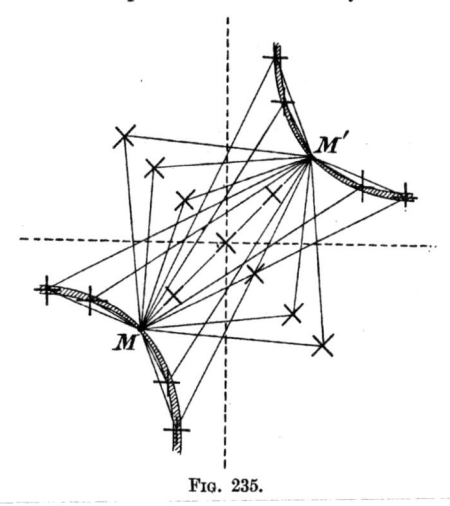

Fig. 235.

isochromatic curves is determined by the double refraction of the section and its thickness. Experiment 6, p. 370, upon plates of muscovite mica of different thicknesses demonstrates this fact. Fig. 236 shows the interference figures given by muscovite mica

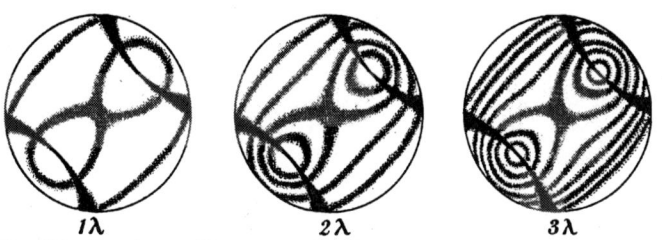

1λ 2λ 3λ

Fig. 236.—Interference Figures given by Plates of Muscovite Mica of Different Retardations (45° Position).

in plates having (for perpendicular light) retardations of 1λ, 2λ, and 3λ (D line) respectively. The number of wave-lengths retardation is given by the number of dark curves between each melatope and the centre of the field, through which rays pass

288   BIAXIAL CRYSTALS

at right angles to the section. The angle between the optic axes is constant for any given wave-length of light and therefore the apparent distance between them remains the same in all sections of whatever thickness. When the acute angle between the optic axes is larger than about 80°, as observed in air (see below), the melatopes may lie outside the field of the microscope, and differentiation between the acute and the obtuse axial angle is difficult. When the axial angle is 90° the distinction between the acute and the obtuse angles disappears.

The angle between the axes which is observed under the microscope with a dry system of lenses is not the actual angle (generally

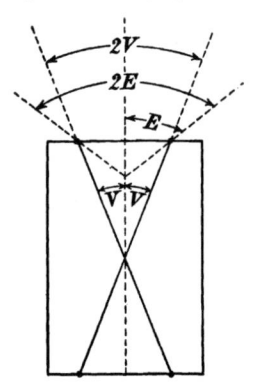

FIG. 237.—Real and Apparent Optic Axial Angles.

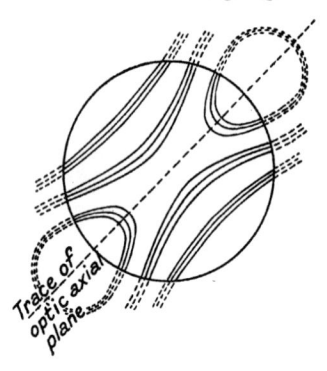

FIG. 238.—Interference Figure given by a Section Normal to the Obtuse Bisectrix (45° Position).

referred to as 2V, Fig. 237) within the crystal but an apparent angle in air (2E) due to the external refraction of rays at the surface of the section. The relation between the real and apparent optic axial angles is given by the formula

$$\sin E = \beta \sin V$$

where β is the intermediate refractive index of the substance.

Sections Normal to the Obtuse Bisectrix. Sections cut at right angles to an obtuse bisectrix usually show a partial figure consisting of lemniscate curves similar in shape to those of the acute bisectrix, but the points of emergence of the optic axes lie well outside the field (Fig. 238). On rotation of the section, the isogyres swing into the field, and when the plane of the optic axes is parallel to the cross-wires, form a cross, and swing out again. They move, however, more rapidly than those of an acute section, and remain in view for only a small rotation of the stage (p. 314). When the optic axial angle is very small this section approaches in character to the optic normal section now to be described, and also to that shown by a vertical section of a uniaxial substance.

Sections Parallel to the Optic Axial Plane. These sections display an interference figure very like those given by vertical sections of uniaxial substances, in that the equal retardation bands are hyperbolic in shape and lie in four quadrants as shown in Fig. 239. Diffuse dark brushes move rapidly into the centre of the field on rotation of the stage, momentarily darken the field and pass out again in the direction of the acute bisectrix. In parallel light between crossed Nicols the section is characterised by displaying the highest double refraction for the substance, α and γ lying in the plane of the section.

The two quadrants in which the acute bisectrix lies show lower colours at a given distance from the centre of the figure than do the other two quadrants, but when the optic axial angle, 2V, approaches 90°, this colour distinction becomes less apparent, until when the crystal is neutral it disappears altogether.

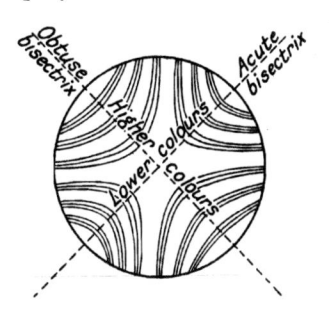

FIG. 239.—Interference Figure given by a Section Parallel to the Optic Axial Plane (Optic Normal Section). Depicted in 45° Position.

The three types of biaxial sections described above lie normal to the planes of optical symmetry and are characterised by the formation of central dark crosses when the traces of these planes are parallel to the cross-wires. They are characteristically presented by pinacoidal faces of orthorhombic crystals.

Sections Normal to an Optic Axis. In these sections an optic axis emerges in the centre of the field, surrounded by nearly circular rings of equal retardation. A single dark brush passes through the centre of the field (Fig. 240), its curvature depending upon the size of the optic axial angle and the position of the optic axial plane with respect to the cross-wires. When the optic axial plane is parallel to one of these, the brush is also parallel to it and is straight, but if the section be rotated 45° from this position, the

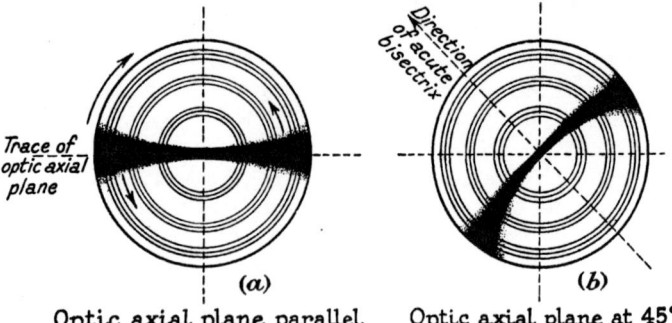

(a) Optic axial plane parallel to a cross wire.

(b) Optic axial plane at 45° to a cross wire.

FIG. 240.—Interference Figure given by a Section of a Biaxial Crystal Normal to one Optic Axis (Medium or Large Optic Axial Angle).

isogyre rotates in the opposite direction and becomes, in general, curved, the amount of curvature being greater the smaller the optic axial angle (Fig. 261, p. 312). It now lies across the optic axial plane and its convex side is directed towards the acute bisectrix. When 2V is 90°, the isogyre remains straight in all positions. Usually only one dark brush is seen in the field, but if the optic axial angle is small, a second optic axis may appear within the field towards the margin with its attendant isogyre (Fig. 241). In parallel light between crossed Nicols such sections are characterised by retaining a uniform faint illumination through a complete revolution of the stage, as already stated (Chap. VII, p. 250).

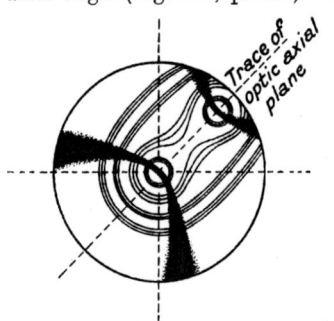

FIG. 241.—Single Optic Axial Figure given by a Crystal with a Small Optic Axial Angle.

Sections Orientated Obliquely to the Bisectrices. Crystals having well-developed prism or pyramidal faces usually lie so as to present sections which are oblique to the symmetry planes of the

SECTIONS ORIENTATED OBLIQUELY TO BISECTRICES 291

indicatrix of the crystal, and so, in convergent light, more or less inclined interference figures are seen. For example, in Figs. 242 (i), (ii), (iii), and (iv), is illustrated an orthorhombic crystal with well-developed prism faces (110), the optic axial plane being coincident with the ac plane. It would lie on the stage as shown in Fig.

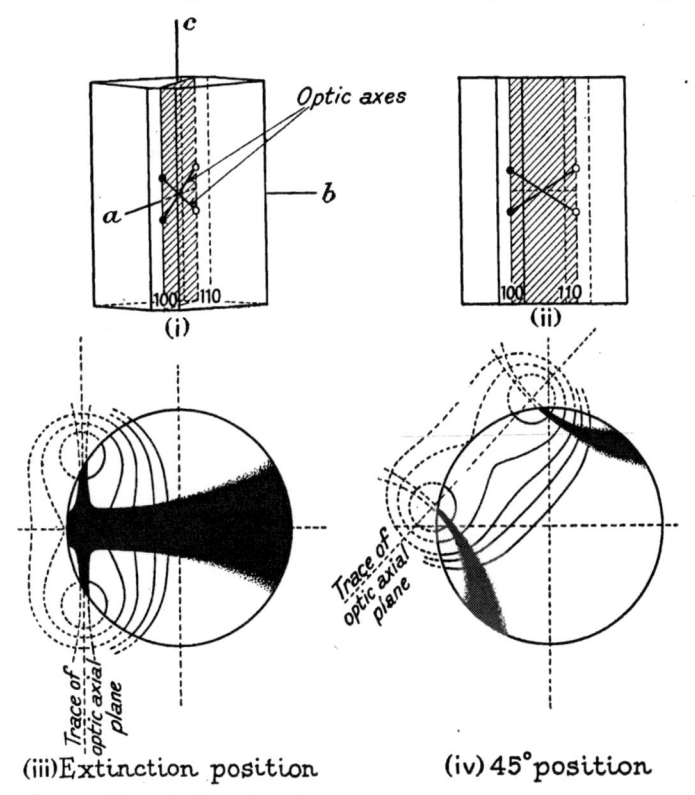

FIG. 242.—Lateral Displacement of Acute Bisectrix Figure.

242 (ii), the plane containing the optic axes being tilted to right (or left) as shown. In convergent light the interference figure would show the lateral displacement of the "eyes" illustrated in Fig. 242 (iii), in which the orientation is such that the trace of the optic axial plane is parallel to the cross-wires. A rotation of the stage through 45° will give a figure similar to that of Fig. 242 (iv).

292   BIAXIAL CRYSTALS

To take another example, should an orthorhombic crystal have well developed macro-domes, (101) (Fig. 243 (i)), it would tend to lie as in Fig. 243 (ii), when, if the optic orientation were the same as

(iii) Extinction position   (iv) 45° position

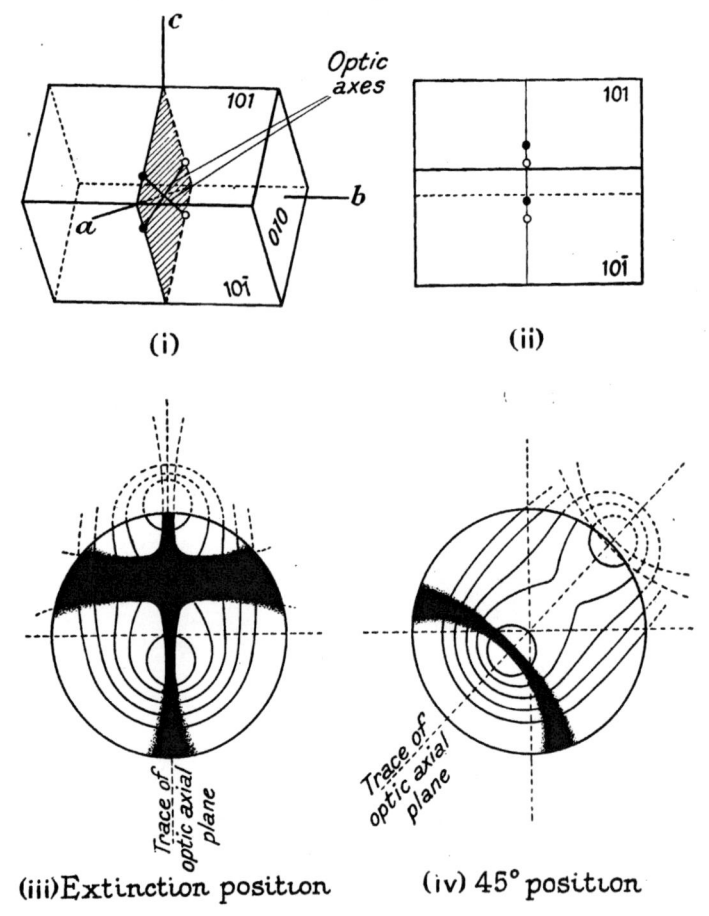

Fig. 243.—Displacement of Acute Bisectrix Figure in the Direction of the Optic Axial Plane.

in the preceding case, the optic axial plane would remain vertical, but the acute bisectrix would not emerge in the centre of the field. The resulting interference figure would bear the aspect shown in

SECTIONS ORIENTATED OBLIQUELY TO BISECTRICES 293

Figs. 243 (iii) and (iv). In cases like those described above, where the section is normal to an optical symmetry plane, *when the crystal is in the extinction position for parallel light, one isogyre passes through the centre of the field and is straight and coincident with a cross-wire.* Where the section is not normal to an optical symmetry plane the isogyres do not pass through the centre of the field in the position where they became straight and parallel to a cross-wire.

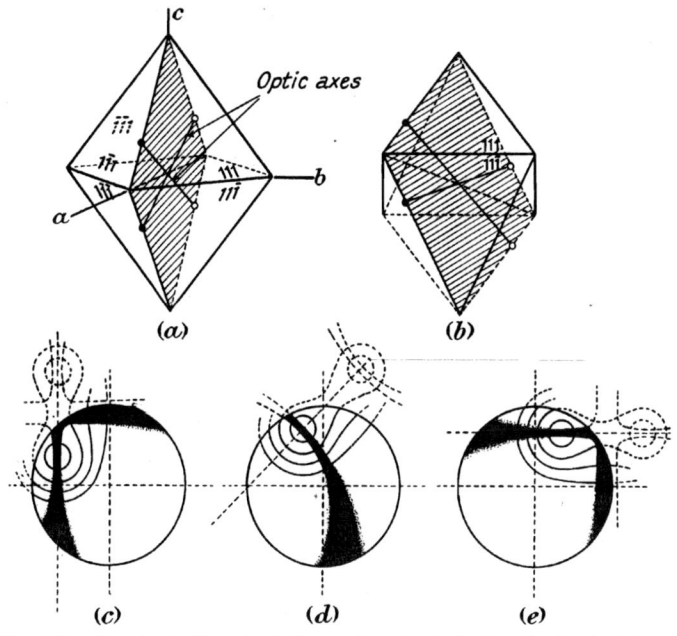

Fig. 244.—Completely Uncentred Figure given by Orthorhombic Crystal lying on (111).

In pyramidal crystals the section presented by the substance lies inclined to all three principal planes of the indicatrix and the isogyres will assume a curved shape when intersecting the centre of the field (*i.e.* in the extinction positions for parallel light). This character distinguishes oblique biaxial interference figures in which only one brush is seen, from inclined uniaxial figures of which the emergence of the optic axis is outside the field of view.

Fig. 244 (*a*) shows a pyramidal orthorhombic crystal with the optic axial plane lying parallel to (010), the acute bisectrix emerging normal to (100). If such a crystal lies as shown in Fig. 244 (b), the resulting tilt of the acute bisectrix is seen in the interference figures at (c), (d), and (e). The decentralised position of the straight isogyre should be noticed.

In general it may be said that the method of procedure in the study of oblique interference figures is to place an isogyre in a position in which it is parallel to a cross-wire (if it can be seen in this position). If the isogyre now occupies the centre of the field the section is normal to an optical plane of symmetry. The stage should be rotated through 45°, the isogyre being watched closely to notice any change of shape, and the direction of the acute bisectrix and therefore the trace of the optic axial plane may be found from the direction of convexity of the brush. The complete figure may then be imagined by completing the outlines of the isochromatic curves. If a change to parallel light is now made the relation of the interference figure to the crystallographic directions may be found.

DISPERSION. EFFECT IN CONVERGENT LIGHT

Hitherto, the discussion of interference phenomena in convergent light has been pursued without emphasising the effects due to light of different colours. In Chapter IV, pp. 120–3, the changes in optical properties of crystals with light of different colours were explained and the effects of these changes with respect to interference phenomena in convergent light will now be dealt with.

In uniaxial crystals, although the values of ε and ω, and therefore the size and shape of the indicatrix, vary with the wave-length of the light, the direction of the optic axis remains unchanged and the crossed isogyre maintains the same position for all colours.

In biaxial crystals, on the other hand, dispersion results in a variety of changes which, though frequently small, are large enough in many substances to have marked effects upon the interference figures. In the first place, dispersion of the refractive indices α, β, and γ, alters the size of the optic axial angle to a greater or less extent, because the relative values of the indices are changed. This variation in the optic axial angle is called *dispersion of the optic axes* and may occasionally be so large as to make the acute bisectrix for one colour (say red) the obtuse bisectrix for another (say blue). Secondly, in addition to the change in both size and shape which the indicatrix undergoes in all uniaxial and biaxial crystals with variation in the wave-length of light used, in the monoclinic and triclinic systems the orientation of the indicatrix relative to the crystal axes varies also. This results in *dispersion of the bisectrices*.

Dispersion in Orthorhombic Crystals. In all substances belonging to this system the three axes of the indicatrix and those of the crystal coincide for light of all colours (Chap. IV, p. 120). The optic axial plane (containing α and γ) may be parallel to any one of the pinacoids (100), (010), or (001). The acute bisectrix may therefore emerge parallel to either a, b, or c, the optic axial plane having two possible positions in each case.

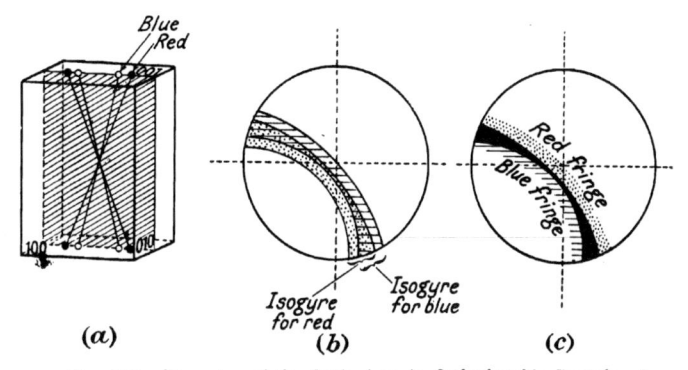

FIG. 245.—Dispersion of the Optic Axes in Orthorhombic Crystals. 1.

An example of moderate dispersion is shown in Fig. 245 (*a*), where the optic axial plane is parallel to (100) and the axial angle is greater for red light than for blue. It is evident that in white light, the isogyres of the interference figure in the 45° position will be composite, consisting of the partly superposed isogyres for different kinds of light as shown in Fig. 245 (*b*) for red and blue.

The composite isogyre of Fig. 245 (*b*) would have an appearance like that illustrated in Fig. 245 (*c*), namely, a dark brush fringed on its convex side with red, and on its concave side with blue light. This is due to the fact that the isogyre for any colour marks the positions in which that colour is extinguished, and the other components of white light reach the eye. In cases where the optic axial angle is greater for red light (ρ), than for violet (v), as in Fig.

245, the fact is summarised briefly by the formula $\rho > \nu$, and the opposite case by $\rho < \nu$.

Dispersion of the optic axes also affects the distribution of colour in the colour bands of the interference figure. In cases of moderate dispersion this effect is most evident in the immediate neighbourhood of the melatopes, where also the coloured fringes of the isogyres

(a) 45° position

(b) Extinction position

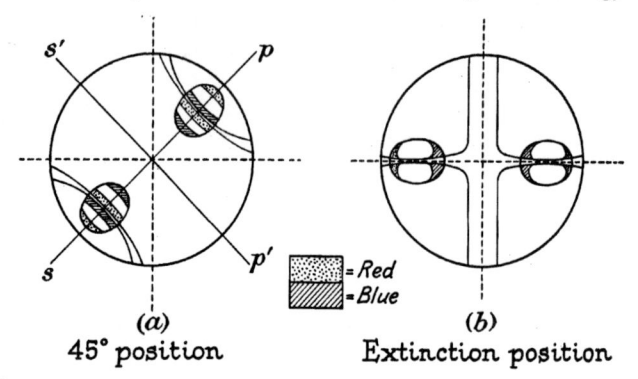

= Red
= Blue

FIG. 246.—Dispersion of the Optic Axes in Orthorhombic Crystals ($\rho > \nu$). 2.

show up most clearly. In Figs. 246 (a) and (b), the arrangement of colours around these two points is shown diagrammatically for the case $\rho > \nu$. The colours would be reversed for the opposite case. It should be noted that in the extinction position the coloured fringes of the isogyres disappear; and also that the figure is symmetrical about the trace of the optic axial plane, sp, and about the optic normal, $s'p'$, this following from the fact that only the optic axes and not the bisectrices are dispersed.

Dispersion in Monoclinic Crystals. In crystals belonging to the monoclinic system one of the symmetry planes of the indicatrix lies parallel to $b(010)$, *i.e.* to the one possible symmetry plane of the system, whatever the wave-length of the light, and it follows that one of the axes of the indicatrix, either α, β, or γ, must be coincident with the direction of the b axis for all colours of light. Three cases are possible, and are illustrated in Figs. 247 (i), (ii), and (iii), the ortho-axis b being in turn acute bisectrix, obtuse bisectrix and optic normal. In (i) and (ii), the optic axial plane is normal to $b(010)$ and the plane of symmetry, and in (iii) it is parallel to $b(010)$. The cases figured represent a fairly large degree of dispersion but not so

DISPERSION IN MONOCLINIC CRYSTALS

extreme as to lead to the phenomenon of crossed-axial-plane dispersion (see p. 122), examples of which will be given below.

In addition to angular dispersion of the optic axes which is always present in some degree, dispersion of the bisectrices which happen to lie in the plane parallel to (010) takes place, as the result of a rotation of the indicatrix as a whole about the fixed axis b.

In the first case (Fig. 247 (i)), b is the acute bisectrix (α or γ), the obtuse bisectrix (γ or α) together with the optic normal lying in

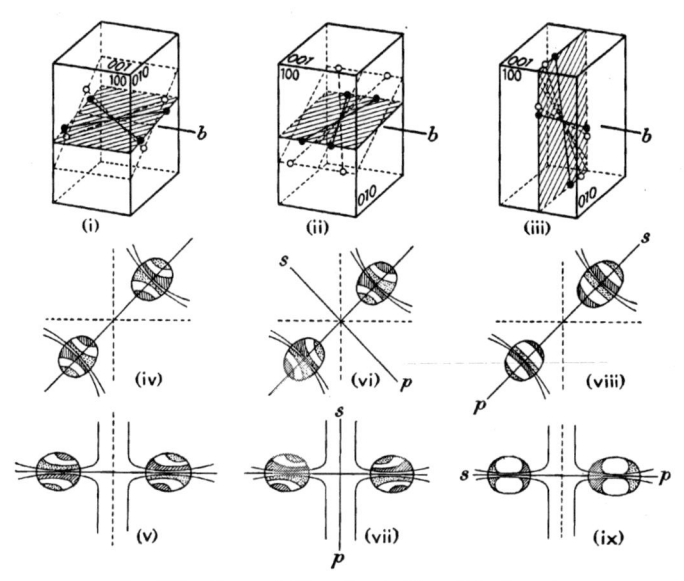

Fig. 247.—Types of Dispersion in Monoclinic Crystals.
Lines about which the interference figures are symmetrical are marked s p.

the symmetry plane (010). Dispersion takes place by a rotation of the optic axial plane around b, the melatopes of the acute axial angle rotating around the fixed bisectrix, and the obtuse bisectrix (together with the optic normal) being dispersed in the plane (010). This type of dispersion is called *crossed* or *rotated*. Viewed in convergent light the interference figure is seen to be centro-symmetrical (Fig. 247 (iv) and (v)).

In the second case (Fig. 247 (ii)), b is the obtuse bisectrix, the acute bisectrix and the optic normal lying in the symmetry plane.

298 DISPERSION. EFFECT IN CONVERGENT LIGHT

A similar rotation of the indicatrix displaces the acute bisectrix normal to a line joining the melatopes of the figure. The positions of successive interference figures for different wave-lengths of light are thus parallel to one another (the distance between the melatopes in successive figures being different, however, owing to dispersion of the optic axes). The dispersion is therefore called *horizontal* or *parallel*. The interference figure in white light will only be symmetrical with respect to a line normal to the trace of the optic axial plane. This type of dispersion is best recognised in the position in which the isogyres are crossed when the narrow bar joining the melatopes will be seen to be bordered by different colours above and below (Fig. 247 (vii)).

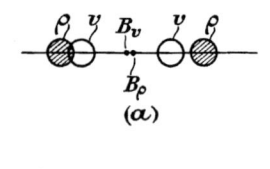

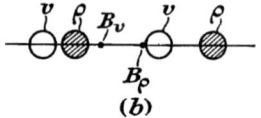

FIG. 248.—Inclined Dispersion in Monoclinic Crystals.

In the remaining case (Fig. 247 (iii)), the optic normal β coincides with the b axis and the optic axial plane lies parallel to the clinopinacoid (010). The optic axes and the acute and obtuse bisectrices are dispersed parallel to the clinopinacoid. With changing colour of light the position of the melatopes changes along the line joining them, and the interference figure will be symmetrical only with respect to the trace of the optic axial plane (Fig. 247 (viii)). Such dispersion is called *inclined*. The colour bands round one melatope will be more elongated than around the other, the actual order of the colours around the melatopes depending upon the amount of dispersion of the optic axes and bisectrices. Thus, in the case shown in Fig. 248 (a) the isogyres in the interference figure will be fringed as shown in Fig. 247 (viii). Should, however, the dispersion be as great as that shown in Fig. 248 (b) the colours around one melatope would be reversed.

Dispersion in Triclinic Crystals. In the triclinic system, there being no planes nor axes of symmetry by which the orientation of the indicatrix may be controlled, dispersion of both optic

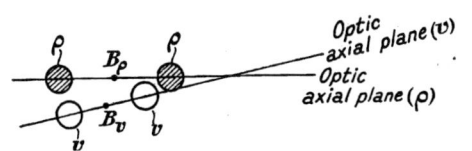

FIG. 249.—Dispersion in Triclinic Crystals.

axes and bisectrices takes place irregularly. A possible case is shown in Fig. 249.

Crossed-Axial-Plane Dispersion. In certain biaxial substances in which the intermediate refractive index β lies near to either α or γ, the variation in size of the optic axial angle for a comparatively small change in the wave-length of light is very marked. If the acute interference figure of such a substance is observed in red light, and the wave-length is changed towards blue (by the use of a monochromator or a series of filters), it will be seen that the axial angle will become smaller as the wave-length is shortened, the substance becoming uniaxial for some intermediate colour of the spectrum. Further progressive change in the colour of the light will be followed by an opening of the optic axes in a direction at right angles to the former trace of the optic axial plane, the maximum size of angle being reached for violet light. The optic axial planes for red and blue light are normal to one another. This type of dispersion, called *crossed-axial-plane* dispersion, has already been briefly treated in Chapter IV, p. 122. The case of brookite mentioned there is illustrated in Figs. 250 (*a*), (*b*), and (*c*). The interference figure of this substance in white light (like that of most substances which show the phenomenon) is symmetrical but anomalous and without isogyres.

A very accessible example is afforded by *p*-nitrobenzaldehyde.[*] If a little of this substance is melted between a slide and a cover slip and allowed to crystallise slowly, many of the crystals (probably orthorhombic) show acute bisectrix figures consisting of two strongly coloured isogyres (Fig. 251 (*a*)). In red, green, and blue lights (obtained simply by means of colour filters) the isogyres appear in the positions shown in Figs. 251 (*b*), (*c*), and (*d*), respectively. The

[*] E. S. Davies and N. H. Hartshorne, *J. Chem. Soc.*, 1934, 1830.

reason why definite isogyres appear at all in white light is probably that the isogyres for wave-lengths between red and yellow are much more crowded together than those for other parts of the spectrum.

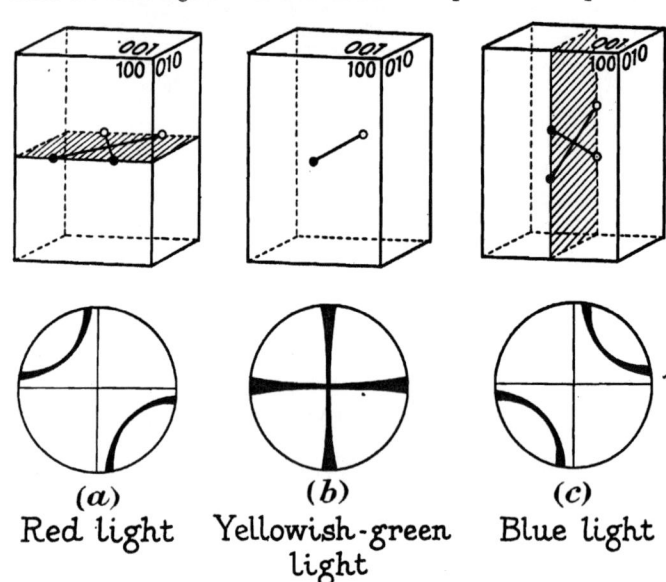

(a) Red light (b) Yellowish-green light (c) Blue light

Fig. 250.—Crossed-Axial-Plane Dispersion of Brookite.

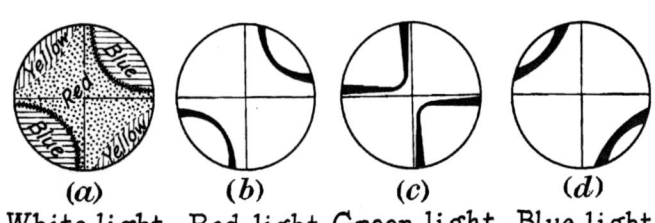

(a) White light (b) Red light (c) Green light (d) Blue light

Fig. 251.—Crossed-Axial-Plane Dispersion of p-Nitrobenzaldehyde.

Other orthorhombic substances * which show dispersion of this type are rubidium sulphate, Rb_2SO_4; cæsium selenate, Cs_2SeO_4; and saccharo-lactone, $C_6H_{10}O_5$, which is also optically active.

* A. E. H. Tutton, *Crystallography and Practical Crystal Measurement* (MacMillan, 1922), Vol. 2, 1058.

In monoclinic crystals, crossed-axial-plane dispersion involves a change in the orientation of the optic axial plane from (010) to a plane normal to this, or *vice versa*, *i.e.* between the orientations shown in Fig. 247 (ii) and (iii). It will be apparent that this change may be accompanied by the types of dispersion illustrated in these figures. An example is afforded by *2,2'*-dipyridyl [*] in which at 3650 Å the optic axial plane is parallel to (010), whilst at 5461 Å it is normal to this plane. At 4150 Å the crystal is uniaxial. Between 3650 and 4150 Å, inclined dispersion is shown, and between 4150 and 5461 Å, horizontal dispersion.

Other examples of this type of dispersion in monoclinic crystals are shown by cæsium magnesium sulphate, $Cs_2Mg(SO_4)_2, 6H_2O$, and the corresponding selenate, and by ethyl triphenylpyrrholone.[†]

Dispersion Resulting in Change of Optical Sign. In certain uniaxial crystals in which the birefringence is small, dispersion may result in a change of sign. The example of torbernite has already been mentioned in Chapter IV, p. 122. An organic example is afforded by benzil,[‡] which is uniaxial and positive at all wavelengths down to the visible violet, in which it becomes isotropic, and at shorter wave-lengths is negative.

In some biaxial crystals also a change of optical sign has been observed with change of wave-length, that is to say, the dispersion of the optic axes results in the optic axial angle passing through a value of 90°. This is shown by the mineral danburite, and a number of organic examples have been described by W. M. D. Bryant, *viz.*, *trans*-stilbene (see p. 122) ; and the picrates of dimethylamine, diethylamine, triethanolamine, which become optically neutral at approximately 4820, 4660, and 4900 Å respectively.[§]

Measurement of Dispersion. Accurate measurements of the amount of dispersion should be made in monochromatic light of different wave-lengths. To measure the amount of dispersion of the bisectrices in monoclinic and triclinic crystals, the method employed is to determine the maximum extinction angles of the substance in light of different colours, instead of making observation upon the interference figures.

---

[*] W. M. D. Bryant, *J. Amer. Chem. Soc.*, 1941, **63**, 511.
[†] A. E. H. Tutton, *op. cit.*
[‡] E. E. Jelley, *Phot. J.*, 1934, **74**, 514.
[§] J. Mitchell, Jr., and W. M. D. Bryant, *J. Amer. Chem. Soc.*, 1943, **65**, 128.

Determination of the Optical Sign of a Crystal by Observations upon Interference Figures

The differences between positive and negative crystals have already been explained in Chapter IV. They may be summarised briefly for the purpose of this section by means of the four diagrams in Fig. 252 which give the traces of the directions of vibration and relative velocities of convergent rays in basal sections of uniaxial

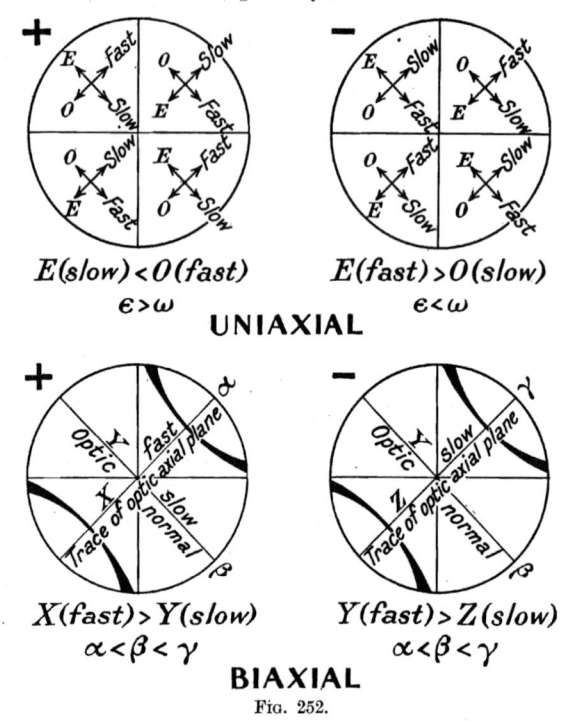

+ $E(slow) < O(fast)$
$\epsilon > \omega$

− $E(fast) > O(slow)$
$\epsilon < \omega$

UNIAXIAL

+ $X(fast) > Y(slow)$
$\alpha < \beta < \gamma$

− $Y(fast) > Z(slow)$
$\alpha < \beta < \gamma$

BIAXIAL

Fig. 252.

crystals, and the principal vibration directions in sections of biaxial crystals normal to the acute bisectrix.

The problem of determining sign, therefore, resolves itself into finding out (1) in uniaxial interference figures, whether the radial or tangential direction of vibration is the faster, and (2) in biaxial crystals, whether in the acute interference figure the optic normal direction giving the refractive index β (*i.e.* the normal to the line joining the melatopes) is faster or slower than the ray vibrating parallel to the trace of the optic axial plane, this ray giving either α in positive, and γ in negative crystals.

The accessories used for this purpose are the mica plate ($\frac{1}{4}\lambda$ retardation); the gypsum or selenite plate (unit retardation plate); and the quartz wedge. The principles underlying the use of these accessories have already been explained in Chapter VII, pp. 260–3, and need not be repeated here. It will only be necessary to describe their application to the present problem. In general, it may be said that the gypsum plate is better adapted for use with sections having low double refraction, the mica plate and quartz wedge being more convenient for sections with higher birefringences.

It cannot be too strongly emphasised that in the following discussion, the phenomena illustrated in the diagrams are true only for the conditions described, and care must be exercised to see that the orientation of the interference figure, plate or wedge, and microscope slot, are identical with those given in the text before using the diagrams as a guide in practical work. It is always best to work out each case from first principles.

Uniaxial Crystals. If a $\frac{1}{4}\lambda$ retardation mica plate is placed in the slot of the microscope between crossed Nicols so that its vibration directions lie at 45° from the cross-wires, the appearance of a uniaxial interference figure is characteristically changed. The isogyres disappear, the colour bands are slightly moved radially towards the centre in one pair of diagonally opposed quadrants of the field, and outwards in the other diagonal pair; at the same time there appear two dark spots near the intersection of the cross-wires on a line at 45° to them (Figs. 253 (a) and (b)). These spots

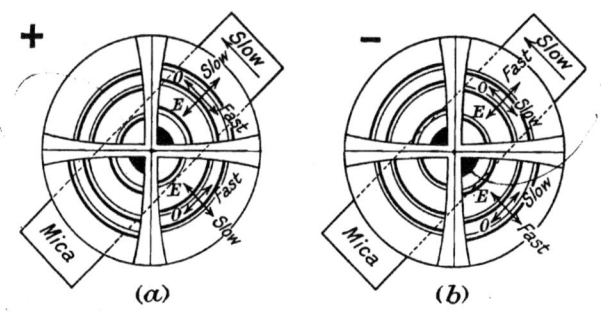

FIG. 253.—Effect of the $\frac{1}{4}\lambda$ Mica Plate on + and − Uniaxial Interference Figures.

are really segments of circles. The diagonal upon which the spots lie is determined by the direction in which the relative retardation is reduced. For example, in Fig. 253 (a) is shown the effect produced by a mica plate with fast length superposed NE–SW upon a positive uniaxial interference figure. The fast directions of vibration in crystal and plate are at right angles to each other in the NE–SW quadrants and so the relative retardation in these is reduced, the dark spots marking the positions at which exact compensation results. In the NW–SE quadrants the relative retardation is increased. The reason why the colour bands move in the way shown is as follows. In the diagram the position of the colour bands before the insertion of the plate is represented by the fine concentric circles around the centre. Consider the first one marking the locus of points at which retardation is exactly one wavelength for a certain colour of light. In the NE–SW quadrants this circle has now only $\frac{3}{4}\lambda$ retardation and the 1λ retardation circle will lie more remote from the centre. In the other quadrants (NW–SE) the position of the first fine ring now marks the locus of points having $1\frac{1}{4}\lambda$ retardation, and therefore the 1λ band will lie nearer the centre of the figure. All the colour bands will naturally be affected in the same way. Discontinuity of the rings will therefore occur at the isogyres. The negative case is shown in Fig. 253 (b).

A unit wave-length retardation plate of gypsum giving a red of the first order between crossed Nicols may be used in the same way as the mica plate. The relative retardation of the plate is not affected by the isogyres, which appear red. Reference to the birefringence diagram, p. 248, will show that a small increase in relative retardation will change the first order red to blue of the second order, whilst a slight reduction in relative retardation will give a yellow of the first order. In the quadrants of the interference figure in which the relative retardation is increased, the margin of the isogyres near the cross will be fringed with blue and similar positions in the opposite quadrants will be occupied by yellow (Fig. 254). The colour bands suffer alteration in position as in the case already described when the mica plate is used, but the relative retardation of each ring is altered by exactly 1λ instead of $\frac{1}{4}\lambda$ for light of a definite wave-length. The positive and negative uniaxial cases with a fast length gypsum plate are shown, the various bands being labelled with the retardation they represent.

When the quartz wedge is used, the results are exactly analogous to those seen when the mica and gypsum plates are used, except that the movement of the colour bands is continuous as the wedge is pushed through the microscope slot thin end first, the movement

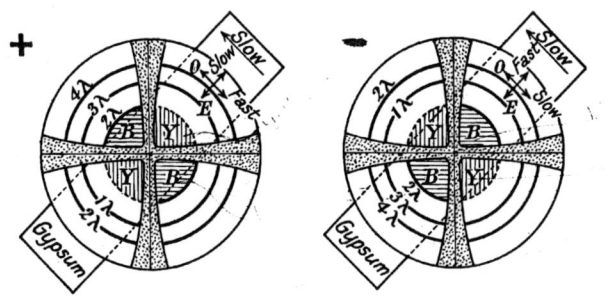

Fig. 254.—Effect of Unit Retardation Gypsum Plate on + and − Uniaxial Interference Figures.

being towards the centre in the quadrants in which the colours are raised (increased relative retardation), and away from the centre in the opposite quadrants where the colours are lowered in response to a decrease in relative retardation. In Fig. 255 the movement

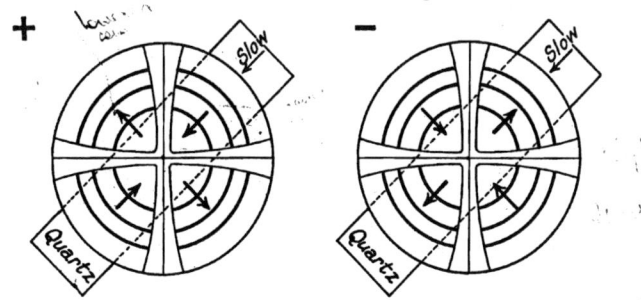

Fig. 255.—Effect of Quartz Wedge on + and − Uniaxial Interference Figures.

of the colour bands in positive and negative uniaxial interference figures is indicated by the small stout arrows, a quartz wedge with its slow direction of vibration parallel to its length being used, orientated NE–SW with respect to the cross-wires.

When inclined interference figures are presented, such as those shown in Fig. 232, the same methods may be adopted as those just detailed. The stage is revolved, the movement of the straight isogyres is noted and a quadrant is selected the orientation of which in relation to the complete figure is known. The test is then made with one of the plates or a wedge, the latter being used upon very inclined figures of which only the colour bands may be seen.

Sections parallel to the optic axis may be recognised by their interference figures (p. 282), and the direction of the optic axis deduced. The sign of the substance may then be determined in parallel polarised light by use, e.g., of the quartz wedge.

Biaxial Crystals. The methods for determining sign in biaxial crystals are conducted in exactly the same way as those for uniaxial crystals, and any of the three accessories already mentioned may be used. The mica plate is, however, not in general so satisfactory in the case of biaxial crystals as are the gypsum plate and quartz wedge. When an acute interference figure is examined by means of the $\frac{1}{4}\lambda$ retardation mica plate the dark spots in diagonally opposite quadrants are seen near the melatopes if these are placed parallel to a cross-wire before inserting the plate in the 45° slot. The effect of a plate with a fast length orientated NE-SW above a positive interference figure is shown in Fig. 256. For a negative substance the spots would appear in the opposite quadrants.

When the gypsum plate or quartz wedge is used to determine sign, the interference figure should be orientated so that the melatopes make an angle of 45° to the cross-wires, so that the directions of vibration in the plate lie parallel to the principal vibration directions in the crystal section.

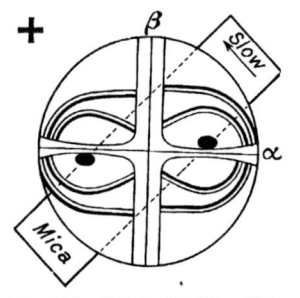

Fig. 256.—Effect of $\frac{1}{4}\lambda$ Mica Plate on a + Biaxial Interference Figure (Acute Bisectrix).

Consider the case of a positive biaxial section normal to the acute bisectrix. If this section is orientated so that the melatopes lie NE-SW, the fast direction of vibration (α) will be parallel to the line joining the melatopes and the slow direction (β) will lie normal to it. If now a quartz wedge with slow length is inserted parallel to the trace of the optic axial plane of the crystal (fast), then the colours will be lowered. As the wedge is pushed in the lemniscate colour bands will move out in the direction of the optic normal, bands of smaller replacing those of higher retardation (Fig. 257 (i)). The phenomena for positive and negative crystals in different orientations is given in Figs. 257 (i)–(iv).

When a unit retardation plate is used, the phenomena for biaxial crystals are analogous to those given by uniaxial crystals. The isogyres, which become red on insertion of the plate, are fringed near

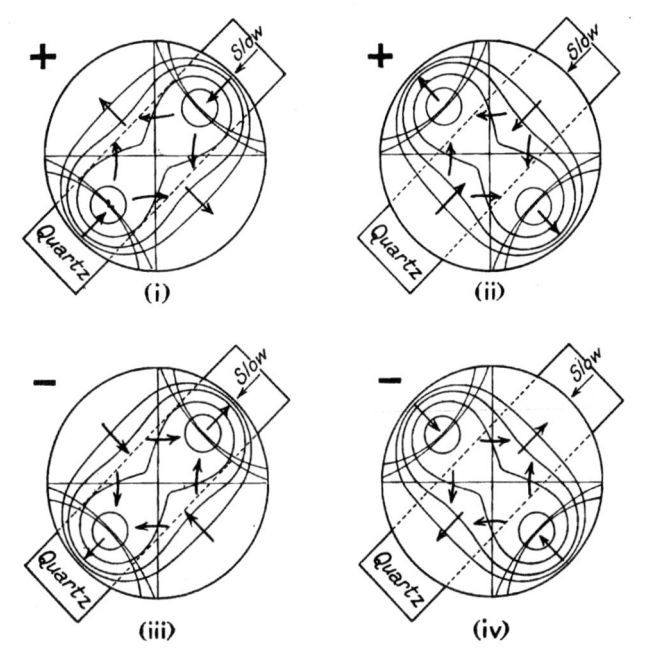

Fig. 257.—Effect of Quartz Wedge on + and − Biaxial Interference Figures (Acute Bisectrix).

the melatopes with yellow on the sides facing those portions of the field in which the retardation has been reduced, and blue on the opposite sides. Figs. 258 (a) and (b) show the disposition of the coloured fringes for positive and negative biaxial interference figures, using a 1λ retardation plate with fast length, in a NE–SW direction. These diagrams should be carefully compared with those for uniaxial crystals (Fig. 254), when it will be seen that the disposition of the colours with respect to the 45° diagonals is the same, *i.e.* for the conditions stated, in positive crystals yellow NE–SW, blue NW–SE, and *vice versa* for negative crystals.

It should be remembered that the optical effects (directions of movement of colour bands, etc.), obtained for any given orientation of an acute

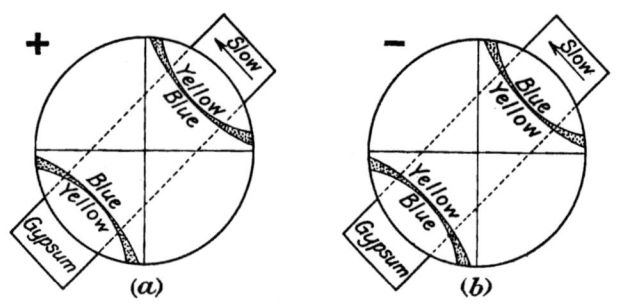

Fig. 258.—Effect of Unit Retardation Gypsum Plate on + and − Biaxial Interference Figures (Acute Bisectrix).

interference figure and plate or wedge, will be reversed if the obtuse figure is examined.

On partial biaxial interference figures the same procedure as for complete figures is followed after the direction of the acute bisectrix has been determined by turning the isogyre 45° from its

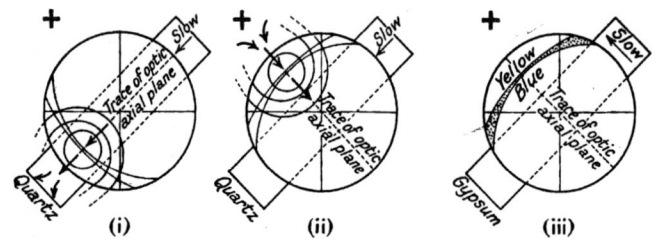

Fig. 259.—Effect of Quartz Wedge, (i) and (ii), and Gypsum Plate, (iii), on an Inclined Biaxial Interference Figure (+).

straight position (if it can be seen), and noting the direction of its convexity, which is the required direction. Figs 259 (i), (ii), (iii) show a typical case for an inclined positive biaxial substance with slow quartz wedge and fast gypsum plate. These diagrams should be compared with Figs. 257 and 258. Should the isogyre leave the field in the diagonal position its curvature cannot be determined with certainty and another section should be sought.

On sections normal to the optic axial plane (see p. 289, Fig. 239), the optical character may be determined with the quartz wedge as described for optic normal sections of uniaxial substances, after deducing the direction of the acute bisectrix as given on p. 289, the acute bisectrix direction being slow (γ) for positive, and fast (α) for negative substances.

Measurement of the Optic Axial Angle

The exact measurement of optic axial angles by the crystallographer involves the cutting of accurately orientated sections which are examined in convergent polarised light upon some form of rotation apparatus which enables each melatope of the interference figure in turn to be brought into coincidence with the cross-wires in the ocular of the instrument. If the internal angle, 2V, is to be measured directly, the section is immersed in a liquid having a refractive index equal to the intermediate index β of the crystal. Under the polarising microscope, this method may be adopted by using a small stage goniometer such as that of the Miers pattern (Fig. 167, p. 177), or the modified stage goniometer by Swift (p. 214). Another method is to observe the angular distance between the optic axial sections of the crystal, upon the universal stage (p. 413), the latter being used with parallel light and low-power objectives. These sections are distinguished by their uniform illumination on rotation between crossed Nicols in a plane normal to the axis of the microscope.

It is possible to calculate the value of 2V if all three refractive indices are known. These may be determined under the microscope as described in pp. 316–18. The formulæ given below show the relation between the refractive indices and half the acute angle for positive and negative crystals respectively, Vα being half the negative acute axial angle and Vγ that for positive crystals :

$$\cos^2 V\alpha = \frac{\gamma^2(\beta^2 - \alpha^2)}{\beta^2(\gamma^2 - \alpha^2)}$$

$$\cos^2 V\gamma = \frac{\alpha^2(\gamma^2 - \beta^2)}{\beta^2(\gamma^2 - \alpha^2)}$$

Alternatively it is possible to calculate any principal refractive index from a knowledge of V and the other two refractive indices, and this may be useful when γ is too high to be measured directly. This problem is discussed more fully on pp. 319 *et seq.*

The most rapid and convenient methods of measuring the optic axial angle under the microscope are those in which observations are made upon interference figures of which one or both of the melatopes appear in the field. Several factors militate against absolute accuracy. In the first place the primary interference figure at the principal focus of the objective is formed upon a curved (nearly spherical) surface, and what is really observed by the eye is an orthographic projection of the figure. Secondly, in thin sections the isogyres may be so diffuse as to make it difficult to determine very accurately the points of emergence of the optic axes. It must be remembered that the size of the optic axial angle depends upon the wave-length of light, and that therefore a precise determination can only be made in monochromatic light. The dispersion of the optic axial angle in many substances is, however, so small that a mean value, sufficiently definitive for diagnostic purposes, can be obtained in white light. It is possible in the most favourable cases to approach within 1° of the true value of 2E (the apparent angle in air, p. 288) by microscopic methods using monochromatic light. Usually, however, the accuracy is of the order of $\pm$ 4°. It goes without saying that the optical system of the microscope must be carefully centred.

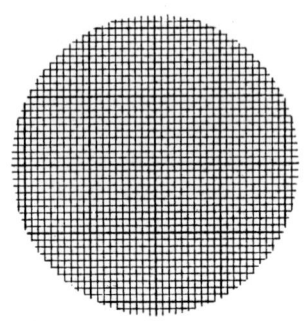

Fig. 260.—Co-ordinate Micrometer Scale.

Measurements upon the figures are made by means of a scale placed in the focal plane of the ocular (or at a conjugate focus where it may be viewed simultaneously with the interference figure). A simple linear scale which can be placed at 45° to the cross-wires will do, but a 0·1 mm. co-ordinate micrometer scale (Fig. 260) is much better and makes the plotting of the positions of the isogyres upon diagrams (p. 313 below) a simple matter. For great accuracy in measuring the figures a double-screw-micrometer eyepiece as recommended by Wright * may be used, in which cross-wires are moved in two directions at right angles, the amount of movement being read on a drum fitted to each screw.

* F. E. Wright, *Methods of Petrographic-Microscopic Research*, p. 155 (1911).

The Mallard Method (Two Melatopes in the Field).

E. Mallard showed that half the apparent distance between the melatopes of a biaxial interference figure, as measured on an eyepiece scale, is proportional to the sine of the angle which the optic axes make with the bisectrix, and a constant factor which depends upon the lens system of the microscope. The relation between the apparent axial angle, 2E, and half the apparent distance D, between the melatopes is given by Mallard's formula

$$D = K \sin E.$$

K being the *Mallard constant*. In order to find K for any lens system of the microscope, D is measured in terms of the eyepiece scale upon an interference figure the optic axial angle of which is known. An example will make the method clear. A plate of barytes cut normal to the acute bisectrix with 2E = 63° 10′ gave D = 12, in the 45° position, using a Reichert No. 5 objective and a Reichert No. III ocular. K therefore works out to 22·9. One good determination will usually suffice for ordinary work, but two or three determinations should be made upon substances having very different values of 2E for greater accuracy. Once the Mallard constant (sometimes denoted by M), is known, the value of 2E for any section showing a central (or slightly inclined figure) having two melatopes in the field may be found, and the apparent angle reduced to the true angle (2V) by means of the formula

$$\sin E = \beta \sin V$$

where β is the intermediate refractive index of the crystal. Graphical methods for solving the formulæ have been devised and reference should be made for these to the works of Johannsen and Wright already quoted.* In microscopes where an arrangement exists for altering the position of the Bertrand lens, the Mallard constant must be determined not only for a fixed combination of ocular and objective but also for a definite position of the Bertrand lens.

Instead of working out the angular equivalents of D by means of the Mallard formula each time, the divisions of the microscope scale may be calibrated once for all for any fixed lens system by means of an apertometer † which is placed upon the stage, the

---

* A. Johannsen, *op. cit.*, pp. 469 *et seq.*, p. 491; F. E. Wright, *op. cit.* (plates).

† A. Johannsen, *op. cit.*, p. 133.

angular equivalents of the scale being then read off and noted. Three or four substances of known 2E may be used for the same purpose and a graph may be constructed showing the relation of E and D.

Estimation of 2V on Optic Axial Sections. It has already been stated that when the angle 2V is 90° the single isogyre of the optic axial section is straight and pivots around the centre of the field (the " compass figure "). In uniaxial crystals (2V = 0°), the basal section gives a figure in which the angles between the isogyres are 90°, and when 2V is very small the acute figure approximates to that of the uniaxial one, each separate isogyre in the 45° position

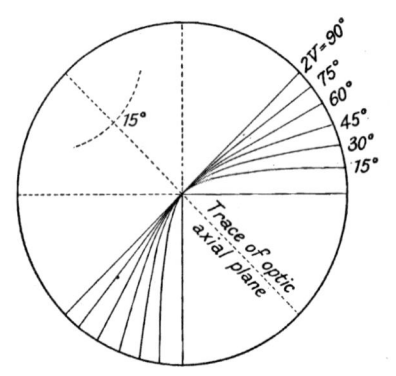

FIG. 261.—Curves showing approximately the Relation of the Shape of the Isogyre to the Optic Axial Angle (2V), in a Section Normal to one Optic Axis (45° Position).

being sharply curved, the two arms of each brush forming an angle of nearly 90° between them. The amount of curvature of the isogyres in the 45° position is clearly a function of the size of the optic axial angle, though it depends somewhat upon other factors such as the mean refractive index of the crystal. It may therefore be used to estimate the size of the optic axial angle. Fig. 261 shows approximately the course of the middle line of the isogyre in optic axial sections for values of 2V between 0° and 90°, at intervals of 15°. The field of view shown is that given by an average microscope using an objective of about 0·80 N.A., and it will be noted that when 2V is less than about 25° the other isogyre of the interference figure is also seen. The curves are based on those

MEASUREMENT OF THE OPTIC AXIAL ANGLE 313

given by F. E. Wright * for different cases, and may be used to estimate optic axial angles to perhaps $\pm 10°$. Comparison of an actual isogyre with the curves is simplified and at the same time made more accurate by plotting it upon squared paper with the aid of a co-ordinate micrometer scale.

Sections with Large Axial Angle Normal to a Bisectrix.
When the apparent angle between the optic axes in air is greater than about 85°, the melatopes in a section normal to a bisectrix will lie outside the field unless the numerical apertures of condenser and objective are exceptionally large, and it will not be possible to discover by inspection whether the acute or obtuse interference figure is being examined. Michel-Lévy used, as a measure of the angle, the amount of rotation of the stage necessary to bring the isogyres of such a figure from the crossed position to become tangential to a given circle. The circle (or circles) used were engraved upon a glass plate and placed in the focal plane of the eyepiece or at a conjugate focus of the interference figure, thereby being viewed simultaneously with it.

To use the method it is necessary to know the angular value of the circle of reference, and Michel-Lévy obtained this by using a substance of known 2E. The formula he gave involves a knowledge of the refractive index of the glass in the objective, and as this is not usually known, a modified formula by F. E. Wright, using Mallard's constant (see above), is given here:

$$\sin E = \frac{K}{\sqrt{\sin 2\phi}}$$

where E is half the required unknown angle, K the Mallard constant determined for the circle of reference by means of a substance of known 2E (as described on p. 311), and ϕ is the angle of revolution of the stage from the position of crossed isogyres until they become tangent to the circle of reference.

The margin of the field may be substituted for a circle of reference engraved on glass, but difficulty is then experienced in deciding the exact point of tangency and the difficulty is increased when the isogyres are diffuse.

The method is not nearly so accurate as that of Mallard and a relatively large error is introduced should the section be slightly inclined to the bisectrix. As a means of deciding whether an acute

* *op. cit.*

or an obtuse figure is under observation the method is useful, and Dana * uses a rough approximation of the method to differentiate between acute and obtuse sections, giving the amount of rotation of the stage necessary to bring the isogyres from the crossed position to the margin of the field as roughly $> 30°$ and $< 15°$ respectively.

A more accurate modification of the Michel-Lévy method has been devised by Johannsen and Phemister † in which a substance of known 2E, the isogyres of which fall inside the field, is viewed simultaneously with the figure of the unknown substance to be measured. The known substance is placed upon the upper lens of the condensing system, and the unknown substance on the stage

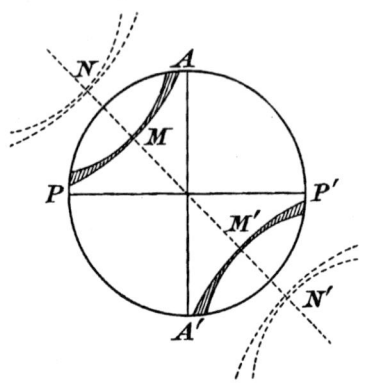

FIG. 262.—Determination of the Optic Axial Angle by the Method of Johannsen and Phemister.
MM' = melatopes of standard section.
NN' = melatopes of unknown section.

as near as possible to the lower crystal. When both sections have been arranged in the 45° position with the directions of their optic axial planes coinciding (Fig. 262), a confused figure is seen. The upper section is rotated, and when its isogyres moving inwards fall over those of the lower substance, the dark brushes come into view because both lower and upper isogyres mark the locus of points of extinction. The isogyres are made to coincide at the melatopes of the lower crystal, the lower section is then removed, and the stage reading noted. The unknown substance is then turned until the crossed position of the isogyres is reached. The reading of the

* E. S. Dana, *Text Book of Mineralogy*, revised by Ford, p. 310 (1932).
† *J. of Geol.*, 1924, **32**, 81.

stage is again taken and the difference gives the angle ϕ of the previous formula, which now becomes

$$\sin \text{E of the unknown substance} = \frac{\sin \text{E of known substance}}{\sqrt{\sin 2\phi}}$$

In this method the circle of reference is replaced by the distance between the melatopes of a section having a known value of 2E. The advantage of the method is that there is no ambiguity about the position at which the two sets of isogyres become superposed and errors in the estimation of the Mallard constant are eliminated.

A plate of muscovite having a relative retardation of 2λ is suggested as a standard section of reference. An accuracy of $\pm 1°$ is claimed if care is taken to place the standard section and ocular scale exactly at 45°.

The Objective Diaphragm. C. B. Slawson * devised a method of measuring interference figures by means of a variable diaphragm placed in the rear focal plane of the objective, where it is superimposed directly upon the interference figure itself, and calibrated to give the angular distance from the centre of the field of any point in the interference figure. This method eliminates certain sources of error inherent in the Mallard and other methods, because calibration is dependent only upon the objective and not upon a plurality of factors such as a fixed lens system, tube length, size of eyepiece scale, etc. Another advantage is, that observation of the interference figure may be made with the eyepiece removed or by any of the methods previously described.

Fig. 263 shows the objective with its diaphragm D, the movement of which is controlled by means of the milled ring B, the angular rotation of the latter being shown on an engraved scale. This scale is calibrated either by means of an apertometer viewed through the diaphragm, or by using sections of which the value of 2E is known.

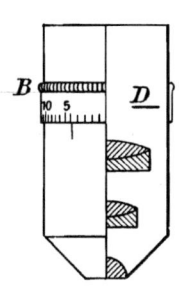

Fig. 263.—Objective with Iris Diaphragm in the Focal Plane for Slawson's Method of determining Optic Axial Angles (diagrammatic).

In use, the diaphragm is closed until the melatopes are on the edge of the field, and the scale reading gives a value representing E, one half of the apparent optic angle.

* *Amer. Min.*, 1934, **19**, No. 1, 25.

316   THE DETERMINATION OF THE REFRACTIVE INDICES

On inclined figures, *if the trace of the optic axial plane passes through the centre of the field*, 2E may be measured without great loss of accuracy by finding the angular distance of each melatope from the centre by taking a reading for each in turn and adding together the two angles thus obtained.

Ordinary objectives which are used for convergent light work, having a focal length of about 4 mm., may be modified by the addition of such a diaphragm, but larger interference figures are given by an objective specially made, having a focal length of about 8 mm. and an aperture of 0·85.

The Determination of the Refractive Indices of Birefringent Crystals

The immersion method of determining refractive indices has already been described in Chapter VII with special reference to isotropic crystals. Now that the study of crystals in parallel and convergent polarised light has been dealt with, the reader is in a position to understand how the method may be applied to birefringent crystals.

In the first place, by using the method in ordinary unpolarised light as for isotropic crystals, the *mean* refractive index of a birefringent substance may be determined, and in certain cases this may be of diagnostic value, especially if the birefringence is low. It is, however, usually necessary, and always more satisfactory, to determine the principal refractive indices of the substance, or, at least, the indices shown by a particular section of it, and this can be done by making use of the fact that in polarised light the two refractive indices shown by any crystal (not presenting an optic axial section) may be separately compared with the index of the immersion liquid in the following way. Crossed Nicols are inserted and the stage is turned until the crystal is extinguished. In this position, one of the vibration directions of the crystal is parallel to that of the polariser (which may be ascertained as described in Expt. I, p. 362, if not already known), and the light transmitted by the crystal vibrates entirely in this direction (see p. 101). On removing the analyser therefore, the index for this direction may be compared with the index of the liquid by either the Becke, or Schroeder van der Kolk test. The stage is then turned through

OF BIREFRINGENT CRYSTALS 317

90°, when a comparison of the other index of the crystal with that of the liquid can be similarly made.

To determine the principal refractive indices, crystals lying so as to present these indices must be examined in the above way, such crystals being variously recognised by their outward form, polarisation colours, the nature of their extinction, or their interference figures as will now be described.

Uniaxial Crystals. Every section has ω as one of its indices, so that the determination of this index may be readily made. ω is also the sole index shown by optic axial (basal) sections, which may be recognised by their isotropism in parallel light, and by their characteristic interference figure consisting of concentric rings of colour and a black cross (p. 278).

ε is shown only by sections containing the optic (and c) axis. Crystals with a well-developed prism form (parallel to c) usually lie on a vertical face and so present $\varepsilon\omega$ sections. Such crystals can often be recognised by their appearance (see e.g. Figs. 83, 98, and 99, Chap. III); they give straight extinction and optic normal interference figures. It may also be recalled that the birefringence is a maximum for crystals so orientated, and that they therefore show the highest polarisation colours in relation to their thickness.

The determination of ε on irregular material, which is too finely divided to give satisfactory interference figures, may be made by examining a number of grains showing the highest polarisation colours, taking as ε the value found which differs most widely from ω. Obviously the correctness of the result depends upon the grains being orientated at random, so that some present $\varepsilon\omega$ sections. A marked rhombohedral or basal cleavage might result in an overwhelming proportion of the grains being bounded by faces parallel to these directions, when examples lying so as to show ε would be rare or absent. If the appearance of the material suggests this possibility, it should be mixed with glass powder to induce it to adopt other orientations (see p. 213). When this method is employed, the determination may often by simplified by making use of the fact that basal plates (recognised by their isotropism in parallel light) give ε directly if they are caused to stand on edge.

Biaxial Crystals. Sections giving centred interference figures show principal refractive indices as follows:

Figure	Optical Sign	Indices shown
Acute bisectrix	+	α, β
,, ,,	−	γ, β
Obtuse bisectrix	+	γ, β
,, ,,	−	α, β
Optic normal	+ or −	α, γ
Single optic axis	+ or −	β

The sections giving the above results are the principal and circular sections of the indicatrix. In addition to these, any section which is perpendicular to a principal section of the indicatrix, *i.e.* to an optical symmetry plane of the crystal, shows one of the principal refractive indices, namely that for light vibrating at right angles to the symmetry plane concerned. Such sections give the type of inclined interference figure described on pp. 290–2, in which, in the extinction position, one isogyre is straight and passes through the centre of the field. In this position, the principal index shown is that for light vibrating at right angles to the direction of this isogyre. The optical sign and the relation of the vibration direction to the optic axial plane determine whether the index is α, β, or γ.

The principal indices may be determined on finely divided irregular material by a method similar to that described for uniaxial crystals namely by examining a large number of grains showing the highest polarisation colours and taking as α and γ respectively the lowest and highest indices found. β is given by those grains showing a uniform illumination between crossed Nicols on rotating the stage. If the shape of the grains suggests that their orientation is not sufficiently random, they should be mixed with glass powder, and in any case they should be tested in various positions, whenever it is possible to roll them about by shifting the cover slip with a needle.

Acicular and fibrous crystals present all the indices of the substance during rotation about the axis of elongation, a fact which is useful if the crystals will "stay put" in many different positions of rotation. If the extinction is straight along the axis of elongation at all positions of rotation, the index along the length must necessarily be constant and a principal index.

A point of interest and importance is that no section of a biaxial crystal can have indices which are both above, or both below β. In some cases this affords a means of determining β, for if one grain is found with a lower index of, say, 1·635, and another with an upper index of this same value, then β is 1·635.

OF BIREFRINGENT CRYSTALS 319

Indirect Determination of γ. It frequently happens that the value of the refractive index γ is so high that it cannot be determined directly by immersion methods. As stated on p. 309 its value may be calculated if the values for V, α, and β are known. The experimental errors in the determination of these constants are, however, often large enough to make such a calculation very inaccurate. Wood and Ayliffe * have proposed a method which is a combination of experimental and graphical methods, and makes possible an approximation to the value of γ with a fair degree of accuracy.

A crystal is mounted on the needle of a rotation apparatus as described in Chapter VI, p. 216, so that its optic axial plane is normal to the axis of rotation. It is evident that in this orientation

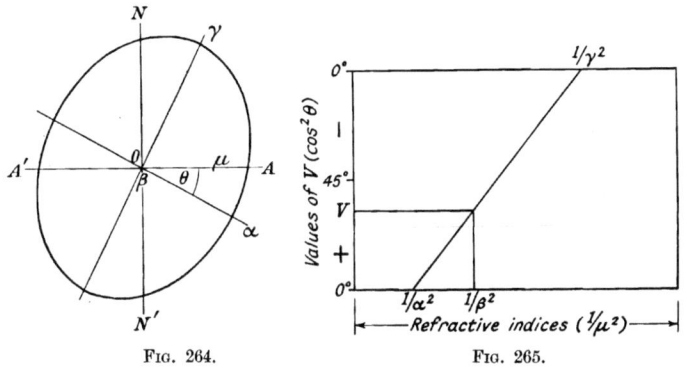

FIG. 264.   FIG. 265.

the values of α, β, and 2V can be found directly by immersing the crystal successively in liquids of increasing refractive index, because as it is rotated, it presents sections giving a constant refractive index β along the axis of rotation, and, at right angles to this, values which will vary between α and γ. Fig. 264 represents a section of the biaxial indicatrix of a crystal mounted in this way, the axis of rotation being β, normal to the paper, and NN' being the axis of the microscope. AOA' is the trace of an elliptical section of the indicatrix parallel to the microscope stage, having β as one axis and μ as the other. Light passing through the section normal to AOA' will be polarised, and two rays will be produced having refractive indices β and μ. θ is the angle through which the crystal has been turned from a position in which the $\alpha\beta$ section was horizontal (and in which the refractive indices α and β may be measured).

* R. G. Wood and S. H. Ayliffe, *Phil. Mag.*, 1936, (7), **21**, 324.

The mathematical relationships between α, β, μ, and θ, are expressed by the equation:

$$\frac{1}{\mu^2} = \frac{1}{\gamma^2} + \left(\frac{1}{\alpha^2} - \frac{1}{\gamma^2}\right) \cos^2 \theta.$$

In such an equation, $\frac{1}{\mu^2}$ varies linearly with $\cos^2 \theta$. If now values of $\frac{1}{\mu^2}$ and $\cos^2 \theta$ are plotted along the abscissa and ordinate respectively $\frac{1}{\alpha^2}, \frac{1}{\beta^2}$, and $\frac{1}{\gamma^2}$ plot on a straight line as shown in Fig. 265.

The method of Wood and Ayliffe was to immerse the crystal mounted as described in liquids of successively greater refractive index, and turn it until a match was obtained. By this means a number of values for θ and corresponding refractive indices were obtained. $\cos^2 \theta$ was plotted against $\frac{1}{\mu^2}$ for each pair of values, and the straight-line graph thus obtained gave $\frac{1}{\alpha^2}$ on the lower line, $\frac{1}{\gamma^2}$ on the upper line, and a value of $\frac{1}{\beta^2}$ when $\theta = V$, the latter being the angle measured in a liquid of refractive index β.

The fact that a number of measurements is necessary to get the value of γ by extrapolation by this method makes for greater accuracy.

It is easy to construct such a diagram in which the values of the refractive indices themselves are plotted against values of V. This is facilitated since F. E. Wright * (in a paper which contains graphical solutions for many crystallographic formulæ) has given tables of $\cos^2 \theta$, and values (to six decimal places) for $\frac{1}{\mu^2}$ from $\mu = 1\cdot40$ to $2\cdot48$.

A diagram of this kind has been used by the authors for many years, and is included as a folder (Fig. 313) at the end of this book. Mertie † published a similar one in 1942 but did not mention the paper here quoted of Wood and Ayliffe, though he gives other references to various types of graphical solutions to the same problem.

* F. E. Wright, " Methods in Microscopical Petrography," *Amer. J. Sci.* 1913, **36**, 509.
† J. B. Mertie, Jr., *Amer. Min.*, 1942, **27**, 538.

OF BIREFRINGENT CRYSTALS 321

Such a diagram has a number of uses in addition to that of finding the value of γ by extrapolation. It can be used to test the compatibility of experimentally determined results for the optical constants of a crystal. Further it can be used to demonstrate that small errors (of the order of ± 0.002 in refractive indices, and $\pm 4°$ in the value for 2V) will result in large errors in the determination of γ by extrapolation, especially if α and β are very near to one another, 2V small, and the optical sign positive. A little intelligent experimentation with the diagram will demonstrate the size of the errors to be expected when calculating one constant from a knowledge of the other three, using formulæ such as those given on p. 309. The testing of the compatibility of his own results will also provide a very wholesome corrective to the student who has an exaggerated idea of the accuracy of his own work.

CHAPTER IX

LIQUID CRYSTALS

When certain solid organic compounds are heated, they do not pass directly into the liquid state, but first adopt a state which has properties intermediate between those of a true crystal and those of a true liquid. Thus on reaching a certain temperature, the solid undergoes transformation into a turbid phase which is both birefringent and fluid, the consistency varying in different cases from that of a paste to that of a freely flowing liquid. At a higher temperature, this phase " melts " to form a clear isotropic liquid of normal type. On cooling, these changes take place in the reverse order, though some supercooling (p. 18) may occur when the transition temperatures are reached, as in the case of ordinary crystallisation.

O. Lehmann,[*] who was the first to make a systematic study of such substances, proposed the name *liquid crystal* to describe the intermediate state. This name is still commonly used, though objection has been raised to it on the ground that a substance in this condition is not in fact a crystal. A more satisfactory designation, proposed by G. Friedel [†] is the *mesomorphic state* (Gk. *mesos*, intermediate ; *morphe*, form), from which follows the term *mesophase* to denote a phase which is in this state. Other terms in use are *crystalline liquid, anisotropic melt*, and *paracrystal*.

For a given compound, the temperature limits between which the mesomorphic state is stable are definite transition points, which vary slightly with pressure in accordance with the Clausius-Clapeyron equation (Chap. I, p. 17) like true melting-points and the transition points between enantiotropic polymorphs. At the upper limit the mesomorphic and liquid states can exist in equilibrium with one another ; at the lower one the mesomorphic and

---

[*] O. Lehmann, *Z. physikal. Chem.*, 1889, **4**, 468.
[†] G. Friedel, *Ann. Phys.*, 1922, **18**, 273–474.

crystalline states can similarly exist together. With some compounds, however, it is possible, *e.g.*, by rapidly cooling the melt, to obtain a mesophase which is metastable with respect to the crystalline state of the substance under all conditions, and therefore stands in a monotropic relation to it.

The reason for the existence of the mesomorphic state is suggested in a general way by the molecular structures of the compounds which assume it. All such compounds have molecules which are much elongated, and in some cases flattened as well, and which possess one or more active (polar) groups. The shape of the molecules predisposes them to set themselves parallel to one another, like logs in a river, and in the crystalline state they are arranged in this way and are held together by local attachments due to the polar groups, as well as by the unspecific van der Waals attraction. It is not surprising therefore that on raising the temperature the transition to the random arrangement characteristic of the true liquid state should be achieved in stages, *i.e.* that the weaker linkages snap first, leaving the molecules with some freedom of relative movement, before they have acquired sufficient thermal energy to overcome in any marked degree the tendency to set themselves parallel to one another. Thus the medium acquires the ability to flow but remains birefringent owing to the preferred orientation of the molecules. It is not, however, yet wholly clear why many compounds which might be expected from their molecular structures to form liquid crystals do not in fact do so. The explanation is probably that in these cases the melting-point is too high for the mesomorphic state to be realised. Thus for example if the substance possesses more than one very strongly polar group, the local attachments formed between these in the crystal are likely to be so strong that they will break only at temperatures at which the molecules have sufficient thermal energy to form the isotropic liquid phase directly.

For more detailed studies and suggestions regarding the dependence of mesomorphic properties upon molecular structure, the reader is referred to the researches of Vorländer, Oseen, Bernal and Crowfoot, and others.* Examples of the gradation of mesomorphic properties in homologous series are given later in this chapter.

Mesomorphic phases are of two main types, which are designated

* See for example the contributions to a " Discussion on Liquid Crystals and Anisotropic Melts," *Trans. Faraday Soc.*, 1933, **29**, 1032 *et seq.*

respectively as *smectic* and *nematic*. The characteristics of these two types are discussed in detail later in the chapter, but for the moment it may be noted that the essential difference between them is that smectic phases have a stratified structure, the molecules being arranged in layers with their long axes approximately normal to the planes of layers, whilst in nematic phases the only restriction on the arrangement is that the molecules preserve a parallel or nearly

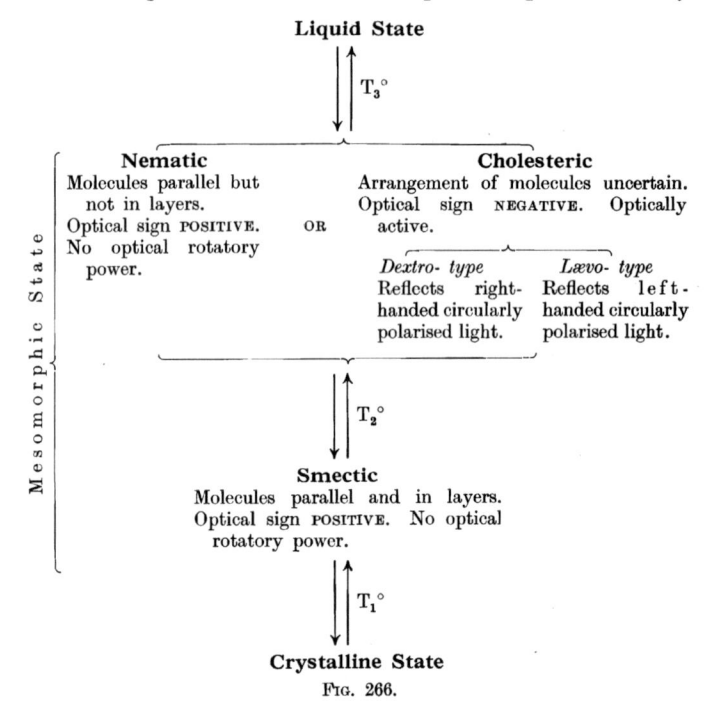

Liquid State

$\updownarrow T_3°$

Mesomorphic State {

Nematic
Molecules parallel but not in layers.
Optical sign POSITIVE.
No optical rotatory power.

OR

Cholesteric
Arrangement of molecules uncertain.
Optical sign NEGATIVE. Optically active.

Dextro- type
Reflects right-handed circularly polarised light.

Lævo- type
Reflects left-handed circularly polarised light.

$\updownarrow T_2°$

Smectic
Molecules parallel and in layers.
Optical sign POSITIVE. No optical rotatory power.

$\updownarrow T_1°$

Crystalline State
Fig. 266.

parallel orientation. Most mesomorphic substances which have been studied are either exclusively smectic or exclusively nematic, but some can exist as both types of phase, and in these cases there is always a definite transition temperature below which the more ordered smectic phase is stable, and above which this changes to the nematic modification. A few substances have been found to possess more than one smectic phase, and here also the temperature ranges of stability of the different phases are sharply defined.

EXAMPLES OF MESOMORPHIC COMPOUNDS

A third type of phase, the *cholesteric*, so called because it is shown mainly by cholesteryl derivatives, is considered by some workers to be a special type of nematic phase, for reasons that will appear later. Substances having a cholesteric phase may exist also as a smectic phase at lower temperatures, but never as a nematic phase of the ordinary type. The relationship between the stability ranges of the various phases of mesomorphic substances is shown diagrammatically in Fig. 266, together with their more important structural and optical characteristics. Some examples are given below.

Examples of Mesomorphic Compounds (with stability ranges of phases.)

I. COMPOUNDS SHOWING A SINGLE MESOMORPHIC PHASE

Smectic

Ammonium oleate: $CH_3.(CH_2)_7.CH=CH.(CH_2)_7.COONH_4$
Smectic at ordinary temperatures.

Ethyl *p*-azoxybenzoate:

$$C_2H_5OOC-\langle\ \rangle-N=N-\langle\ \rangle-COOC_2H_5$$
$$\downarrow$$
$$O \qquad 114\text{--}120°.$$

Ethyl *p*-azoxycinnamate:

$$C_2H_5OOC.CH=CH-\langle\ \rangle-N=N-\langle\ \rangle-CH=CH.COOC_2H_5$$
$$\downarrow$$
$$O \qquad 140\text{--}ca.\ 250°.$$

Nematic

p-Azoxyanisole: $CH_3O-\langle\ \rangle-N=N-\langle\ \rangle-OCH_3$
$$\downarrow$$
$$O \qquad 116\text{--}133°.$$

p-Azoxyphenetole: $C_2H_5O-\langle\ \rangle-N=N-\langle\ \rangle-OC_2H_5$
$$\downarrow$$
$$O \qquad 135\text{--}165°.$$

Anisaldazine: $CH_3O-\langle\ \rangle-CH=N-N=CH-\langle\ \rangle-OCH_3$
$$160\text{--}180°.$$

Cholesteric

Cholesteryl benzoate :

$$\text{structure of cholesteryl benzoate}$$

146–178·5°.

Cholesteryl acetate : Cholesteric phase metastable. Produced by rapid cooling of the liquid.

Amyl cyanobenzalaminocinnamate :

$$CN\langle\ \rangle CH{=}N\langle\ \rangle CH{=}CH.COOC_5H_{11}$$

92–105°.

II. Compounds Showing more than one Mesomorphic Phase

Cholesteryl pelargonate :

$$CH_3.(CH_2)_7.COOC_{27}H_{45}$$

Crystalline $\underset{78°}{\rightleftharpoons}$ Smectic $\underset{79°}{\rightleftharpoons}$ Cholesteric $\underset{90·5°}{\rightleftharpoons}$ Liquid.

Ethyl anisal-p-aminocinnamate :

$$CH_3O\langle\ \rangle CH{=}N\langle\ \rangle CH{=}CH.COOC_2H_5$$

Metastable crystal $\underset{83°}{\rightleftharpoons}$ Smectic I $\underset{91°}{\rightleftharpoons}$ Smectic II $\underset{118°}{\rightleftharpoons}$

Stable crystal, m.p. 108°   Nematic $\underset{139°}{\rightleftharpoons}$ Liquid

(On heating the stable crystalline modification, smectic phase II is obtained directly. Smectic phase I and then a metastable crystalline modification may be formed on cooling smectic phase II if all traces of the stable crystalline form are absent.)

According to a widely accepted theory (E. Bose *), the molecules in a mesophase in bulk, which is not being subjected to external

* E. Bose, *Physikal. Z.*, 1909, **10**, 32, 230.

orientating forces, are not orientated in the same direction throughout the whole medium but are grouped into clusters or "swarms" in each of which they lie parallel or approximately so, but in a direction which is random with respect to those in other swarms. On this view, therefore, the medium resembles a mass of small crystals rather than a single crystal, but with this difference that, owing to the mobility of the molecules, the swarms do not remain constant in size but are continually exchanging molecules with one another, and the arrangement of the swarms is not a rigid one. This theory explains the turbid appearance of a liquid crystal as being due to the scattering of light by the swarms. It also accounts for the fact that the orientating effects produced by electrical and magnetic fields on nematic phases (see later) are much larger than would be expected if the fields acted on the molecules individually; for if the medium is grouped into swarms, each will be orientated as one unit, and since its moment (length × charge, or pole strength) will be much greater than that of a single molecule, the effect will be much increased. Calculations of the average diameter of a swarm on the basis of this effect give a value of the order 10^{-6} cm., which is equivalent to a content of about 10^5 molecules.

The optical properties of liquid crystals, in which we are primarily interested here, have necessarily been studied for the most part on thin or comparatively thin films mounted between glass or other transparent surfaces. The orientation of the molecules in such small quantities of material is profoundly affected by the capillary and other forces associated with the presence of these supporting surfaces. To such cases the swarm theory would appear to have little application, and we shall not consider it further.

Smectic Mesophases. The word "smectic", *i.e.* "soaplike", is used because the soaps, *e.g.* ammonium oleate, form liquid crystals of this type (Gk. *smektis*, fuller's earth; *smegma*, soap). As mentioned above, the essential feature of the structure of a smectic phase is that the molecules are arranged in layers with their long axes approximately normal to the planes of the layers (Fig. 267). The spacing of the molecules within each layer is not uniform, however, as it would be in a true crystal. The fluidity of the phase is due to the fact that the layers are flexible and can glide over one another.

The existence of this stratified structure was originally inferred from certain phenomena which smectic phases exhibit, and which can be observed under the microscope, notably the formation of "stepped drops" and focal conic patterns (see below). It has since been confirmed beyond reasonable doubt by X-ray analysis. For example, in the case of ethyl p-azoxybenzoate (Friedel, 1925), a diffraction pattern is obtained which corresponds to a single repeat distance in the structure, equal approximately to the length of the molecules. This is just what is to be expected from the arrangement illustrated in Fig. 267 (a), the repeat distance being the thickness of the layers, and the absence of other repeat distances showing that there is no uniformity in the spacing of the molecules

(a) Single layers.   (b) Double layers.

Fig. 267.—Arrangement of Molecules in Smectic Phases.

within the layers. In some substances, e.g. thallous stearate and oleate (Hermann *) it appears that the layers are two molecules thick (Fig. 267 (b)). As far as is known, all substances which form a smectic phase have their molecules arranged in layers in the crystalline state as well.

When a small quantity of a substance in the smectic condition is supported on a plane surface with which it does not form strong local attachments, the natural tendency is for the strata to arrange themselves parallel to the surface, and they will either do this spontaneously, especially if the viscosity of the medium is low, or they can be readily made to do so. The resulting *homeotropic*

* K. Hermann, Faraday Society's "Discussion on Liquid Crystals and Anisotropic Melts ", *loc. cit.*

structure, as it is called, is optically homogeneous and shows the optical characters of a *positive uniaxial* crystal with the optic axis parallel to the mean direction in which the long axes of the molecules lie, *i.e.* at right angles to the strata. This is to be expected by analogy with the optical properties of crystals in which long molecules are arranged parallel to one another and have an isotropic distribution along the axes at right angles to the optic axis (see p. 149). In the smectic phase the side to side spacing of the molecules is random, and this must also result in uniaxial character.

A simple demonstration of the homeotropic structure may be made as follows. A little ammonium oleate, which is smectic at ordinary temperatures and has a pasty consistency, is pressed into a thick layer between a clean glass slide and cover slip. On examination between crossed Nicols the preparation shows a confused structure due to the fact that the layers of molecules are crumpled in an irregular manner; the viscosity of the medium is too high for them to flatten out spontaneously. If now the cover slip is moved gently round and round with a slight downward pressure and the preparation be examined between crossed Nicols from time to time, it will be observed that the polarisation colours gradually disappear and eventually the medium becomes isotropic for parallel light. On changing to convergent light, however, a centred uniaxial cross will be obtained (but probably no rings, since the preparation is likely to be rather thin by this time), and by the use of the first order selenite plate the optical sign may be shown to be positive. The manipulation of the cover slip has " ironed out the creases " as it were, and caused the strata to arrange themselves parallel to the glass surfaces.

A very striking example of the homeotropic structure, which also provides evidence of its stratification, is afforded by the " stepped drops " (" gouttes à gradins ") discovered by Grandjean, which are formed by many smectic phases under suitable conditions. To obtain these, a few crystals of the substance are heated on a surface which must be one which does not orientate the molecules, and must be perfectly clean. It is difficult to clean glass sufficiently for this purpose, but a freshly cleaved surface of muscovite mica serves very well for many substances, *e.g.* ethyl *p*-azoxybenzoate and -cinnamate. When the transition point is reached, the crystals " melt " to form one or more perfectly flat drops, built up of a number of horizontal layers, which slide about on one another as the drops are forming, or when the mica slip is disturbed (Fig. 268). (To produce and observe this phenomenon most conveniently, as for many other experiments with liquid crystals, a microscope hot stage is required, and a suitable one of simple type is described at the end of the chapter.) Under sufficient magnification the layers are seen to be fringed by a chain of focal conic structures (see later), indicating that here the layers are contorted, but elsewhere the drops are optically homogeneous and show a centred positive uniaxial figure. The thickness of the layers is variable, but from such measurements as have been made, appears to be a multiple of the length of the molecules, in agreement with the accepted structure of a smectic phase. The reason why the medium splits up into

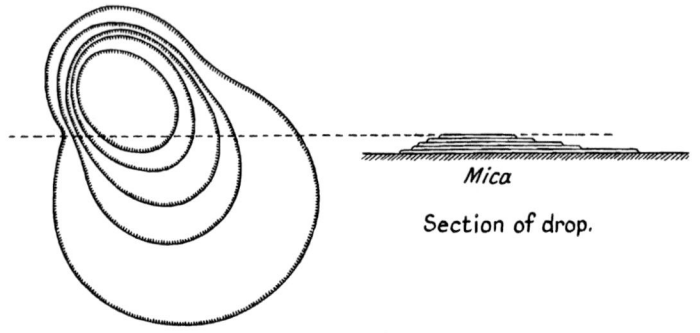

Mica

Section of drop.

Fig. 268.—Stepped Drop ("Goutte à Gradins").

these multiple layers is probably that traces of impurity in the crystals establish dividing surfaces when the drop is in course of formation.

Another type of homeotropic structure in which the strata do not necessarily lie parallel to the supporting surface may be obtained when large crystal plates grown on a microscope slide from the melt are slowly heated to the smectic region. Thus, if ethyl *p*-azoxybenzoate is melted between a slide and cover slip on a hot plate and then slowly cooled and inoculated with a trace of the crystalline form at the edge of the cover slip, it crystallises as large elongated plates. If the temperature of the preparation is then cautiously raised to the smectic range, each plate forms a homeotropic structure in which the single optic axis appears always to have the same or nearly the same direction as that of the acute bisectrix in the original (biaxial) crystal. The optical homogeneity does not, however, extend over the whole area which was occupied by the crystal, but is found only in the central part. For some distance on either side of the original crystal boundary, there is a region in which the orientation changes gradually into those of the neighbouring homeotropes by way of the focal conic type of structure. If the temperature is raised to the upper part of the smectic range where the medium is less viscous, it may happen that the focal conic structure invades the whole preparation, *i.e.* that the homeotropic structures break down. The same thing may occur at lower temperatures if the cover slip is displaced a little. If further movements are imparted to the slip, however, the medium may become homeotropic again, but this time it does so as a whole with the layers parallel to the glass surfaces, as in the experiment with ammonium oleate described above.

Still another way of producing homeotropy is to melt a little of the substance across a small hole in a glass or mica plate. This produces a film supported only at its edges and on entering the smectic range of temperature it adopts the homeotropic structure spontaneously with the optic axis at right angles to the surface, *i.e.* with the strata parallel to the plane of the film.

When a smectic phase is supported between surfaces with which it forms strong local attachments, the strata become contorted. Around each centre of attraction the molecules are compelled to adopt a radiating or fan-like arrangement, in which, however, the constant thickness and parallelism of the strata must be preserved, and which must conform with the arrangements around neighbouring centres. From microscopic study it may be inferred that the strata actually form series of parallel curved surfaces of the kind known in geometry as the cyclides of Dupin, and of the possible structures satisfying the above conditions, this is in fact the one having the lowest potential energy. It is known as the *focal conic* structure for reasons which will presently appear. An example of a Dupin cyclide is shown in Fig. 269. It may be described as a closed hollow circular ring, having a circular cross-section the diameter of which varies from a maximum PQ to a minimum RS. PQ and RS may be called the principal cross-sections and they occupy diametrically opposite positions on the cyclide as shown in the figure.

The geometrical basis of the focal conic structure is shown in

Fig. 270. It consists of a family of parallel, equally spaced cyclides, shown in principal section in (a) and in plan in (b). The common centres of the principal cross-sections are at A and B. Lines in the figure which belong to the same cyclide bear the same numbers, thus 1,1 ; 2,2 ; etc. Cyclide No. 1 may be taken as the starting-point in the description. It is indicated by thicker lines than the others. In this cyclide the radii of the principal sections are AO and OB, *i.e.* the central hole which appears in the example shown in Fig. 269 has here shrunk to a point, O. KL and MN in (a) are the " spine " and " belly " respectively of the cyclide. (These features are omitted from the other cyclides in the figure to avoid confusion.)

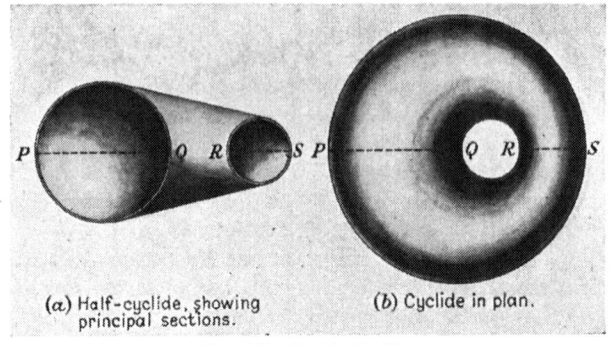

(a) Half-cyclide, showing principal sections.   (b) Cyclide in plan.

FIG. 269.—Dupin Cyclide.

The cyclides inside No. 1 are of the type shown in Fig. 269 up to No. 4. In No. 5, however, the radius of the minor principal cross-section has shrunk to a point (B), and above this number the cyclides are incomplete, the cross-sections shrinking to points, TT', UU', VV' (see (b)) before the full circuit of the hollow ring has been made. In other words instead of being rings they are crescents. (Mathematically a cyclide of this type opens out beyond the points of the crescent to form another crescent, *i.e.* the arcs defining the plan of the crescent, *e.g.* VXV', are continued to form two intersecting circles. In the smectic phase, however, the second crescent can have no existence, since it would take up space already occupied by other cyclides.) It may be shown that *the points B, T, U, V, etc. lie on an ellipse of which O is one of the foci and A and B are the vertices.*

Considering now the cyclides outside No. 1, viz., 9, 10, 11, etc., it will be seen that they have no central hole, but only a " dimple " above and below, e.g. E and F for cyclide No. 9, and C and D for cyclide No. 13. It may be shown that *these points D, F, E, C, etc., lie on a hyperbola of which O is a vertex and B is a focus.*

An ellipse and a hyperbola related to one another in this way are termed a pair of focal conics, and this gives the name to the whole structure. More precisely the relationship between the curves may be expressed as follows. Writing the equations for the curves as:

$$\frac{x^2}{a^2} + \frac{y^2}{b^2} = 1 \text{ for the ellipse}$$

and
$$\frac{x^2}{a'^2} - \frac{y^2}{b'^2} = 1 \text{ for the hyperbola}$$

then it may be shown that
$$a' = \sqrt{(a^2 - b^2)}$$
$$b' = b$$

A property of a pair of focal conics is that the hyperbola is the locus of the apices of cones of revolution of which the ellipse is a common section. Thus C is the apex of a cone of revolution ACB, of which the lines AC and BC lie on diametrically opposite sides. Or to put it another way, either AC or BC, or any other line drawn from C to the ellipse, will generate the cone ACB if it be imagined to be rotated about the cone axis, which is the tangent to the hyperbola at C. Similarly E, F, and D are the apices of cones of revolution AEB, AFB, and ADB respectively, and of all these cones the ellipse is a section.

A particular case is that in which the cyclides have the same cross sectional diameter all the way round, and are thus " anchor rings ". (A familiar example of this type of surface is afforded by the inflated inner tube of a motor or bicycle wheel.) The ellipse then becomes a circle and the hyperbola a straight line.

The most direct evidence that the focal conic structure exists in a smectic phase with crumpled strata is that ellipses and hyperbolæ can actually be seen as dark lines when the phase is examined under the microscope, and that these curves appear in pairs, each consisting of an ellipse and a hyperbola bearing a focal conic relation to one another. Their visibility is to be expected, for, as seen from Fig. 270, they mark structural, and therefore optical discontinuities

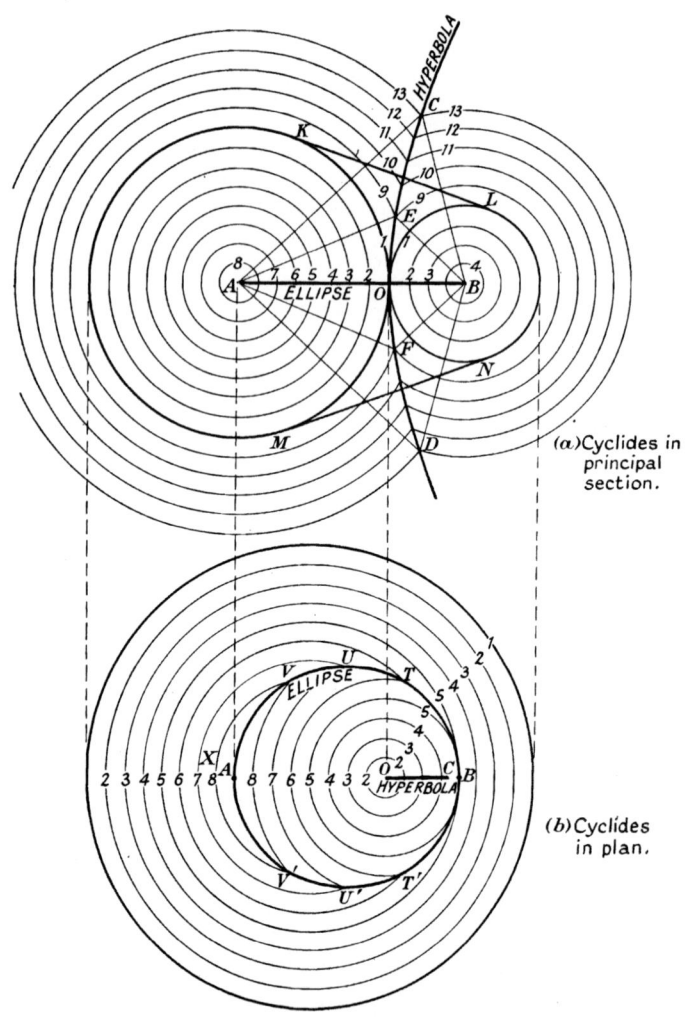

Fig. 270.—Focal Conic Structure.

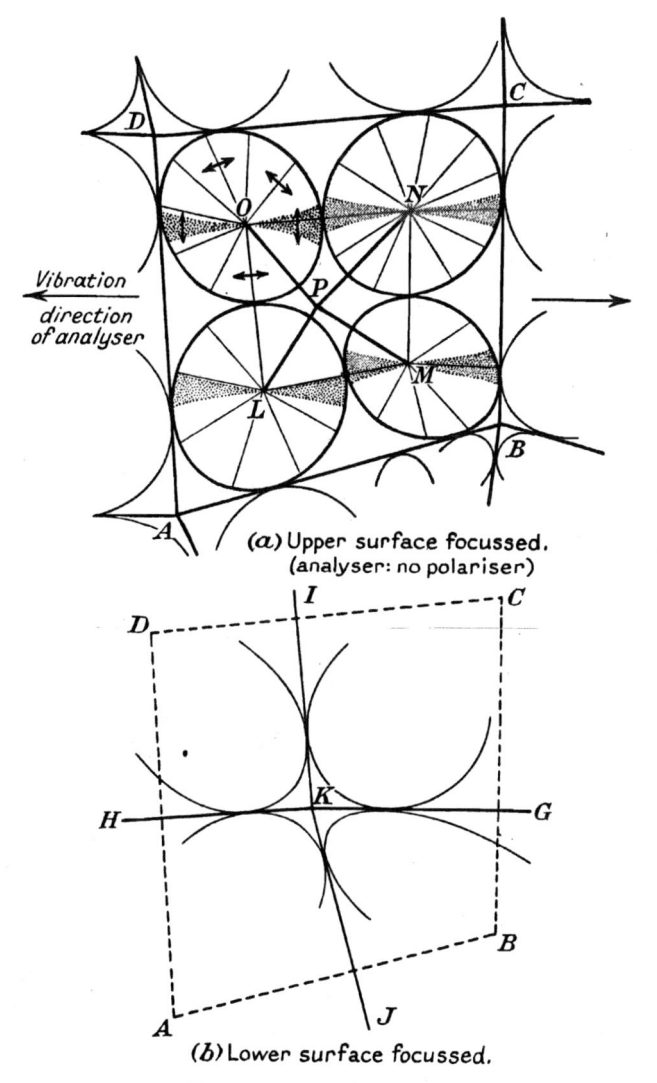

(a) Upper surface focussed.
(analyser: no polariser)

(b) Lower surface focussed.

FIG. 271.—Polygonal Structure.

in the medium; they are the loci of points at which the strata undergo a sharp change in direction. It may not be possible to see both curves of a pair at one and the same position of the microscope tube, since they are not co-planar, but in such cases they can often be traced by altering the position of focus. Furthermore, one or other of the curves may be incomplete, depending on their position and orientation with respect to the supporting surfaces. Thus if the ellipse rests on one of these surfaces, only one branch of the hyperbola can exist, while if the hyperbola lies on a surface, there can only be one half of the corresponding ellipse. A complete ellipse together with both branches of the hyperbola can appear where the crumpling is due to nuclear associations inside the medium, instead of to local attachments to the supporting surfaces.

The simplest manifestation of the focal conic structure and the one which can be studied in most detail, is the so-called *polygonal structure* (Figs. 271 and 272). This is obtained in general when the preparation is fairly thick and not too viscous. G. Friedel * (to whom the study and explanation of the focal conic structure are largely due) gives the following method of inducing ethyl p-azoxy-benzoate to adopt the polygonal type. A thick film of the substance containing a trace of resin, which seems to stabilise the structure, is heated between a slide and a cover slip (which have previously been lightly treated with hydrofluoric acid), until the isotropic liquid phase is just beginning to appear. The temperature is then lowered a few degrees into the smectic region, and the cover slip is meanwhile gently agitated. The object of the treatment with hydrofluoric acid is to promote local attachments to the glass by roughening its surface, and so avoid the homeotropic structure being formed. If the microscope be then focussed on either surface of the film the field will be seen to be divided into a number of irregular polygonal areas which may have any number of sides, and in each of which there appears a family of ellipses which are tangential to the sides of the polygon and to each other. The spaces between the larger ellipses are occupied by smaller ellipses, and the spaces between these with still smaller ones and so on. (See Fig. 271 (a), an imaginary case showing the larger ellipses in a four-sided polygon ABCD, and portions of the ellipses in neighbouring polygons.) The sides of the polygons are either straight or slightly curved. Two other important observations may be

* *loc. cit.*

SMECTIC MESOPHASES 337

made. The first is that the major principal axes of all the ellipses in a given polygon pass (when produced if necessary) through one point (*e.g.* P in the polygon ABCD in the figure). The second is that if the upper surface be focussed and the analyser (without the polariser) be inserted, or, alternatively, if the lower surface be focussed and only the polariser be used, each ellipse is seen to be crossed by a straight brush the direction of which is parallel to the vibration direction of the Nicol. This brush is similar in appearance to that seen in the single optic axial figure (90° position) of a biaxial crystal, and its narrowest part coincides with one of the foci of the ellipse (L, M, N, etc. in the figure).

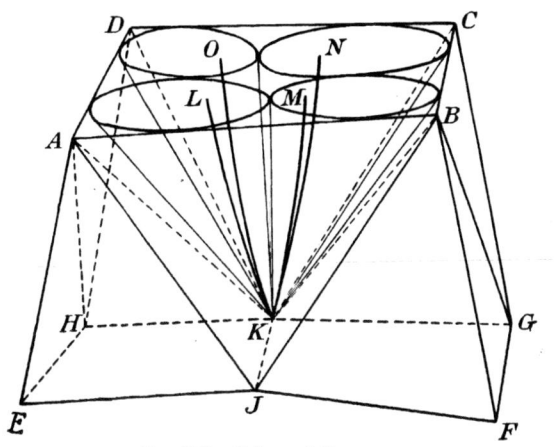

FIG. 272.—Polygonal Structure.

If now the focus of the microscope be gradually altered from the upper to the lower surface of the preparation, it will be possible to make out that (*i*) each ellipse is partnered by one branch of a hyperbola, which lies in the vertical plane containing the major axis of the ellipse and which meets the ellipse at the focus crossed by the dark brush (L, M, N, etc. in the figure), and that (*ii*) the hyperbolæ belonging to ellipses in the same polygon all meet on the lower surface at one point (K, Fig. 271 (*b*)) which is immediately below the point of intersection of the major axes of the ellipses (P, Fig. 271 (*a*)). This point of intersection of the hyperbolæ turns out to be a common corner in the system of polygonal areas on the

338   LIQUID CRYSTALS

lower surface, which has now come into view. The polygon edges which radiate from this point intersect at right angles the projections of the edges of the polygon on the upper face to which the hyperbolæ belong. Thus in the figure, the edges KG, KH, KI, KJ respectively intersect at right angles the projections of the edges BC, AD, DC, and AB of the polygon ABCD.

The above observations can be explained as follows. The medium is divided into pyramids and tetrahedra, the bases of the pyramids being the polygons observed on the upper and lower surfaces. For example, the block of material represented in Fig. 272 consists of an inverted pyramid KABCD, two erect pyramids AEJKH and BJFGK, and three tetrahedra ABJK, ADKH, and BCKG. The

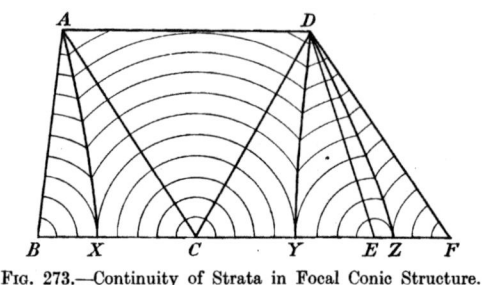

FIG. 273.—Continuity of Strata in Focal Conic Structure.

ABC and DCE are inclined cones belonging to different pyramids. Their elliptical bases touch at C. ADC is therefore a section of a tetrahedron.
DCE and DEF belong to the same pyramid, and touch along the common generator DE. AX, DY, and DZ are the hyperbolæ inside the three cones. For simplicity these hyperbolæ have been assumed to be coplanar, *i.e.* the major axes of the three associated ellipses, BC, CE, and EF, are in line.

pyramid base ABCD may be taken to correspond to the polygon shown in Fig. 271 (*a*), and the apex of the pyramid, K, to the point lettered similarly in Fig. 271 (*b*). Each pyramid contains a family of cones of revolution, which are in general inclined, and which touch one another along common generators. The elliptic bases of the cones rest on the base of the pyramid. The hyperbolæ which bear a focal conic relation to these ellipses meet at the apex of the pyramid. To illustrate these points the four largest cones in the pyramid KABCD, together with their hyperbolæ LK, MK, NK, OK, are sketched in in Fig. 272. Each cone corresponds to the domain lettered ABC in Fig. 270 (*a*) (assuming that C is the termination of the hyperbola), and the portions of the cyclides which this domain contains are continuous with those of neighbouring tangent cones. Thus in Fig. 273, DCE and DEF are two cones inside the same pyramid, which touch at the common generator DE. DY and DZ are their respective hyperbolæ. It will be seen that the cyclides of one cone pass continuously into those of the other. The figure, to be sure, only represents the situation at a single line of contact, but assuming that each pyramid is completely filled with cones, small ones in the crannies between the larger ones, at least down to the limit set by the intermolecular distance, then the strata of each cone will at every point on its surface be continuous with the strata of some other cone, except at the surface of the pyramid, where, as shown below, it may be inferred that they are continuous with those in the tetrahedral regions. It is not possible of course to prove by direct observation that the pyramids are completely filled with cones, since ellipses smaller than the limit of microscopic resolution cannot be seen. It may, however, be concluded with some confidence that this is in fact the case, for otherwise there would have to be some kind of intermediate structure filling the smallest spaces, and this would involve discontinuities in the strata, additional to those which are inevitable at the hyperbolæ, with a consequent increase of potential energy.

The structure within the tetrahedra must now be considered. It should first be noted that the pairs of polygon edges exemplified in Fig. 271 (b), which are at right angles to one another, e.g. KG and BC, KH and AD, are each made up of an upper and lower edge of a tetrahedron. Reference to Fig. 272 will make this clear. From this and from the fact that these edges are often seen to be slightly curved, it may be inferred that they are parts of focal conic curves, i.e. that in each pair one edge is part of an ellipse and the other a part of a hyperbola. The Dupin cyclides corresponding to these curves will then be continuous with those of the cones on the surfaces of neighbouring pyramids. For a single section this is shown in Fig. 273. ADC is a section of a tetrahedron, and its cyclides, based on the edges AD and C are seen to be continuous with those of the cone DCE (DC being the line of contact between the cone and the tetrahedron), and similarly with those of the cone ABC belonging to a different pyramid from that containing DCE.

The reason for the dark brushes seen across each ellipse when the polygonal structure is observed in polarised light is as follows. It should first be recalled that the long axes of the molecules are everywhere at right angles to the strata, i.e. to the cyclide surfaces, and so lie along all lines joining the hyperbola to the ellipse, *e.g.* along AC, BC, OA, OB, and OV in Fig. 270. In the plane of the elliptic base of a cone these lines radiate from the focus from which the hyperbola springs, and some of these lines are shown for each ellipse in Fig. 271 (*a*). Suppose first that the analyser only is inserted, and the microscope is focussed on the upper surface of the preparation. Ordinary light then enters the medium and is resolved into extraordinary and ordinary rays. Every extraordinary ray vibrates in the plane containing the ray and the optic axis of the medium, *i.e.* the direction of the long axes of the molecules. Owing to the curvature of the strata, however, the inclination of this axis varies continuously from point to point, and it can be shown that as a result of this the ray will be deviated, and will follow a curved path.* Consequently, few of these rays reach the objective of the microscope. The direction of propagation of the ordinary rays is, however, not affected by the curvature of the strata, since they vibrate at right angles to the long axes of the molecules. These rays therefore are mainly responsible for the formation of the image, and their vibration directions on emergence from the medium are at right angles to the radial lines mentioned above. The vibration directions for some of the rays are shown in the top left-hand ellipse in the polygon in Fig. 271 (*a*). From this it will be clear that rays emerging on those radial lines which are parallel to the vibration direction of the analyser will be extinguished.

If now the lower surface of the preparation be focussed and the polariser (but not the analyser) inserted, extraordinary and ordinary rays will be produced when the light enters the medium as in the previous case, except that there will be no ordinary rays where the long axes of the molecules lie parallel to the vibration direction of the polariser (for the resolved light then has no component normal to this direction), and no extraordinary rays where these axes are at right angles to this vibration direction. The extraordinary rays are deviated and get " lost " as before, whilst for the reason just indicated, the ordinary rays only appear where the molecules are not parallel to the vibration direction of the polariser. Thus each ellipse shows a brush parallel to this direction.

* The path of an extraordinary ray moving in such a medium was calculated by Grandjean : *Bull. Soc. Min.*, 1919, **42**, 42. See also *Proc. Roy. Inst.*, 1934, **28**, 89.

When a smectic phase is deposited from the isotropic melt or from a solution, it frequently first appears in the form of particles having characteristic elongated shapes, and showing evidence of focal conic structure. These particles have been termed *bâtonnets* (little rods, or batons) by the French workers. When two of these particles meet they fuse together.

Nematic Mesophases. The term "nematic" means thread-like (Gk. *nema*, thread) and is used because of the mobile thread-like lines which are observed in nematic phases which have been prepared by rapid cooling of the melt. The lines, like the ellipses and hyperbolæ seen in smectic phases, are due to discontinuities in the structure.

The properties of nematic phases indicate that the molecules are approximately parallel to one another, but that they are not in layers (see Fig. 274). The parallel arrangement is shown by the optical character, which is uniaxial positive, and the absence of layers by the fact that X-ray diffraction reveals no periodicity in the structure. It must be concluded that the molecules can be drawn past one another in the direction of their long axes, a degree of freedom which is excluded in smectic phases by the comparatively strong attractions between the ends of the molecules, or at least between corresponding parts of the molecular chains. This agrees with the fact that nematic phases are in general more fluid than smectic phases. Particles of dust, for example, can wander quite freely through a nematic phase without destroying its structure.

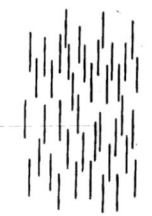

Fig. 274. Arrangement of Molecules in Nematic Phase.

Electric and magnetic fields orientate the molecules in a nematic phase, but have no effect on a smectic phase once it is established. (They do, however, influence the orientation of the molecules in a smectic phase which is in process of formation from the melt.) This again indicates a greater freedom of movement of the molecules in the nematic type of phase. The molecules set themselves with their long axes at right angles to the lines of force in an electric field, and parallel to the lines of force in a magnetic field. Somewhat intense fields are necessary to produce complete orientation.

Nematic phases exhibit neither focal conic structures nor "gouttes à gradins", for, as we have seen, these are the results of a stratified structure. Also they separate from the melt as spherical drops, and not as "bâtonnets". Between crossed Nicols the drops display a black cross, the arms of which are parallel to the vibration directions of the Nicols, thus indicating a spherulitic type of structure.

Bernal and Crowfoot * have studied the crystal structures of p-azoxyanisole, p-azoxyphenetole, and anisal 1 : 5 diaminonaphthalene which form nematic phases, and find that the molecules are not arranged in layers but are interlocked, or to use the authors' term, imbricated (Fig. 275). Moreover in the anisole and phenetole, the positions of the molecules along the directions of their long axes are somewhat variable, as is shown by certain of the spots in the X-ray diffraction pattern being "smeared". Thus the crystal structure gives a forecast as it were of the arrangement and degrees of freedom of the molecules in the nematic state. Not all nematic substances which have been studied in this way show an imbricated structure in the crystalline state, however; in some the molecules are arranged in layers. But it seems that where this is the case the substance can exist as a smectic phase as well as a nematic one.

FIG. 275.—Imbricated Structure.

In a nematic preparation between glass surfaces, the molecules in contact with the glass have a striking tendency to attach themselves sideways to it, and the orientation of those in the bulk of the phase appears to be governed by those sticking to the glass. This may be shown by the following experiments. A slide and cover slip are treated with hydrofluoric acid and are carefully washed and dried without touching the surfaces. Molten p-azoxyanisole or p-azoxyphenetole is then allowed to run in between the slide and the slip by capillary action. On cooling and inoculating the edge of the film with a grain of the crystalline form, fairly large crystal plates are obtained. If the preparation is then heated to the nematic range of temperature, each plate assumes the nematic state without changing its outlines, and it is found that although it is quite fluid, for dust particles can often be seen wandering quite

* See Faraday Society's "Discussion on Liquid Crystals and Anisotropic Melts", *loc. cit.*

freely through the preparation, yet it is birefringent (uniaxial positive) and uniform in optical properties. Each plate has its own special optic orientation, which has presumably been determined by that of the original crystal. If now the cover slip be slightly shifted sideways, the borders of the uniform plates become doubled (Fig. 276). This is because the molecules in the top layer are attached to the cover slip and move with it, while those in the bottom layer remain stuck to the slide. The molecules in the bulk of the preparation have to accommodate themselves to the shearing effect. If the cover slip be turned, instead of being shifted in one direction, then the medium is forced to adopt a twisted structure, and this causes it to *rotate* the plane of polarisation of the light,

(a) Before   (b) After

← Direction of Shift

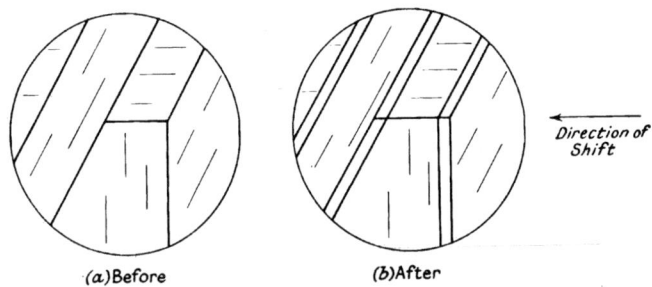

FIG. 276.—Effect of Shifting the Cover Slip on a Nematic Preparation consisting of Uniform Plates (diagrammatic).

i.e. it becomes optically active (*cf.* Reush's experiment with mica plates, p. 127).

Now let the temperature be raised to just inside the isotropic liquid region, and then lowered so that the nematic phase forms again. It will be found that practically the same pattern of nematic plates is produced as before. The reason for this is that when the preparation was melted to the isotropic phase, many of the molecules attached to the glass remained there without change of orientation, and on cooling they directed the molecules in the bulk of the medium into their original arrangements.

We must now consider further the thread-like lines which were referred to at the beginning of this section, and from which nematic phases derive their name. They appear, as already stated, in phases which have been prepared by the rapid cooling of the melt, and more particularly if the film is thick, and if no special care has been taken in the preparation of the glass surfaces. They have the form of serpentine curves, much intertwined, and of no definite geometrical type. Some appear to be partly or wholly anchored to the glass, but for the most part they follow freely all the movements of the liquid, continually changing their length and shape the while, or in some cases disappearing without leaving any trace. It seems certain that these lines are due to discontinuities in the structure, *i.e.* that they correspond to the ellipses and hyperbolæ in smectic phases, but that unlike these curves they are not compelled to conform to a definite geometrical law because there is no stratification in the medium. The nature of the discontinuity is a matter of some doubt. It may be of the same nature as in a smectic phase, that is to say the long axes of the molecules may be directed towards the lines in a radial manner. Alternatively, the lines may denote the presence of vortices, perhaps hollow and vacuous, thus giving a surface at which total reflection occurs, around which the molecules are circulating. In this case the long axes of the molecules would be tangential to circles normal to, and centred on, the lines. Possibly both types of discontinuity exist. In either case the lines may be regarded as axes around which the medium is structurally " rolled up " (" axes d'enroulement ", to use G. Friedel's term).

If the substance has been allowed to crystallise between the glass surfaces before the thread structure described above is established, and the latter is observed through the analyser (no polariser), it will be seen that parts of the field are more brightly illuminated than others. In the brighter parts the threads appear as sharply defined thin lines, whilst in the darker parts they are blurred. If we accept the threads as being " axes d'enroulement ", this may readily be explained as follows. Extraordinary rays passing near a thread are deviated by the strong curvature of the structure, as in the case of the focal conic structure of smectic phases, and so for the most part do not reach the objective. The ordinary rays are not deviated since they vibrate at right angles to the long axes of the molecules. The vibration directions of the rays emerging from the preparation are, however, determined by the molecules stuck to the upper glass surface, and the orientation of these at any place was determined by the orientation of the crystal which was there when the substance was in the solid state. If the long axes of the molecules on a particular area of the glass are at right angles to the vibration direction of the analyser, then ordinary rays will reach the eye, and since these were not deviated in passing near the threads, the area will be brightly illuminated and the images of the threads in it will be sharp. If, however, the long axes of the surface molecules are parallel to the vibration direction of the analyser, only extraordinary rays will be permitted to pass, and since this component of the light suffered deviation in passing near the threads, its intensity will be low and the images of the threads will be blurred.

Cholesteric Mesophases. Cholesteric phases have properties which are markedly different from those of smectic and nematic phases. In the first place they are *optically negative*, a property which, in the case of cholesteryl derivatives, is no doubt connected with the broad flat sterol residue in the molecule (see for example the formula of cholesteryl benzoate given on p. 326). They are also *optically active*, and often very strongly so; the rotation may amount to a whole turn for a layer only a small fraction of a millimetre thick. But to the naked eye, their most striking property is that of scattering vivid colours, recalling those of a peacock's feather, when they are illuminated with white light. These colours are studied to the best advantage when the substance is in the form of a thin uniform film between glass, or other transparent surfaces. The colour of a ray scattered at a particular angle to the surface of the film depends on (i) the substance, (ii) the temperature, and (iii) the angle of incidence of the illuminating beam. For a fixed set of conditions this colour comprises a band of wave-lengths having a strongly accentuated maximum of intensity, so that it is approximately monochromatic. For a given substance and temperature, the mean wave-length of the band has a maximum value when the directions of illumination and of observation are both normal to the surface of the film, and a minimum value when these directions make grazing angles with the surface.

A mixture of about 55% of cholesteryl benzoate and 45% of cholesteryl acetate is convenient for studying this phenomenon, since it gives a cholesteric phase which can be preserved in the supercooled condition at ordinary temperatures for several hours or even much longer. When a film of this mixture is melted and allowed to cool so as to enter the cholesteric region, it first displays a deep violet tint when the directions of illumination and observation are normal to the surface, but appears dark when these directions are oblique, because shorter wave-lengths than the violet lie outside the visible spectrum. As the temperature falls the light scattered for normal directions of illumination and observation changes successively to blue, green, yellow, and red, and at every stage light of shorter wave-length is seen when these directions are oblique. With other substances the sequence of colours with falling temperature is the reverse of that just given, and in some cases only a small portion of the spectrum is traversed. In all cases, however, the wave-length falls as the directions of illumination and observation become more oblique.

It should perhaps be emphasised that the colours shown by cholesteric films in ordinary diffused daylight are the result of the scattering of rays coming from many different directions, and that the results stated above can only be observed without ambiguity when the specimen is illuminated with a parallel beam.

Another characteristic of the scattered light is that it is *circularly polarised*, the direction of the circular motion being right-handed with some substances (*dextro-* type) and left-handed with others (*lævo-* type). A further very remarkable feature is that if the incident light is already circularly polarised in the same sense as that normally scattered by the substance, *e.g.* if right-handed circularly polarised light is directed at a dextro- substance, it is scattered *without change of sense*. In all other cases when circularly polarised light is reflected, the sense of the rotation is reversed, *i.e.* right-handed is changed into left-handed and *vice versa*. If, however, right-handed circularly polarised light is directed at a lævo- substance, or the reverse, the light is *transmitted* without change of sense.

The optical activity of cholesteric phases, for normally incident rays, has been found to conform to the following rule : *in a dextro-substance the plane of polarisation of the incident light is rotated to the right when its wave-length is less than that of the light scattered (at right angles to the surface) with maximum intensity ; it is rotated to the left when its wave-length is greater than this.* For a lævo- substance, lævo-rotation occurs when the light is of shorter wave-length than that scattered at maximum intensity and dextro-rotation when the wave-length is greater.

The remarkable combination of optical properties outlined above has so far not been fully explained. The iridescent colours suggest some kind of layer structure, in which the thickness of the layers is of the order of the wave-length of light, and the existence of such a structure has in fact been demonstrated by Grandjean as follows: If a little of a cholesteric substance is melted into a cleavage crack in a plate of mica, so that it takes the form of a thin wedge of narrow angle, fine parallel lines following contours of equal thickness appear on cooling. These are the edges of parallel layers ("plans de Grandjean"), the thickness of which (as measured by reference to the Newton's colours displayed by the crack before it was filled) is of the order of 2000 Å.

G. Friedel (*loc. cit.*) has expressed the view that cholesteric phases constitute a special type of nematic state. His main arguments are as follows : (1) Substances are known which can exist in both a nematic and a smectic phase. Others are known to have both a cholesteric and a smectic phase. But no substance has been observed to exist in both a nematic and a cholesteric phase. (2) When certain mixtures of dextro- and lævo-cholesteric substances are heated, the typical cholesteric properties gradually disappear and are replaced by nematic properties. There is no discontinuity in the change.

For a full discussion of the optical properties of cholesteric phases, the reader is referred to the above paper by Friedel.

Gradation of Mesomorphic Properties in Homologous Series. Bennett and Jones * have studied the mesomorphic properties of the *p-n*-alkoxy-benzoic and *-trans*-cinnamic acids. The lower members of the series are nematic, but with increasing length of the aliphatic chain smectic properties appear as shown below:

I. ALKOXY-BENZOIC ACIDS

Mesophase(s)

Methoxy-	CH_3O-⬡-$COOH$	None, but a *mixture* of the two gives a nematic phase
Ethoxy-	C_2H_5O-⬡-$COOH$	
Propoxy-	C_3H_7O-⬡-$COOH$	
n-Butoxy-	C_4H_9O-⬡-$COOH$	A single nematic phase
n-Amyloxy-	$C_5H_{11}O$-⬡-$COOH$	
n-Hexyloxy-	$C_6H_{13}O$-⬡-$COOH$	

* *J. Chem. Soc.*, 1939, 420.

		Mesophase(s)
n-Heptyloxy-	$C_7H_{15}O\text{-}C_6H_4\text{-}COOH$	
n-Octyloxy-	$C_8H_{17}O\text{-}C_6H_4\text{-}COOH$	One nematic and one smectic phase, the temperature range of the latter increasing at the expense of that of the former with increasing length of the aliphatic chain
n-Nonyloxy-	$C_9H_{19}O\text{-}C_6H_4\text{-}COOH$	
n-Decyloxy-	$C_{10}H_{21}O\text{-}C_6H_4\text{-}COOH$	
n-Dodecyloxy-	$C_{12}H_{25}O\text{-}C_6H_4\text{-}COOH$	
n-Cetyloxy-	$C_{16}H_{33}O\text{-}C_6H_4\text{-}COOH$	

II. *Trans*-Alkoxy-cinnamic Acids

A similar gradation is found, though smectic properties appear somewhat later in the series. Thus from the methoxy- to the nonyloxy- acids inclusive a single nematic phase is formed. Thereafter (decyl-, dodecyl-, and cetyl-) a smectic phase makes its appearance in addition to the nematic phase, with an increasing temperature range in relation to that of the nematic phase.

Apart from the general trend from nematic to smectic properties which these series show, there are two points of special interest. The first concerns the fact that the methoxy- and ethoxy-benzoic acids show no mesomorphic properties singly, but do yield a nematic phase when they are mixed. This may be attributed to their high melting-points (184° and 195° respectively) as compared with the other acids in the series, which results in the region of stability of the crystalline state completely overlapping as it were the (theoretical) range of existence of the mesomorphic phase (see p. 323). When the melting-point is lowered by admixture, nematic properties appear. The second point of interest is that mesomorphic properties are shown by such an apparently short molecule as propoxy-benzoic acid. That it does so, is taken as strong evidence that it is in fact double molecules of the acid which are concerned. As the formulæ below show, the double molecule (I) has a very similar structure to that of *p*-axoxy phenetole (II), which is a typical nematic substance.

$$C_3H_7O\text{-}C_6H_4\text{-}C\genfrac{}{}{0pt}{}{OH\text{---}O}{O\text{---}HO}C\text{-}C_6H_4\text{-}OC_3H_7$$

(I)

MESOMORPHIC PROPERTIES IN HOMOLOGOUS SERIES 349

$$C_2H_5O\langle\bigcirc\rangle N{=}N\langle\bigcirc\rangle OC_2H_5$$
$$\downarrow$$
$$O$$

(II)

It is probable that all the acids exist as double molecules in the mesophases, for the crystal structures of simple long chain fatty acids are known to consist of double layers in which the polar —COOH groups point inwards (owing to their strong attractions for each other) with the aliphatic chains on the outside. It seems reasonable to suppose that the acids in the present series have a similar crystal structure and that this is connected with the appearance of smectic properties in the higher members.

Friedel (*loc. cit.*) has drawn attention to a somewhat similar gradation of properties in the cholesteryl esters of the fatty acids, as shown below:

	Ester		Mesophase(s)
A.	Formate	$H.COO.C_{27}H_{45}$	⎫
	Acetate	$CH_3.COO.C_{27}H_{45}$	⎬ A single *cholesteric* phase
	Propionate	$C_2H_5.COO.C_{27}H_{45}$	
	Butyrate	$C_3H_7.COO.C_{27}H_{45}$	⎭
B.	Caprylate	$C_7H_{15}.COO.C_{27}H_{45}$	⎫
	Pelargonate	$C_8H_{17}.COO.C_{27}H_{45}$	⎬ Two mesophases, one *cholesteric*
	Caprate	$C_9H_{19}.COO.C_{27}H_{45}$	and one *smectic*
	Laurate	$C_{11}H_{23}.COO.C_{27}H_{45}$	
	Myristate	$C_{13}H_{27}.COO.C_{27}H_{45}$	⎭
C.	Stearate	$C_{17}H_{35}.COO.C_{27}H_{45}$	⎫ A single *smectic* phase
	Oleate	$C_{17}H_{33}.COO.C_{27}H_{45}$	⎭

It will be seen that when the chain is short (group A), the effect of the cholesteryl residue predominates; when it is very long (group C), it itself determines the properties and only a single smectic phase is formed; whilst when it is of intermediate length (group B), both parts of the molecule influence the mesomorphic behaviour, a smectic phase being stable in the lower part of the mesomorphic temperature range and a cholesteric phase in the upper part.

A Simple Microscope Hot Stage for the Study of Liquid Crystals. The stage illustrated in Fig. 277 is easily made and has the advantage of being so thin (about $\frac{1}{16}''$) that the slide is not raised appreciably above its normal position in relation to the microscope condensing system, so that good interference figures can still be obtained. From top to bottom in the order shown the components are as follows:

(i) A copper plate not more than $\frac{1}{32}''$ thick, about $3\frac{1}{2}'' \times 2\frac{1}{2}''$, and with a $\frac{1}{4}''$ hole in the centre. The plate should be as smooth as possible, and, for the reason given below, should be rolled to a slightly cylindrical contour with the concave

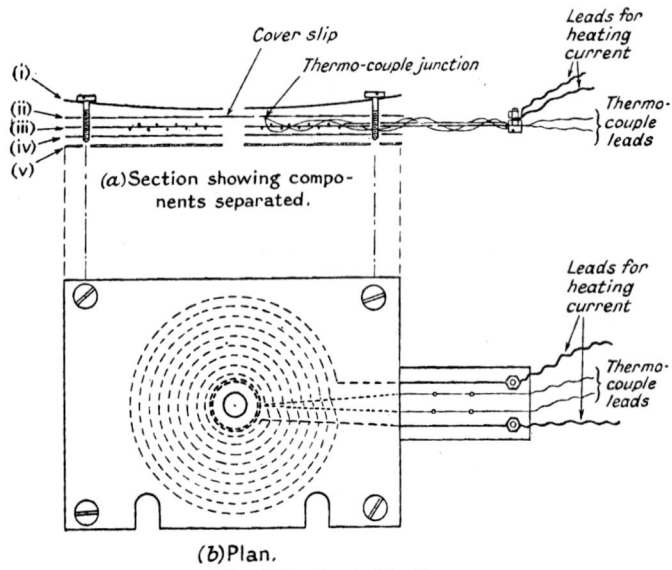

FIG. 277.—Simple Hot Stage.

side uppermost. This plate and the components below are slotted at one edge, as shown in (b) to allow the stage clips to be used.

(ii) A thin sheet of mica with a central hole, slightly larger than $\frac{1}{2}''$, in which is placed a $\frac{1}{2}''$ circular cover slip, resting on—

(iii) A sheet of mica with a tongue at one side to carry the binding posts for the leads to the heating element and to the thermo-couple (see below). The central hole is $\frac{1}{4}''$. On this sheet is threaded either thin nichrome wire as shown in the figure, or, better, nichrome ribbon in zig-zag fashion. The length of wire or ribbon should be such as to pass 1–2 amps. at 36 volts, which is a convenient working voltage. The sheet also carries a copper-constantan thermo-couple, with the junction as near to the centre of the stage as possible. It goes without saying that the heating element and the thermo-couple leads must be so threaded on the sheet as to avoid short circuits.

(iv) A thin mica sheet with a $\frac{1}{4}''$ central hole.

(v) A sheet of asbestos paper with a $\frac{1}{4}''$ central hole.

The microscope stage must be drilled and tapped to take the four small bolts which, as shown in the figure, hold the whole assembly together. These bolts are cautiously tightened until the copper plate is just pulled flat, as is shown by testing with a microscope slide laid on the top; if adjustment is correct the slide will show no tendency to rock. The slight spring in the plate results in a gentle pressure being applied at the centre, and this makes for a compact assembly. If the plate is made flat in the first instance, the tightening of the bolts necessary to give minimum thickness causes it to adopt a slightly domed shape.

The thermo-E.M.F. is read by means of a sensitive millivoltmeter, the scale being calibrated to give temperatures by making observations on the stage of the melting of substances of known melting-points. The cold junction of the circuit can be immersed in ice and water in a vacuum flask, or if the experiments are not too prolonged, in water at a temperature near that of the laboratory, say 18°. This latter alternative has the advantage that the whole of the voltmeter scale becomes available for useful readings.

The heating current is controlled by means of a rheostat, and in the absence of draughts, it is quite easy to maintain the temperature constant to within one or two degrees.

It is perhaps inadvisable to use a stage of this type on a microscope of high grade, for a considerable amount of heat is conducted to the microscope and substage, including of course the polariser. The latter can if desired be protected by surrounding it with a jacket through which cold water is circulated. The authors have, however, used the stage for periods of about an hour at temperatures up to 150° without jacketing the polariser, and no ill effects to this component seem to have resulted.

Recommended for Further General Reading

The extremely comprehensive paper by G. Friedel referred to in the text above (*Ann. Phys.*, 1922, **18**, 273–474).

W. H. Bragg, " Liquid Crystals ", *Nature*, 1934, **133**, 445.

CHAPTER X

METHODS OF ATTACK AND EXPERIMENTS

The optical examination of crystalline material must be carried out systematically if the maximum amount of information is to be gained expeditiously. It is impossible to lay down hard and fast rules of procedure, because so much depends upon the nature of the material and the requirements of the problem in hand, but for the benefit of the beginner we have attempted in this chapter to afford some guidance by giving a method of carrying out the examination of a single compound, followed by examples indicating points of importance and methods of attack in some typical cases of the chemical application of the polarising microscope. The chapter concludes with a series of simple experiments.

THE SYSTEMATIC EXAMINATION OF A SINGLE COMPOUND

The optical examination of a single compound may be termed the fundamental operation in chemical work with the polarising microscope, since even when the problem is to identify the constituents of a mixture, it is necessary, in the absence of published data, to determine first the optical properties of authentic specimens of the compounds which may be present.

The following method has been found satisfactory by the authors in a variety of cases. (In giving this method, it is assumed that the substance is to be examined as completely as possible. Where this is unnecessary, the examination will naturally be concluded as soon as sufficient data for the purpose in hand have been obtained.)

The substance should be prepared and mounted for examination using one or more of the methods described in Chapter VI. The methods selected will depend upon both the properties of the substance and the nature of the problem.

Monochromatic light is used where necessary, according to the dispersion of the substance and the degree of precision of the measurements required, *e.g.* for determining the refractive indices, and where there is marked dispersion, extinction angles and interference figures also.

The crystals are first examined between crossed Nicols to see whether they are **birefringent** or belong to the **cubic system**. Crystals that appear to be isotropic in parallel light must not be passed as cubic until it has been shown that they do not give interference figures; they may be birefringent crystals presenting optic axial sections.

Cubic Crystals

The **colour, habit, cleavage,** and **refractive index** should be determined on several crystals.

Birefringent Crystals

The procedure depends upon whether the crystals are well formed, in which case they will adopt a limited number of definite orientations owing to their tendency to lie on their larger faces; or poorly formed, in which case their orientation is likely to be more or less random.

Well-formed (including acicular) crystals

An isolated crystal is selected and the following properties, so far as they are applicable, are determined in the order given:

Colour and **habit**;

Edge angles;

Cleavage;

Vibration (extinction) directions in relation to the crystal edges;

Pleochroism in relation to the vibration directions;

Directions of fast and slow vibration (sign of elongation, if elongation is shown);

Polarisation colours (whether normal or abnormal);

Birefringence (estimated by the order of the polarisation colours, having regard to the probable thickness of the crystal);

Interference figure and its orientation with regard to the crystal outline. Also properties deducible from the interference

SYSTEMATIC EXAMINATION OF A SINGLE COMPOUND 355

figure, if suitable, namely **orientation of the optic axes and optic axial plane, optic axial angle, optical sign,** and **dispersion of the optic axes and bisectrices.**

The above determinations must be repeated on several other crystals on the slide. A sketch should be made of each crystal studied, showing its outline, edge angles, cleavage lines, vibration directions, pleochroism, and interference figure (correctly orientated). A survey of the data and sketches will show whether the crystals present one, two, or more sections. Thus crystals of the monoclinic double salt schönite, $K_2SO_4,MgSO_4,6H_2O$, grown from a drop of solution on a slide present two sections, due to the development of plates parallel to (001) and (110) respectively. These sections and, for reference, an ideal crystal of schönite are shown in Fig. 278.

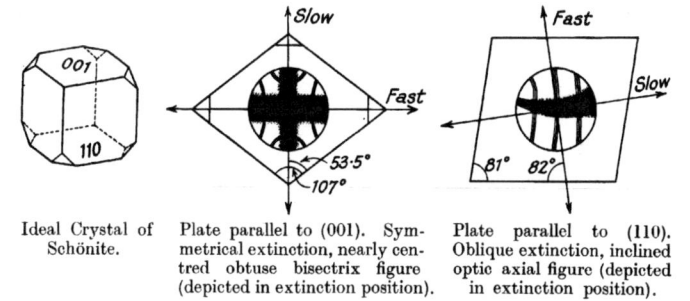

| Ideal Crystal of Schönite. | Plate parallel to (001). Symmetrical extinction, nearly centred obtuse bisectrix figure (depicted in extinction position). | Plate parallel to (110). Oblique extinction, inclined optic axial figure (depicted in extinction position). |

FIG. 278.—Sections presented by Schönite ($K_2SO_4,MgSO_4,6H_2O$) grown from a Drop of Solution on a Microscope Slide.

The refractive indices of the sections presented are now determined. This involves the mounting of further portions of the material in liquids of known refractive index. The sections can usually be recognised very rapidly in these new mounts by means of some distinctive property, e.g. their edge angles.

The properties so far determined are likely to be of great value for identifying the substance, and are particularly appropriate for this purpose as they are the most readily observed (except, of course, when the substance adopts some quite different crystal habit, which presents other sections).

The complete optical examination of a crystal, however, includes the determination of the principal refractive indices, and these may not have been shown by the above sections. Thus in the example of schönite given above, neither of the vibration directions of the (110) section corresponds to a principal index. On the other hand, the (001) section, which happens to be nearly normal to the obtuse bisectrix, does show one principal refractive index. The fast direction, which is normal to the optic axial plane, corresponds to the index β; the slow direction gives a value very near to γ. The sections may also have given such uncentred interference figures as to make the determination of the optical sign, optic axial angle, etc., uncertain or impossible. If this is so, it will be necessary to complete the examination by studying other sections of the substance. This may be done either by tilting the crystals into other positions (p. 213), by cutting sections (p. 216), or by using a rotation apparatus (p. 213). Alternatively, irregular grains of the substance, prepared by crushing a large crystal to a powder of about 0·05 mm. particle size, may be mounted on a slide, and since their orientation will be more or less random (unless they are platy), some of them will present the sections required (p. 318; see also below).

Poorly formed Crystals

If the crystals exhibit marked elongation, flattening, or cleavage, an examination on the lines detailed under " well-formed crystals " may show that they present a limited number of definite sections, and can therefore be treated under that heading, using, perhaps, cleavage lines in place of edges to describe the orientation of the vibration directions, etc.

If, however, the crystals are granular and not too large (*i.e.* not much above 0·05 mm.), so that they may be freely rolled about on the slide, they will present, or can be made to present, almost any desired section. The examination of such material will naturally be confined to determining those properties which are unconnected with external form, namely :

Colour ;

Pleochroism (in relation to the principal vibration directions) ;

Polarisation colours (whether normal or abnormal) ;

Birefringence ;

Optical sign ;

Optic axial angle ;

Dispersion of the optic axes and bisectrices ;

Principal refractive indices.

SYSTEMATIC EXAMINATION OF A SINGLE COMPOUND 357

The grains should not be mounted in liquids which differ widely from them in refractive index (see p. 124). The examination is made on those grains which present, or which can be turned over so as to present, principal optical sections as shown by their behaviour between crossed Nicols in parallel and in convergent light. If the interference figures are poor, as they may be if the grains are very small, γ and α (in the biaxial case) should be sought on those grains showing the highest polarisation colours, taking the highest and lowest values so found as these indices respectively; β will be given by those grains showing a uniform illumination on rotating the stage.

PRESENTATION OF OPTICAL CRYSTALLOGRAPHICAL DATA

The results of the microscopic examination of a compound must be reported accurately and unambiguously. It is also highly desirable that they should be presented in as concise a form as possible. Recommendations on this subject have been published by a committee appointed by the Division of Analytical and Micro Chemistry of the American Chemical Society,* with the object of standardising the presentation of optical crystallographic data in the publications of that Society. With this report the authors are in entire agreement, since it emphasises many points which they have long felt to be important, and which are embodied in the following paragraphs.

The description of the properties of a crystal is greatly improved by annotated drawings on which may be shown the *habit, edge angles*, directions of *cleavage*, and the relations of these to the optical vectors. Ambiguities are thereby avoided and the presentation gains in conciseness. *Descriptions of habit should always be accompanied by a statement of the conditions of crystallisation, e.g.* whether obtained from solution or by sublimation, and if the former, the solvent used. The most prominent faces developed should be named, if possible, by means of Millerian indices. This presupposes that the directions of the crystallographic axes can be recognised, as in uniaxial crystals, or are deliberately assigned. If this is not possible, the forms present should be described in general terms as pinacoidal, prismatic, or pyramidal. The directions of *cleavage* in relation to the crystallographic axes, or to prominent faces, should be given

* *J. Ind. Eng. Chem. (Anal.)*, 1945, **17**, 603.

by means of a statement of the real or projected angles between them, and a qualitative estimate made as to whether the cleavage is perfect, good, or imperfect. The *crystal system* should be stated if possible, and also any indication of the *symmetry class* as deduced from, say, a hemihedral or hemimorphic habit.

Pleochroism should always be related to the axes of the indicatrix, or to the vibration directions of the section or sections commonly presented, *e.g.* X = yellow, Z = red.

Drawings illustrating the *optical orientation* in biaxial crystals may be done as in Figs. 278 and 286, in which the interference figure in a stated position (at extinction or at 45°) for the section is shown in a central circle, correctly orientated with respect to the crystal outline. The number of isochromatic rings in the figure actually depends on the thickness and birefringence of the section; in the drawing it is only necessary to show a sufficient number to illustrate their general direction.

In uniaxial crystals drawings to show the optical orientation are not necessary, since the direction of the optic axis is always parallel to the c axis.

It is possible to state the optical orientation of a biaxial crystal unambiguously without the use of a drawing if the crystal morphology is completely known. In the orthorhombic system it is only necessary to state (i) the crystallographic axis which is parallel to the acute bisectrix, and whether the latter is X or Z, (ii) the orientation of the optic axial plane. Thus in the case of potassium sulphate, Z (acute bisectrix) = c; the optic axial plane is (100). Another method used by many authors is to state the orientation of the optic axial plane, the optical sign, and the crystallographic axis to which X or Z is parallel. Thus in the case just given the description would run : optic axial plane is (100) ; (+) ; Z = c.

In the crystals of the monoclinic system, it is necessary to state (*i*) the indicatrix axis which is parallel to b, (*ii*) the extinction angle measured on (010) between c and the nearest indicatrix axis, and whether this lies in the acute or obtuse angle β between the a and c axes, and (*iii*) the optical sign. It is conventional for greater clarity also to state whether the optic axial plane is parallel or normal to (010), though it will be evident that this information is contained in statements (*i*) to (*iii*). It should be noted that if Y = b, the orientation of the optic axial plane is completely defined by this statement alone ; it must be (010). Some authors use a

TYPICAL APPLICATIONS OF THE POLARISING MICROSCOPE 359

$+$ or $-$ sign to denote whether the indicatrix axis referred to in (*iii*) lies respectively in the obtuse or acute angle β. This gives rise to no ambiguity, but other authors use these symbols to denote whether the extinction is measured in a clockwise or anticlockwise direction from c. This is ambiguous unless it is clear which is the positive and which the negative end of the b axis. It is therefore better to state definitely whether the direction concerned lies in the obtuse or acute angle. *Example*: Gypsum, $CaSO_4,2H_2O$, $Y = b$; $X \wedge c = 37\frac{1}{2}°$ in the acute angle β; $(+)$.

In the triclinic system it is necessary to state the angular relationships of two of the indicatrix axes to the three crystallographic axes, and the optical sign. Microscopically this information cannot be obtained in quantitative terms without the use of a universal stage. For most purposes, however, it is sufficient to describe the orientation of the optical vectors in relation to the sections commonly presented. See for example Fig. 281, in the experiment on $CuSO_4,5H_2O$.

The probable accuracy of values of *refractive indices* given should always be stated. This depends upon a number of factors, such as whether white or monochromatic light is used, the precision of the method of comparing the indices of crystal and liquid, and whether the habit of the crystal allows the measurement of the principal indices while it lies in stable positions.

In the above treatment of the presentation of optical crystallographic data, only points needing special emphasis have been dealt with. A complete description of the crystal would of course include all the properties listed in the previous section, for example the value of the optic axial angle and the nature of the dispersion.

POINTS OF IMPORTANCE AND METHODS OF ATTACK IN SOME
TYPICAL APPLICATIONS OF THE POLARISING MICROSCOPE

1. *The determination of the optical properties of a compound for record purposes*

The examination should be as complete as the material will allow. Since the main object is, presumably, to obtain data which may assist other workers in recognising the compound, due attention should be paid to any characteristic habits developed, and the properties of the optical sections which these habits commonly present when they are mounted on a slide.

The following are fundamental optical constants which are

independent of external crystalline form, and are therefore of particular importance for record purposes: principal refractive indices, optic axial angle, optical sign, dispersion of the optic axes and bisectrices.

2. *The optical examination of a material of entirely unknown composition, as a preliminary to its chemical analysis*

Here, the examination will be directed towards determining the general nature of the material, e.g. whether it consists of one compound or more. It is unlikely that, in the complete absence of information, identification of the material or any constituent of it will result.

A simple case is that in which all the crystals on the slide show the same optical characters, indicating that the material consists of a single compound. Such a result would be obtained with a compound, the crystals of which were of a single platy habit, and so orientated themselves similarly on the slide.

Other simple cases are those in which some of the crystals are birefringent and others isotropic, or some give uniaxial, and others biaxial interference figures. These results show that at least two crystal species are present.

Another certain indication that more than one kind of crystal is present depends upon the facts that in uniaxial crystal sections one index is always ω, whilst in biaxial crystal sections, the two indices cannot both be greater or less than β (see p. 318). Consequently if two crystals are found on the slide, one of which has two indices greater than a certain value, and the other has two indices below this value, it proves that the crystals are different.

In other cases the conclusions may be less certain. It may, for example, be impossible to be sure whether two crystals showing different optical characters are different species or different sections of the same species. Sometimes it is possible to resolve the uncertainty by rolling the crystals into other positions. Otherwise, marked differences in habit, colour, refractive index, and birefringence should be sought. On occasion it is worth while to attempt a partial separation of the material by fractional crystallisation, e.g. by allowing a drop of its solution to evaporate on a slide. If one kind of crystal separates before the other, it indicates that they are different. It should be noted, however, that the fractional crystallisation of salt mixtures may cause confusion, owing to reciprocal salt action and the appearance of salts not in the original mixture at all.

3. *To determine whether two crystalline bodies are identical or not*

The following are cases falling under this heading : (*i*) a substance, *e.g.* the product of a reaction or the contents of an unlabelled bottle, is suspected of being a certain compound, and it is required to find out whether this is so or not by comparing it with an authentic specimen of the compound ; (*ii*) two substances known to have the same chemical composition are to be compared in order to see whether they are identical or isomeric. (It should be pointed out that should they prove to have different optical properties, this does not exclude their being *polymorphic* forms of the same compound ; in such a case, however, the more stable form will usually show some signs of growing at the expense of the other, when they are placed in contact, particularly in the presence of a solvent, as in the case of potassium nitrate, Expt. 9, p. 373.)

The specimens should always be examined carefully in their original state, and if sufficient data does not thereby emerge to decide on their identity or otherwise, possible differences in habit (which might mask optical similarities by presenting different sections) should be eliminated, when practicable, by recrystallising the specimens from the same solvent and under the same conditions. A comparison of the two under the microscope will then show, usually in a few minutes, whether they are the same or not.

Should it not be possible to recrystallise the specimens, and the first examination leaves their identity doubtful, they should both be reduced to a powder of about 0·05 mm. particle size (that is, of course, if they are not already so finely divided), and a determination of the principal refractive indices made on the resulting grains (see above, p. 357). Should the values of these indices agree for the two substances, and the other optical properties be not inconsistent, their identity may be taken as reasonably certain.

4. *To identify the constituents of a mixture, it being known that certain compounds only can be present*

For this, a previous knowledge of the optical properties of the possible compounds is necessary, and may be obtained either from published data, or by the examination of authentic specimens. The problem is similar to the preceding one, but more complex.

To eliminate possible variations of habit, the authentic specimens examined should have been crystallised under as nearly as possible the same conditions as the mixture, if these conditions are known and can be reproduced. Sometimes this precaution is unnecessary, because the compounds are readily distinguishable by properties that are quite independent of the orientation of the crystals on the slide. For instance, suppose that the chemical analysis of a mixture of salts shows it to contain the ions K^+, Na^+, Br^-, and Cl^-. The possible combinations of these ions that are stable at room temperature (in the presence of water vapour) are KCl, KBr, NaCl, and $NaBr,2H_2O$, and therefore any of these salts may be present. The first three are cubic, and so are extinguished at all orientations between crossed Nicols. They may be readily distinguished by their refractive indices which differ considerably thus ; KCl, 1·4903 ; KBr, 1·5590 ; NaCl, 1·5442 (all n_D). $NaBr,2H_2O$ is monoclinic and therefore shows polarisation colours between crossed Nicols. These data alone are sufficient to identify the constituents of any mixture of the salts.

Cases of difficulty under this heading may often be simplified (provided the mixture does not consist of salts ; see section 2 above) by partially separating the constituents by fractional crystallisation.

EXPERIMENTS

The following experiments have been designed to give the beginner practical instruction in handling and testing the polarising microscope, and in carrying out the more important optical tests which can be made with it. The experiments may well form the basis of a short practical course for students, and many of them have in fact been used by the authors for that purpose. Before beginning the experiments, the student should prepare the set of apparatus described on pp. 194–8 and illustrated in Fig. 174, and also a set of immersion media covering the range from 1·45 to 1·74 in steps of 0·01, such as the mixtures suggested by F. E. Wright (p. 232).

EXPERIMENT 1. *To check the setting of the Nicol prisms and the cross-wires in the microscope*

It is first necessary to see that the Nicols are accurately crossed when in the " 0 " position. Remove the eyepiece, objective, condenser, and mirror, insert both Nicols (" 0 " position), and, holding the instrument up, point it at the sun or at a distant source of artificial light. On looking along the tube, the field should be

EXPERIMENTS

uniformly dark and should lighten up immediately on rotating the polariser the slightest amount in either direction.

Now find the vibration direction of light transmitted by the polariser. In a calcite prism this may be determined by inspection, if the type of construction is known (see p. 165), but it is as well to check it experimentally. This may be done very readily by observing the pleochroism of a section of biotite mica cut parallel, or nearly parallel, to the optic axis. (Biotite is biaxial, but most specimens have such a small optic axial angle, that for practical purposes they may be regarded as uniaxial.) Any mounted section of biotite-bearing granite, such as may be obtained from a dealer in petrological specimens, will be found to contain numerous flakes of biotite, recognisable by their brown colour and perfect cleavage.* Focus the microscope on such a section, using a low-power objective (e.g. 1"), and find a crystal of biotite showing numerous parallel cleavage lines. These lines, which are the traces of the basal cleavage, run at right angles to the optic axis. Now insert the polariser and rotate the stage. It will be observed that the substance is strongly pleochroic, the colour changing between pale brown and deep chocolate brown. The vibration direction of the polariser may now be deduced from the fact that the colour is deepest, *i.e.* the absorption of light is greatest, when the cleavage lines are parallel to the vibration direction of the light.

Another method of finding the vibration direction of the light passed by the polariser is based on the fact that light is polarised by reflection (see p. 100). For ordinary glass, maximum polarisation is obtained when the angle of incidence is about 57°. The microscope is placed in front of a window, or other extended source of light, and the microscope mirror and a glass plate, *e.g.* a microscope slide, are arranged as nearly as possible at the angles shown in Fig. 279. The surfaces of mirror and plate should be perfectly clean. Under these conditions, the light entering the microscope vibrates in a direction normal to the plane of incidence, *i.e.* parallel to the E–W cross-wire. With the analyser removed, rotate the polariser and note whether the maximum amount of light is transmitted when it is in its normal position (*i.e.* the " 0 " position), or at right angles to this. From this observation the vibration direction of the light transmitted by the polariser may be deduced, *i.e.* whether

---

* The section may contain tourmaline which is also brown, but shows no cleavage.

it is parallel to the N–S or to the E–W cross-wire. A sufficiently good but not so pronounced effect is obtained even without the glass plate, by tilting back the microscope and adjusting the mirror so that light is reflected from it into the microscope at approximately the polarising angle.

Finally, test the setting of the cross-wires. These must be accurately at 90° to one another, and must coincide with the vibration directions of the Nicols when the latter are in the " 0 " position. Focus the microscope on the edge of a microscope slide, using a carefully centred objective of low power. Now move the slide to a position where the image of the edge passes very close to the point of intersection of the cross-wires, and turn the stage until the image is parallel to one of them. Note the stage reading, and turn the stage until the image is parallel to the other cross-wire. The stage reading should now differ from the previous one by exactly 90°. Check this by repeating the experiment two or three times.*

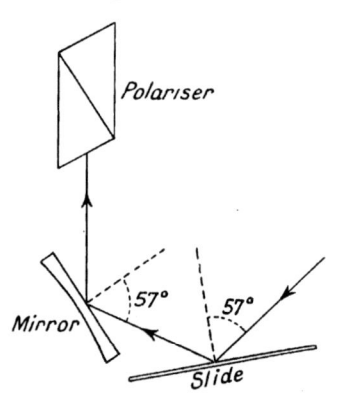

FIG. 279.—Reflection Method of Finding Vibration Direction of Light passed by Polariser.

To test whether the cross-wires coincide with the vibration directions of the Nicols when the latter are in the " 0 " position, a crystal which is known to have straight extinction is observed. Most biotites give straight extinction with reference to the basal cleavage lines, and the section mentioned above is therefore sometimes used for this test. Since, however, biotites are known with an oblique extinction of a few degrees, it is safer to use some substance, the composition and therefore optical properties of which are constant. For example, ammonium nitrate (orthorhombic form) may be used, and suitable crystals can be grown from a drop of the aqueous solution by the method described in Chapter VI,

* It is here assumed that the stage is accurately graduated, and in an instrument of reputable make this may normally be taken for granted. Should there be any doubt on this point, however, it may be resolved by carrying out the experiment with the slide in many different positions on the stage.

p. 200. The temperature must be kept below 32° because above this point the solution deposits another form. The crystals grow as long straight-edged needles, which show straight extinction with reference to their length. When several of these needles have grown to a length equivalent to the radius of the field of view, place a cover slip over the drop to arrest further evaporation. Now select an isolated crystal with perfectly straight edges parallel to the length, and, by moving the slide, bring the image of one of these edges near to the point of intersection of the cross-wires. Centre the objective if necessary, and insert both Nicols in the crossed position.* Rotate the stage and note whether the crystal is extinguished every time the edge becomes parallel to a cross-wire. If so, then the setting of the cross-wires is correct. The exact position of extinction may be determined by approaching it from either side several times and taking the mean, or by using the selenite plate as explained on p. 258.

Should the microscope fail to satisfy any of the above tests, it should be returned to the maker for adjustment.

EXPERIMENT 2. *To examine crystals of sodium chloride, and to observe the effect of urea on their habit*

Add an excess of sodium chloride to 5 c.c. of water in a test-tube, and heat to boiling with constant shaking. Allow to cool, and when the temperature is between 40° and 60° (use a thermometer; do not guess) place a drop of the clear solution on a microscope slide. Watch the crystallisation of the drop under the microscope, using a low-power objective (*e.g.* 1″). Cubic crystals, mostly in the form of square or rectangular plates, are deposited. When they have grown to a convenient size for easy observation, arrest further evaporation of the drop by placing a cover slip over it.

Now examine the crystals between crossed Nicols, and note that they all remain as dark as the rest of the field, no matter into what position the stage be turned. They are isotropic.

To the solution in the test-tube add 1 gm. of urea. Heat to boiling, allow to cool to between 40° and 60°, and remove and observe a drop as before. This time it will be seen that most of the crystals have octahedral facets on their corners (*cf.* Fig. 181 (*a*), p. 221); they are in fact a half-way stage between the cube and the

* The polariser may be already in position. Generally there is no need to remove it for the ordinary observation of specimens.

octahedron. The presence of urea in the solution has resulted in an alteration in the habit of the sodium chloride. Confirm that the crystals are still isotropic by examining them between crossed Nicols.

EXPERIMENT 3. *To determine the refractive index of an isotropic crystal*

Crush a few crystals of potassium chloride to a powder in a small mortar. Avoid using the pestle with a grinding action, as this will produce a large proportion of very finely divided material impossible to study satisfactorily under the microscope. Transfer a minute portion of the powder to the middle of a clean microscope slide by means of a micro-spatula. Cover with a ½" cover slip, pressing down gently to spread out the powder. Bring a small drop of liquid of index about 1·47 to the edge of the cover slip, by means of the dropping rod, when the liquid will run in by capillary action. Compare the refractive indices of solid and liquid by means of the Becke test (p. 225). For this use a medium- or high-power objective (*e.g.* ¼"), and lower the condenser. If a substage diaphragm is fitted, it should be partly closed. Under these conditions, each crystal fragment will be seen to be bordered by a bright ring of light, and if the microscope tube be racked up slightly, this line moves in towards the centre of the fragment, indicating that the crystal has the higher index.

This conclusion is now confirmed by applying the van der Kolk test. Use a 1" objective, open the substage diaphragm, and raise the condenser to its normal position. Illuminate the specimen obliquely by any of the means mentioned on p. 227, so that the field appears half illuminated. Different microscopes vary in regard to the best means of producing oblique illumination for this test, and the most suitable method for the instrument used must be found by trial. Note whether the test is affected by altering the position of the condenser.

Now repeat these tests using a liquid of index about 1·50, and it will be observed that the above phenomena are reversed, showing that the index of the solid is less than that of the liquid.

Repeat the tests using liquids having indices intermediate between those of the liquids already used, until one is found in which the crystals are practically invisible. The index of this liquid will be the index of the potassium chloride, accurate to the second

EXPERIMENTS 367

place of decimals if only white light has been used. If sodium light be used in the final stages of the experiment, and liquids varying in index by 0·002 are prepared by mixing the standard liquids in different proportions, a value of the index for the D line accurate to $\pm$ 0·002, or even less, may be obtained. A knowledge and some control of the temperature contribute to the accuracy of the determination since, as already shown on p. 230, most of the refractive liquids used alter in index by about 0·0005 for a change of one degree in temperature. n_D for potassium chloride is 1·4904.

EXPERIMENT 4. *To study the crystallisation of a solution containing potassium and magnesium chlorides*

A solution containing potassium and magnesium chlorides in equimolecular proportions does not deposit the double salt carnallite, $KCl,MgCl_2,6H_2O$, until a considerable quantity of potassium chloride has first been thrown down. This is the principle underlying the well-known preparation of potassium chloride from natural carnallite. To obtain carnallite as the first product of crystallisation a large excess of magnesium chloride must be present.

Make up a solution containing the two salts in approximately equimolecular proportions by dissolving 3 gm. of potassium chloride and 8 gm. of magnesium chloride ($6H_2O$) in 10 c.c. of boiling water. Place a drop of the solution on a slide, and watch the crystallisation under the microscope. The crystals deposited are all isotropic (test between crossed Nicols), and consist entirely of potassium chloride. Now dissolve 3 gm. of potassium chloride and 26–27 gm. of magnesium chloride hexahydrate in boiling water, making the final volume about 35 c.c. (Owing to the hygroscopic nature of magnesium chloride, an exact weight cannot be specified, and the best concentration for the experiment may have to be found by trial.) Place a drop of the hot solution on the slide, and observe that now, in addition to, or instead of potassium chloride crystals, six-sided plates, mostly elongated, are deposited. These are crystals of carnallite. They give marked polarisation colours and straight extinction with reference to their direction of elongation. The edge angles of the plates are all near to 120°, but the crystals evidently do not belong to the hexagonal or trigonal system, because a plate belonging to one of these systems and showing a regular hexagonal outline would be isotropic to light passing through it normally, whereas the carnallite plates are strongly birefringent. In point of fact carnallite is orthorhombic. By means of the mica plate it may be shown that the crystals have positive elongation.

The distinctive optical properties of potassium chloride and carnallite demonstrated by this experiment could be made use of in the control of a preparation of potassium chloride from carnallite. The first sign of birefringent material in the crystals deposited would indicate that the extraction was complete.

EXPERIMENT 5. *To study a uniaxial interference figure*

A convenient example is afforded by the basal section of biotite, if a specimen with a very small optic axial angle is selected. Biotite cleaves perpendicularly to the optic axis (regarding the substance as essentially uniaxial; *cf.* Expt. 1), so that all that is necessary to obtain a suitable section is to remove a flake from a piece of the substance with a penknife. Biotite is strongly coloured and it is impossible to study thick sections of it, because they are not sufficiently transparent. Cleave, however, first of all the thickest section that will transmit an appreciable amount of light when under the microscope, and bring it to a focus with a high-power objective ($\frac{1}{4}''$ or near). Insert the Nicols (crossed), and note that the section remains dark at all positions of rotation of the stage. This is because the light is passing along the optic axis, or nearly so, in all parts of the field, and along this direction the crystal is isotropic. Now insert the converging lens, remove the eyepiece, and adjust the condenser to give the best illumination.* A black cross, the arms of which are parallel to the vibration directions of the Nicols, and a system of concentric rings will be seen. The colours of the latter are not normal polarisation colours, because they are modified by the brown tint of the substance.

Rotate the stage and note that the arms of the cross do not change their positions. Careful observation will, however, show that at intervals of 90°, the cross separates to a slight extent into the two " brushes " characteristic of the biaxial figure as illustrated in Fig. 280. This is of course due to the fact that biotite is really biaxial.

Now replace the eyepiece and insert the Bertrand lens. The interference figure is again seen and is larger than before, but not so sharply defined. If the instrument is not provided with a

* With the Cooke microscopes fitted with polaroid plates (see p. 180), it is only necessary to remove the eyepiece and open the substage diaphragm.

EXPERIMENTS 369

Bertrand lens, it will probably have a Becke lens (see p. 272) which serves the same purpose, except that it only focusses the figure and not the cross-wires.

Determine next the optical sign of the biotite by inserting the $\frac{1}{4}$ wave mica plate in the slot above the objective. Note that the rings in the two quadrants of the figure through which the " slow " direction of the plate passes are moved outwards, and that there is a black spot near the centre of the figure in each of these quadrants, whilst in the other two quadrants the rings are moved inwards. Biotite is therefore negative. Confirm this result by using (a) the gypsum plate and (b) the quartz wedge (see p. 305).

Now cleave a thinner section of biotite, and observe its interference figure. Note that, although the cross is unchanged, there are fewer

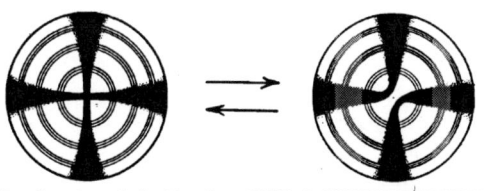

FIG. 280.—Opening of the Pseudo-uniaxial Cross of Biotite on Rotation of the Stage.

rings in the field ; in fact, if the section is thin enough, no rings at all appear.

Having obtained and studied suitable sections, it is advisable to mount them in Canada balsam for future reference (see p. 212).

Mounted sections of quartz (uniaxial, positive) perpendicular to the optic axis and of any desired thickness may be obtained from dealers of petrological specimens, and may be studied in addition to, or instead of, biotite. They have the advantage that they give normal polarisation colours. If thicker than about 1 mm., they reveal the optical activity of the substance, the centre of the cross being replaced by a spot of colour (see p. 281). Another example is afforded by iodoform, which may be crystallised on a slide from xylene, giving hexagonal plates parallel to (0001). The sign is negative.

An example of an inclined uniaxial figure will be met with in Expt. 9.

EXPERIMENT 6. *To study a biaxial interference figure*

Muscovite mica, *i.e.* the mica commonly put to practical use, *e.g.* as an electrical insulator, is almost ideal for this experiment, for it cleaves almost exactly at right angles to the acute bisectrix of the optic axial angle. First place a section 0·1 to 0·25 mm. thick under the microscope, and obtain its interference figure as in the previous experiment (high-power objective, converger, crossed Nicols, Bertrand lens). Rotate the stage and note that the " eyes " and colour rings of the figure rotate with it. Note also that the " brushes " sweep in and out, forming a diffuse cross parallel to the cross-wires whenever the optic axial plane is parallel to one of the latter,* and two hyperbolic curves passing through the " eyes " when the optic axial plane is at 45° to the cross-wires.

Muscovite is monoclinic and possesses horizontal dispersion (see p. 298), but in many specimens this is so slight that the arrangement of colours about the optic axial plane appears to be symmetrical.

Now determine the optical sign of the crystal. Turn the stage until the optic axial plane is parallel to the slot above the objective, and insert the quartz wedge in the slot, thin end first. Note that as the wedge is pushed in (the length direction of the wedge being assumed to be " slow ") the colour rings of the figure in the region between the " eyes " appear to shrink into the latter. Next turn the stage through 90° and repeat. The rings now move outwards as the wedge is pushed in. The acute bisectrix is therefore α and muscovite is negative.

Now study progressively thinner and thinner flakes of muscovite, and note the diminution in the number of rings visible in the field with decreasing thickness. The distance between the melatopes in the 45° position does not vary, however (for a given lens system), because this depends solely on the optic axial angle.

Make permanent mounts of the sections you have studied.

The value of 2E for muscovite is normally between 70° and 80°. An example of a larger optic axial angle is afforded by a cleavage plate of topaz (sign positive), for which 2E varies between about 90° and 120°. With normal equipment this means that the melatopes will lie outside or just on the border of the field. Smaller optic axial angles are shown by biotite (previous experiment), by the

* As explained previously, the cross-wires cannot be seen at the same time as the interference figure, unless the microscope is fitted with a Bertrand lens, but their approximate positions can easily be memorised.

stable form of potassium nitrate (Expt. 9), or by sections of aragonite cut parallel to (001) (2E = 31°; sign negative). The last may be obtained ready mounted from dealers in petrological specimens.

Examples of figures given by biaxial crystals having other orientations than that given above will be encountered in the experiments which follow.

EXPERIMENT 7. *To study crystals of zinc sulphate heptahydrate* $(ZnSO_4, 7H_2O)$

This compound changes to the hexahydrate at 39° and so it must by crystallised below this temperature. Shake up an excess of the salt with 5 c.c. of water, warmed to 25°, until no more appears to dissolve. Place a drop of the clear solution on a slide, and watch its crystallisation under a low-power objective. (If crystallisation does not start after a few minutes, proceed as explained on p. 201.) When the crystals have grown large enough for convenient observation, arrest further evaporation by placing a cover slip over the drop.

The crystals grow as long plates or prisms, which between crossed Nicols give straight extinction with reference to their length direction. A systematic examination of several crystals on the lines given on pp. 354–6 will reveal two types of sections, one in which the extinction is sharp and complete, and another in which it is only partial. In general the polarisation colours of the first type are much lower than those of the second. The reason for this will be explained presently.

The sign of elongation of crystals of the first type, determined by the mica or gypsum plate, will be found to be negative. In convergent light they give an acute bisectrix figure, the optic axial plane being normal to the direction of elongation of the crystal, as may be seen from the relative orientations of the crystal and figure to one another. It is probable that the crystals will be too thin to show more than a single ring of colour in the field of the interference figure, but the behaviour of the brushes leaves no doubt as to its nature. Determine the optical sign from the figure by means of the gypsum plate (the quartz wedge will probably not give clear results with such thin crystals); it will prove to be negative. The direction of elongation is therefore β, and the transverse direction γ, which agrees with the negative sign of elongation found previously.

Crystals of the second type will be found to have positive elongation. In convergent light they give an inclined single optic axial figure. From the direction of curvature of the brush in the 45° position, the direction in which the acute bisectrix lies may be deduced (see p. 290), and this will show that the optic axial plane is normal to the direction of elongation as in the crystals of the first type. From this it may be concluded that the two types are elongated along the same axis, but occupy different positions of rotation about this axis. The optical sign of the substance may be found from the inclined figures, and should agree with the result obtained from crystals of the first type.

The reason for the difference in the sign of elongation between the two types is that crystals of the first present $\beta\gamma$ sections, whilst those of the second present sections having as indices β and a value between β and α. The low polarisation colours of the first type are due to the fact that the difference between β and γ is small, the indices being α 1·4568, β 1·4801, γ 1·4836 (D line).

Now obtain a centred, or nearly centred optic axial figure for the purpose of estimating the optic axial angle from the curvature of the brush (see p. 312). To do this, crush to a coarse powder a large crystal of the substance, obtained either from the stock bottle, or by the slow evaporation of a saturated solution. Mount a little of this powder in a drop of any non-aqueous immersion liquid between a slide and a cover slip. Search between crossed Nicols for those fragments that show a faint uniform illumination on rotation of the stage. These will give the required figures. The mean value of 2V (white light) is 46°.

$ZnSO_4,7H_2O$ belongs to the orthorhombic system.

EXPERIMENT 8. *To study crystals of copper sulphate pentahydrate* $(CuSO_4,5H_2O)$

Recrystallise the salt from a drop of solution on a slide, and examine the crystals in their mother liquor; all, or nearly all will be found to present the section shown in Fig. 281. This section is parallel to $1\bar{1}0$ or $\bar{1}10$. The figure shows the simple outline of the crystal, faces which truncate the edges and corners being omitted. The edge angle given is correct to $\pm 1°$. The extinction angle given is a mean value obtained in white light; the extinction dispersion of this section is considerable, and monochromatic light must be used to get a precise result.

EXPERIMENTS

The directions of fast and slow vibration may be determined on thin crystals (in which the colour of the substance shows but faintly) with the ¼-wave mica plate. On thicker crystals these directions may be determined by noting the effect of the quartz wedge on the colour bands at the wedge-shaped borders of the crystals.

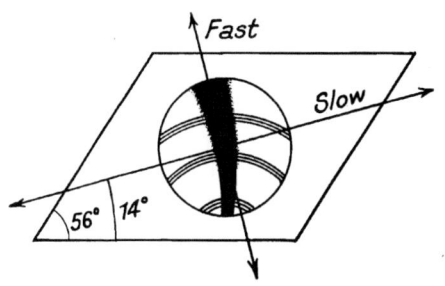

FIG. 281.—Typical Section of $CuSO_4,5H_2O$ (figure in extinction position).

The section shows an inclined single optic axial figure, the emergence of the optic axis lying just outside the field.

Now crush a large crystal of the salt into a coarse powder, mount in a non-aqueous immersion medium, and find crystals giving figures from which the optical sign may be determined, and the optic axial angle estimated. The sign is negative, and 2V (mean value for white light) is 56°.

$CuSO_4,5H_2O$ belongs to the triclinic system.

EXPERIMENT 9. *To study the dimorphism of potassium nitrate*

Saturate 5 c.c. of water with potassium nitrate at about 60°. Let the solution cool to 35°–40°, and then place a drop on a warmed slide. Watch the crystallisation under the microscope with the 1″ objective. Crystallisation usually starts at the edges of the drop in the form of rhombohedra which belong to the trigonal system. This form is unstable at ordinary températures. The stable form usually comes out a little later as long needles parallel to c, which are orthorhombic. Whenever a needle touches a rhombohedron, the latter loses its sharp outline and becomes "fuzzy", because it is converted into minute crystals of the stable form. Generally, however, some of the rhombohedra escape contact with the needles and can be examined at leisure.

Note that the needles give straight extinction with reference to their length, and negative elongation. They do not give satisfactory interference figures, possibly because of twinning on (110), but these may be obtained by crushing large crystals as described for zinc and copper sulphates (Expts. 7 and 8). Acute bisectrix figures may be obtained in this way, and show that the optic axial angle is extremely small; in fact, very careful observation is needed to detect the separation of the brushes in the 45° position. (2V for sodium light is 7°.) The optical sign is best determined with the mica plate, and is negative.

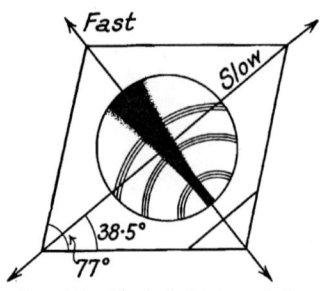

FIG. 282.—Typical Section of the Unstable Form of KNO_3.
The interference figure depicted is that given in the extinction position.

The unstable rhombohedra do not give straight extinction with reference to either of their principal edge directions, but those lying perfectly flat give the symmetrical extinction which is characteristic of the rhombohedral form of the trigonal system. The edge angles and directions of fast and slow vibration of these crystals are as shown in Fig. 282. In convergent light they give greatly inclined uniaxial figures, the emergence of the optic axis being just outside the field of view. Sufficient of the figure is visible for the sign to be determined with the mica plate; it is negative.

EXPERIMENT 10. *To determine the principal refractive indices of a uniaxial crystal*

A convenient example is ammonium dihydrogen phosphate, $NH_4H_2PO_4$, which is tetragonal and crystallises in four-sided prisms parallel to the optic axis, with pyramidal terminations, as shown in Fig. 283. These crystals, almost without exception, lie on a slide on their sides, and so present $\varepsilon\omega$ sections. Occasionally, crystals in which elongation is not marked lie on one of the pyramid faces. When the salt is powdered, many of the fragments lie so as to present $\varepsilon\omega$ sections. It is therefore easy to determine ε and ω either (i) on well-formed crystals, or (ii) on fragments obtained by powdering coarse material (see p. 357).

If method (i) is adopted, the substance should be recrystallised

EXPERIMENTS 375

by rapidly cooling a hot saturated solution while stirring it mechanically (see p. 206). The microcrystalline powder so obtained should be filtered on a Büchner funnel, and dried between wads of filter paper.

Mount a minute portion of the powder in a liquid of index 1·47. With a low-power objective, it will be seen that the majority of the crystals have the shape depicted in Fig. 283, and are lying on their sides. Select such a crystal which is lying perfectly flat, and focus it with a high-power objective (*e.g.* ¼"). Bring the polariser into position, lower the condenser, or partly close the substage diaphragm. Insert the analyser and bring the crystal into an extinction position. Remove the analyser and rack up the microscope tube slightly. The Becke line moves into the crystal. Rotate the stage through 90°, and again carry out the Becke test. The Becke line moves in once more. Both indices of the crystal are therefore greater than that of the liquid. It will be observed that when the light is vibrating in the direction of the optic axis, *i.e.* the direction of elongation, the crystal outline is much fainter than when the light is vibrating across the crystal. ε is therefore much closer to the index of the liquid than is ω. Repeat the above tests on several more crystals of the same type and orientation; the same results should be obtained in every case.

Now mount further portions of the powder in liquids having indices ranging from 1·48 to 1·54, and repeat the above determinations. In white light it will be found that ε is most nearly matched by the liquid of index 1·48, whilst ω falls between 1·52 and 1·53. The use of sodium light and intermediate liquids will enable determinations for the D line to be made, as described in the previous experiment. The correct values for the D line are ε 1·479, ω 1·525.

Fig. 283.—Crystal of $NH_4H_2PO_4$.

The fact that ω is greater than ε may be checked by determining the sign of elongation of the crystals; it will prove to be negative. Also if a few large crystals of the substance be coarsely crushed, fragments may be found giving uniaxial figures from which the optical sign of the substance may be shown to be negative. That the crystals studied above are indeed $\varepsilon\omega$ sections may be proved by their giving optic normal interference figures. These show up particularly well in sodium light.

To carry out method (*ii*), powder a few large crystals of the salt in a small pestle and mortar, until the average diameter of the fragments has been reduced to about 0·05 mm. (see p. 198). Mount minute portions of the powder, each about the volume of a pin's head, in liquids ranging in index from 1·48 to 1·54. In each mount carry out the Becke test in two adjacent extinction positions on several of the fragments that show the highest polarisation colours. The highest and lowest indices so found, ω and ε respectively, should agree with the values given under method (*i*). All fragments, whatever their orientation, should show ω as one of their indices, and if fragments are found which do not satisfy this condition, they must consist of an impurity. Commercial specimens of the salt sometimes contain the diammonium salt, which is biaxial.

In this and other determinations of the refractive indices of anisotropic crystals, the van der Kolk test can be used in place of the Becke test.

Potassium dihydrogen phosphate, KH_2PO_4, may be used for the above experiment instead of ammonium dihydrogen phosphate. It crystallises in the same habit, and its refractive indices (D) are : ε 1·468 ; ω 1·509.

EXPERIMENT 11. *To determine the principal refractive indices of a biaxial crystal*

In biaxial crystals, as prepared for microscopic examination, it rarely happens that the habit is such that all the principal refractive indices can be measured directly on well-developed faces, *i.e.* with the crystals lying in stable positions. In orthorhombic and monoclinic crystals, one or two of the principal indices may be shown by the sections commonly presented, and this will be revealed by their interference figures, but the determination of the remaining index or indices requires that the crystal be supported in an unstable position. An example is afforded by *m*-nitroaniline (Fig. 285). Well-developed triclinic crystals will only present principal refractive indices fortuitously. Method (*ii*) as described in the previous experiment is therefore more generally applicable for biaxial crystals.

A convenient example is copper sulphate pentahydrate, $CuSO_4,5H_2O$. The fragments obtained by powdering large crystals of this salt lie on a slide in a random manner. Crush one or two large crystals of the salt to a powder of the average particle size of 0·05 mm., and mount minute portions of the powder in liquids ranging in index from 1·50 to 1·55. In each mount select several fragments showing the highest polarisation colours, and determine on each fragment the relation of its two refractive indices to that of the liquid by means of the Becke test. Try to roll each fragment studied into other positions by moving the cover slip with a needle, and repeat the test. The lowest and highest indices so found in the whole series of mounts should be 1·51 and 1·54 respectively for white light. To obtain β, search for fragments showing between crossed Nicols a uniform, or nearly uniform illumination on rotating the stage and single optic axial figures. In white light these fragments should be matched by a liquid of index between 1·53 and 1·54.

Results for the D line may be obtained by the use of sodium light and intermediate liquids, the correct values being α 1·514, β 1·537, γ 1·543.

EXPERIMENT 12. *To distinguish a mixture of barium nitrate and lead chloride from a mixture of barium chloride and lead nitrate*

No distinction between these two mixtures is possible by qualitative chemical analysis, and the object of this experiment is to show how this can be effected by the optical method. The optical properties of the four salts which suffice to distinguish them are as follows :

$Ba(NO_3)_2$ Cubic, n_D 1·571.
$Pb(NO_3)_2$ Cubic, n_D 1·7815.
$BaCl_2,2H_2O$ Birefringent (monoclinic); α 1·635, γ 1·660 (D).
$PbCl_2$ Birefringent (orthorhombic); α 2·199, γ 2·260 (D).

Make up a mixture of barium nitrate and lead chloride in approximately equal proportions. Crush to a powder of 0·1 to 0·05 mm. particle size (the lead chloride, if taken from stock, will probably be already in powder form), and mount a little in a liquid of index not exceeding 1·55. Examination between crossed Nicols will show that some of the fragments are isotropic, whilst others are birefringent. If the Becke test be applied, it will be found that all fragments on the slide have indices greater than that of the liquid.

Now mount some of the powder in a liquid of index 1·57. The isotropic fragments will be practically invisible in this, but the birefringent ones will still stand out clearly, the Becke test showing that their indices are all greater than that of the liquid. Repeat with a liquid of index 1·74 ; the indices of the birefringent fragments are still greater than that of the liquid, whilst that of the isotropic fragments is less. The isotropic substance is barium nitrate, and the birefringent substance is lead chloride. (The lead chloride supplied commercially often contains basic chlorides, but these too are birefringent and have very high refractive indices, and so do not interfere with the test.)

Next make and powder a mixture of barium chloride ($2H_2O$) and lead nitrate. In this the isotropic material (lead nitrate) will be found to have an index greater than 1·74 (*cf.* previous mixture), whilst the birefringent barium chloride will show indices between 1·63 and 1·66.

EXPERIMENT 13. *To study crystals of o-, m-, and p-nitroanilines*

This experiment is an example of the use of the optical method to distinguish organic isomerides from one another, and also introduces certain manipulative methods that are of value in dealing with organic substances in general.

o-Nitroaniline exists in three monoclinic modifications, viz. α form, m.p. 68°, β form, m.p. 70°, γ form, m.p. 71°.[*] The γ form, which is the most stable, is deposited from alcohol or alcohol-water solutions as small prisms. Any specimen of the substance which has not been recently melted will probably consist entirely of this form, but for an optical examination it should be recrystallised as follows. Saturate at 40° a mixture of equal volumes of alcohol and water with the substance. (Solutions saturated at higher temperatures may deposit liquid *o*-nitroaniline.) Let the solution cool slowly, meanwhile stirring it mechanically. Filter off the crystal powder, and dry on a porous plate.

The crystals so obtained should be mounted in glycerine. *o*-Nitroaniline is soluble in the common organic liquids and in water, but it may be studied in glycerine because it dissolves in this liquid very slowly.

Note first the pleochroism of the crystals. Some will show none, remaining orange at all positions of rotation of the stage, whilst others will be orange when the light is vibrating parallel to the direction of elongation, and yellow in the position normal to this.

[*] Dippy and Hartshorne, *J. Chem. Soc.*, 1930, 725 ; Hartshorne and Stuart, *ibid.*, 1931, 2583.

Between crossed Nicols the non-pleochroic sections will give straight extinction with reference to their length, and the pleochroic sections will give oblique extinction having a maximum value of 7° in sodium light. In white light the extinction of these pleochroic sections is not perfect, for they only change to a dark green colour. This is due to dispersion of the extinction position, which is very marked in this form of the substance. The whole of the above behaviour indicates that the crystals are monoclinic and are elongated along either the a or the c axis. When lying so that the ac plane is vertical, they give straight extinction, and no dispersion of the extinction position, but for other positions of rotation about the axis of elongation they give oblique extinction and dispersion of the extinction position, both of which properties reach maximum values when the ac plane is horizontal, $i.e.$ normal to the path of the light. Reference to Fig. 153, p. 122, should make this quite clear.

The α and β forms of o-nitroaniline may be obtained as follows. A little of the substance is melted between a slide and a cover slip (see p. 208). The slide is then removed from the hot plate and placed in air heated to about 30°, either by holding it at the right distance from a bunsen burner, or by placing it in an air oven previously adjusted to this temperature. (If the slide is allowed to radiate to the room temperature, an undue amount of the α form is apt to crystallise out.) When solidification is complete, it will be seen, on holding the slide up to the light, that crystallisation has taken the form of fine needles radiating in all directions from a number of centres, $i.e.$ spherulites, and that the needles around some centres are orange, whilst those around the others are yellowish-green. The former are β crystals, the latter α crystals. The needles may be too narrow for the convenient determination of their optical properties, especially in the case of the α form, and if so, broader needles may be obtained by remelting a portion of the film as described on p. 209. A melt is thus obtained which is bordered on one side by a strip of solid consisting of both α and β spherulites, and on cooling, each form inoculates its own species from the melt.

Examine the α crystals first because they change rather quickly into the β form. This change can be readily detected while the optical examination is proceeding; it takes the form of an extension of the areas occupied by the β form, and the appearance of small patches of β form in the areas occupied by the α form. The β form changes into the γ form much more slowly.

The α crystals show both straight and oblique extinction with reference to their direction of elongation, crystals of the latter type being usually encountered the more frequently. The maximum extinction angle is 21°, and those crystals which have this value show weak pleochroism, being pale yellow when the light is vibrating parallel to the direction of elongation, and yellow when it is vibrating normally to this. The pleochroism becomes more marked with decrease in the extinction angle, and finally in the case of crystals giving straight extinction it is pale yellow for light vibrating along the length direction and orange for light vibrating across it. The reason that both straight and oblique extinctions are shown is the same as in the case of the β form, namely that the crystals are monoclinic and are elongated along the a or c axis. In this case, however, it will be noticed that there is no appreciable dispersion of the extinction position when the crystals show oblique extinction.

None of the β needles shows oblique extinction, although they are monoclinic like the other forms. This is because their direction of elongation is the b axis, so that no matter how they be turned about this direction, the ac plane is always vertical. Two types of pleochroism are shown by these needles; for light parallel to the length direction they all show an orange colour, but for light vibrating transversely, some crystals show a darker orange, whilst others show a yellow colour. Crystals of the latter type, which are large enough, give obtuse bisectrix figures in white light. This is distinctive for the β form, since clear figures are only given by the other two forms in monochromatic light owing to marked dispersion.

In the course of several days the β form on the slide becomes transformed more or less completely into the γ form, but the resulting crystals are much too small for optical examination. Between crossed Nicols those areas of the film that have been so transformed merely show confused reds and greens. Measurable γ crystals can, however, be obtained by the transformation of β crystals, if the slide is maintained at about 60° for some time. After a day or two at this temperature there appear large areas of the γ form showing fine multiple twinning, and the green extinction to which reference has already been made. The laminæ in these areas show the maximum extinction of 7° (sodium light) with reference to their direction of elongation, the extinction positions being inclined in opposite directions in adjacent laminæ. Crystals with the ac plane vertical, and therefore showing straight extinction

EXPERIMENTS 381

and no extinction dispersion, may also be found on the slide. Each of these two sections will show the same pleochroism as the corresponding type of γ crystal obtained from solution.

The monoclinic symmetry of the β form can be revealed as follows, and although it is not necessary to do this to distinguish between the nitroanilines, the experiment forms an instructive exercise in manipulation and should be carried out by the beginner. It is first necessary to prepare isolated crystals of the β form, for which purpose use is made of the fact that when o-nitroaniline is melted in a tube and slowly cooled, the crystals first obtained on solidification are almost entirely those of the β form.

Melt completely a few grams of the substance in a lightly corked test-tube by placing the latter in a large beaker of hot water. Then allow the water to cool very slowly, and when about a quarter of

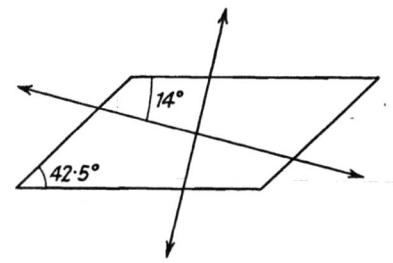

Fig. 284.—*ac* Section of β *o*-Nitroaniline.

the nitroaniline has crystallised, remove the test-tube and quickly pour off the liquid from the crystals. When cold, extract the solid from the tube by breaking the glass around it, and place it on a sheet of white paper. Examine it with a hand lens; it will probably be found to consist of a compact ground mass from which project a number of elongated tabular crystals, 1 to 2 mm. in length. Detach some of the largest of these with a needle, meanwhile holding the lump of solid with a pair of forceps or crucible tongs. If no suitable crystals can be found, the experiment must be repeated with a slower rate of cooling.

Transfer the crystals to a microscope slide, but add neither liquid nor cover slip. In the position which they naturally adopt, they will be found to give the obtuse bisectrix figure and pleochroism characteristic of the second type of β needle found on the slide prepared by melting the substance. Note, however, that the

382   METHODS OF ATTACK AND EXPERIMENTS

figure is not centred, though the crystals are obviously lying on a principal face, the optic axial plane being slightly tilted laterally.

It is now necessary to prepare transverse sections of the crystals. This may be done by means of a razor, the crystal being embedded in hazel pith, as described on p. 217. The sections so prepared are mounted, without removing them from the pith, in glycerine. Those that have been cut successfully will show the outline and oblique extinction of 14° (sodium light) which are depicted in Fig. 284. Dispersion of the extinction position is observed in white light. These results taken in conjunction with the previously determined optical properties of the β form show that it is monoclinic.

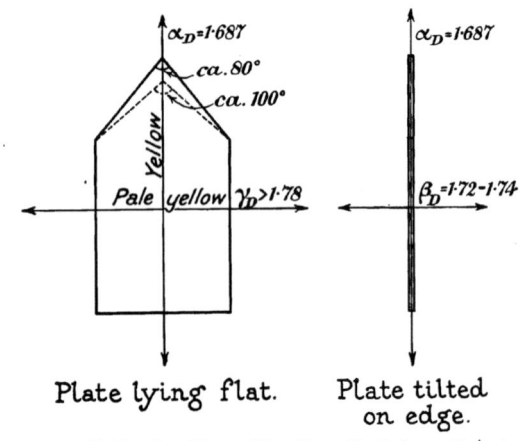

FIG. 285.—Platy Habit of m-Nitroaniline Crystallised from Alcohol or Carbon Tetrachloride.

m-Nitroaniline belongs to the orthorhombic system. Recrystallise a little of the substance from either alcohol or carbon tetrachloride, stirring the cooling solution mechanically. Filter off and dry in the usual way. The crystals may be mounted in the ordinary immersion liquids, since they do not dissolve in these appreciably during the course of the optical examination (a distinction from o-nitroaniline). They are lemon yellow plates having a characteristic hemimorphic shape, being pointed at one end and rectangular at the other. The value of the apex angle of the pointed end is about 100° in some cases and about 80° in others, but all plates show the same optical characters. The pleochroism is weak but

EXPERIMENTS

quite definite, the extinction straight, and the elongation negative. The last property may be determined on plates which are thin enough for their colour not to interfere with the polarisation colours. The plates give centred optic normal figures and are therefore $\alpha\gamma$ sections. β may be determined approximately on plates tilted on edge in mounts containing glass powder (see p. 213). The above properties are conveniently summarised in Fig. 285.

p-Nitroaniline is monoclinic. Crystals obtained from either alcohol, carbon tetrachloride, or benzene by the method described for *m*-nitroaniline or from thymol (see p. 205), develop as yellow six-sided plates, showing straight extinction because they are normal to the *ac* plane. They are weakly pleochroic, transmitting greenish yellow for light vibrating parallel to the length and yellow for the transverse direction. They show positive elongation and inclined single optic axial figures. The brush in the figure is practically straight in all positions of the stage, and the optic axial angle is therefore near 90°. The marked dispersion of the optic axis should be noted. The optic axial plane is normal to the direction of elongation, and the index corresponding to the vibration direction which is parallel to the direction of elongation is therefore β. The above properties together with the edge angles and refractive indices are summarised in Fig. 286.

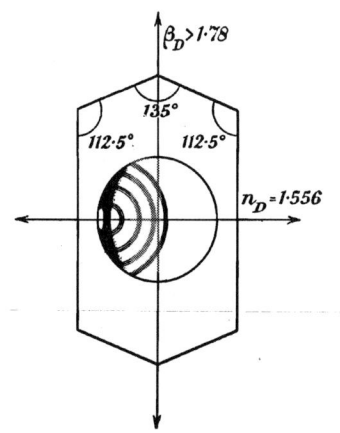

Fig. 286.—Platy Habit of *p*-Nitroaniline Crystallised from Alcohol, Carbon Tetrachloride, Benzene, or Thymol. The interference figure depicted is that given in the 45° position.

Additional substances, suitable for study by the beginner, are given below, with brief notes on their more important morphological and optical characters. The habits described are those developed when they are crystallised from a drop of aqueous solution on a slide.

(1) *Sodium nitrate, NaNO$_3$:* Trigonal. Rhombohedra. Acute profile angle 78°. Perfect rhombohedral cleavage. Isomorphous with the unstable form of KNO$_3$, with optical properties qualitatively similar (see Expt. 9). ε_D 1·336, ω_D 1·585. ε_D' for sections commonly presented, *i.e.* the extraordinary index for this section, 1·467.

(2) *Potassium sulphate, K$_2$SO$_4$:* Orthorhombic. Elongated parallel to *a*, and commonly lying on (021) or (010). Other forms commonly observed are {100}, {001}, {110}, {130}, {011}, and {111}. Z = *c*. Sign positive. 2V = 67°. Optic axial plane (100). α_D 1·493, β_D 1·495, γ_D 1·497. Double refraction weak, $(\gamma - \alpha) = 0·004$, and therefore microscopic crystals rarely show polarisation colours higher than first order white. Interference figures shown by sections commonly presented: (021), inclined single optic axial; (010), obtuse bisectrix.

(3) *Potassium chlorate, KClO$_3$:* Monoclinic. Diamond-shaped tablets parallel to (001), with an acute profile angle of 80°. Perfect (001) and (110) cleavages. Pseudo-trigonal and closely related to NaNO$_3$ (see above). May show lamellar twinning on (001). Z parallel to *b*. Optic axial plane normal to (010). X^*c* = 56° in acute angle β (80°). Sign negative. 2V = 27°. The interference figure on the characteristic (001) section shows the emergence of the acute bisectrix at the margin of the field, with objectives of large N.A. (at least 0·85). With smaller apertures, the appearance is like that of an inclined uniaxial figure, similar to the one given by sodium nitrate (see above), owing to the small optic axial angle. $\rho > \nu$, with horizontal dispersion. α_D 1·408, β_D 1·517, γ_D 1·523.

(4) *Barium chloride dihydrate, BaCl$_2$, 2H$_2$O:* Monoclinic. Tablets parallel to (010), bounded by (100), (001), and (101), with angle β 89° and edge angle (100):(101) = 136°. Commonly twinned on (100) or (001). No good cleavage. Axial plane (010). Z^*c* = 8° in acute angle β. Sign positive. 2V = 84°. α_D 1·635, β_D 1·646, γ_D 1·660.

(5) *Potassium dichromate, K$_2$Cr$_2$O$_7$:* Triclinic. Tablets parallel to (010). The angle $\beta = 91°$, therefore the appearance is sensibly rectangular. Cleavage (010) perfect, (100) and (001) distinct. Commonly twinned on (010). At first sight such crystals appear to be single, but the lower individual can often be seen projecting beyond the upper one; they give imperfect extinction and a confused interference figure. The untwinned crystals present the

EXPERIMENTS 385

characters shown in Fig. 287. Axial plane nearly normal to (001) between (100) and (0$\bar{1}$0). Sign positive. $2V = 52°$. $\rho > \nu$. Markedly pleochroic (see figure). α_D 1·720, β_D 1·728, γ_D 1·820.

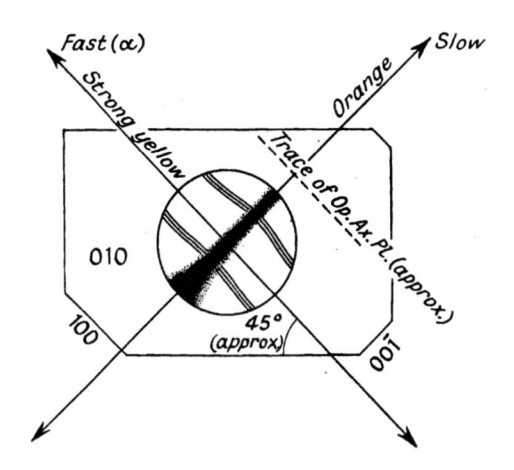

FIG. 287.—Typical Section of $K_2Cr_2O_7$ parallel to (010).
The interference figure depicted is that given in the extinction position.

CHAPTER XI

SPECIAL METHODS

The methods described in the previous chapters are quite adequate for most optical crystallographic work. Occasionally, however, problems occur which need more refined methods for their solution. It may be found, for instance, that the determination of refractive indices to the usual limits of ± 0.002 is not critical enough, and it may then be necessary to resort to *index variation methods* in which either the temperature of the immersion liquid or the wave-length of the illumination, or both, may be varied, when under optimum conditions an accuracy of the order of ± 0.0005 may be possible. Again, the material under examination may, because of an extreme habit, persistently adopt a particular orientation making the determination of principal refractive indices, optic axial angle, and the orientation of the ellipsoid impossible in the absence of any means of turning the crystal into other positions. Simple methods of doing this have already been described in Chapter VI, but the most accurate method involves the use of the *universal stage* by means of which sections of known orientation with respect to the one persistently presented may be examined. Measurements of the dispersion of the refractive indices, and of the dispersion of the birefringence of a substance (a property which appears to be a useful diagnostic character for certain substances *) demand index variation methods for their determination under the microscope. The combination of these methods with the universal stage provides a powerful weapon of attack in many optical problems.

The advantages and disadvantages of each method need to be assessed in relation to the problem in hand, and it in no way follows that the use of more elaborate equipment (which also calls for a long and rigorous apprenticeship) will necessarily be justified in every case.

* A. N. Winchell, *Amer. Min.* 1947, **32**, 336.

Index Variation Methods

General Considerations. The refractive index of any substance is a function of its temperature and of the wave-length of the light. Both solids and liquids have higher refractive indices for shorter wave-lengths, *i.e.* refractive indices are higher for blue light than for red. In general, the change in index resulting from change of wave-length is greater in liquids than in solids. The effect of a rise in temperature is to lower the indices of both solids and liquids, but the change is far greater in liquids than in solids. A rise of, say, 40° C. would affect the refractive index of a liquid in the second place of decimals but would only result in an alteration in the refractive index of the solid in the fourth or fifth place. For all practical purposes, therefore, the limits of temperature within which optical crystallographic work is done leave the refractive indices of solids effectively constant, while producing a marked change in the refractive index of immersion liquids. These facts are illustrated in the dispersion diagram shown in Fig. 288. The heavy lines show the dispersion curves for a flint glass and quartz respectively, while the thinner broken and full lines show the effect upon two liquids of changing both temperature and wave-length. Each broken line, for example, shows the dispersion of *o*-nitrotoluene due to change of wave-length at a definite temperature. The dispersion, $n_F - n_C$, *i.e.* the difference between the refractive index for the F line and that for the C line, is noticeably greater for these liquids than for the solids. It can be seen that at a given temperature a solid may have the same refractive index for a definite wave-length as a liquid in which it is immersed. For example, at X in the figure, the flint glass matches the liquid at a wave-length in the yellow-green, the temperature being 30° C. For wave-lengths less than this value (towards the blue) the solid has a lower refractive index than the liquid, and for longer wave-lengths (towards the red) a higher refractive index. Some idea of the order of the relative effect of temperature and wave-length changes on solids and liquids may also be gathered from the diagram. For the two liquids referred to, *o*-toluidine and *o*-nitrotoluene, a change of 40° C. alters the refractive indices by 0·02, the dispersion at various temperatures being almost constant as is witnessed by the parallelism of the dispersion curves. A change in wave-length from 4861 Å (F line), to 6563 Å (C line), results in a change in refractive index of about the same amount, namely 0·02. Each of these liquids therefore can be made, by temperature and wave-length changes, to cover a range of about 0·04 in refractive index. These

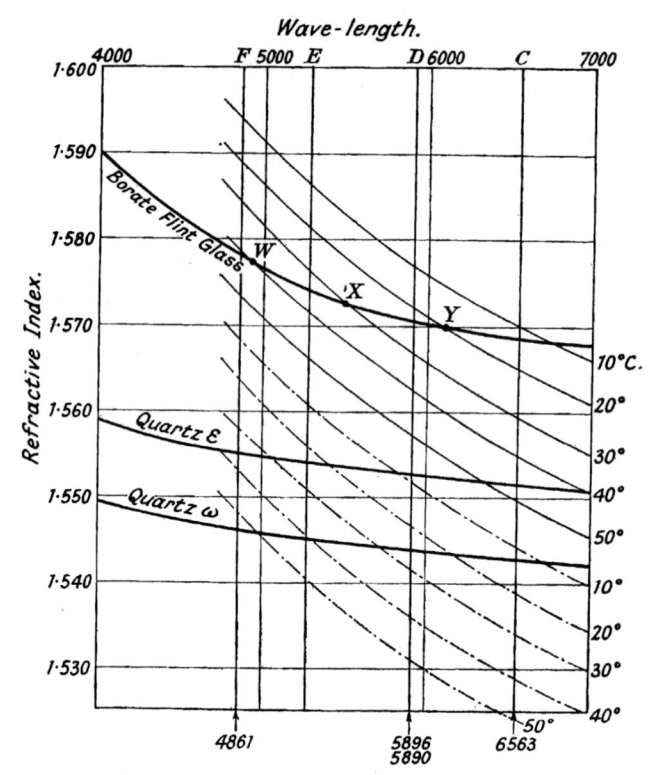

FIG. 288.—Dispersion Curves of Solids and Liquids.
Broken curves refer to *o*-nitrotoluene; solid curves (above) to *o*-toluidine.

effects are about average for the set of immersion liquids recommended by Emmons for variation work (see below).

The application of these principles to the determination of refractive indices under the microscope was first suggested by Merwin [*] and adopted by Tsuboi [†] who used wave-length changes

---

[*] *J. Amer. Chem. Soc.*, 1922, **44**, 1970.
[†] *J. Geol. Soc. Tokyo*, 1925, **32**, and *Min. Mag.*, 1925, **20**, 108.

to effect a match between liquid and crystal. Emmons * varied the temperature of the immersion liquid to achieve the same object. These methods are known as *single variation methods* because one method only is used in order to vary the refractive index. Later, Emmons † combined both means of changing the index in the *double variation method*.

The advantages of variation methods are obvious. Fewer liquids are necessary to cover a given range of refractive indices ; time is saved because fewer mounts need be made ; a greater accuracy in refractive index determinations is possible ; and in the double variation method the dispersion of the refractive indices can be determined. A disadvantage is the added expense of the apparatus. For the single variation method of Tsuboi a direct reading monochromator is required because continuous variation of wave-length is necessary. The cost of this instrument might well be prohibitive for some workers. Again, unless the various units of the variation apparatus (arc, monochromator, refractometer, etc.) can remain set up ready for use, time (which might be sufficient to solve a problem by less elaborate methods) is spent in assembling and "tuning up". It is much cheaper, simpler and just as satisfactory to effect temperature changes rather than those of wave-lengths, and so the single variation method of Emmons will commend itself to most workers, and will be described below. For the double variation method it is not absolutely necessary to have a monochromator, as reasonably accurate results can be got by using a series of "monochromatic" filters (p. 187). These do not give a continuous series of wave-lengths throughout the spectrum, but the adjustment necessary to effect a match at any selected wave-length is made by varying the temperature.

Single Variation Method by Temperature Control. In this method the substance to be examined is mounted in the usual way on the upper thin glass window of a cell through which flows a stream of water, the temperature of which can be controlled, and the crystals are viewed in transmitted light admitted through the lower window of the cell. The essential feature of this cell is that the upper window must be very thin so that the immersion liquid of the mount will very closely approximate to the temperature of the circulating water within the cell. The set-up of the apparatus is

* *Amer. Min.*, 1926, **11**, 115.
† R. C. Emmons, *ibid.*, 1928, **13**, 504 ; 1929, **14**, 415.

Fig. 289.

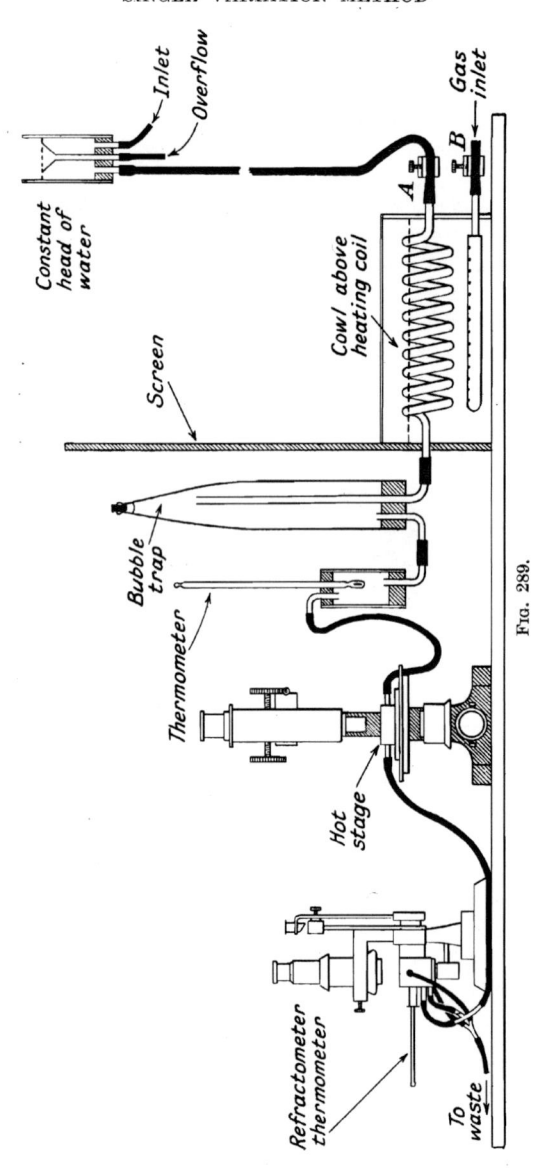

shown diagrammatically in Fig. 289. The water flows from a constant head into a heater made of a coil of copper tube of $\frac{3}{16}''$ bore. Heating is done by means of gas jets under the coil, which has ten turns of 3″ diameter. From the coil the water passes into a bubble trap, conveniently made from an elutriator tube, 18″ long and 2″ diameter at the wide end. This is inverted, narrow end uppermost, the water being led nearly to the top by a glass tube, the outlet being at the bottom. The tap at the upper end of the trap is opened from time to time to release accumulated air. This trap makes a very efficient temperature " buffer ". From the trap the water circulates round a thermometer, then by means of rubber tubing through the cell on the microscope stage, and lastly round the prisms of an Abbé type refractometer, and so to the waste pipe. Emmons suggested controlling the temperature by mixing water

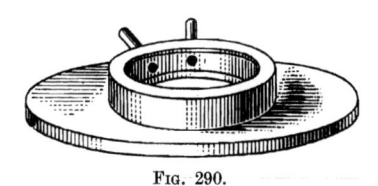

FIG. 290.

from two tanks, one hot and one cold. The above arrangement has been found more convenient and quite satisfactory. The temperature can be adjusted very delicately by varying the flow of gas and water by the screw clips on the rubber tubes as shown at A and B, Fig. 289. An improved heating and circulatory system for use in index variation procedure has been described by Hurlbut.[*] This furnishes water, thermostatically controlled to any desired temperature, and as distilled water is used over and over again by pumping, trouble from air bubbles is avoided.

Various types of cells have been designed. A difficulty at once apparent to the worker who uses a cell of the type illustrated by Winchell [†] (Fig. 290) is that the height of the upper window above the microscope stage makes it impossible to obtain an interference figure in the normal way, and resort has to be made to the Johannsen bulb [‡] (p. 276). To overcome this difficulty, Saylor [§] designed a

[*] C. S. Hurlbut, *Amer. Min.*, 1947, **32**, 487.
[†] A. N. Winchell, *Elements of Optical Mineralogy*, 2nd Ed., 1931, 218.
[‡] A. Johannsen, *Manual of Petrographic Methods*, 1918, 455.
[§] C. P. Saylor, *J. Res. Nat. Bur. Standards*, 1935, **15**, 97.

thin cell for use in variation methods. This cell is about as thick as an object slide, and interference figures may be readily examined. The following description is taken from the paper referred to. Three pieces of brass, A, B, and C, 0·5, 0·9, and 0·2 mm. thick respectively are cut to shape as shown by the thick lines in Fig. 291. From B half the thickness of the plate is chiselled away from the upper side (as shown by the stippling), and from the underside, the

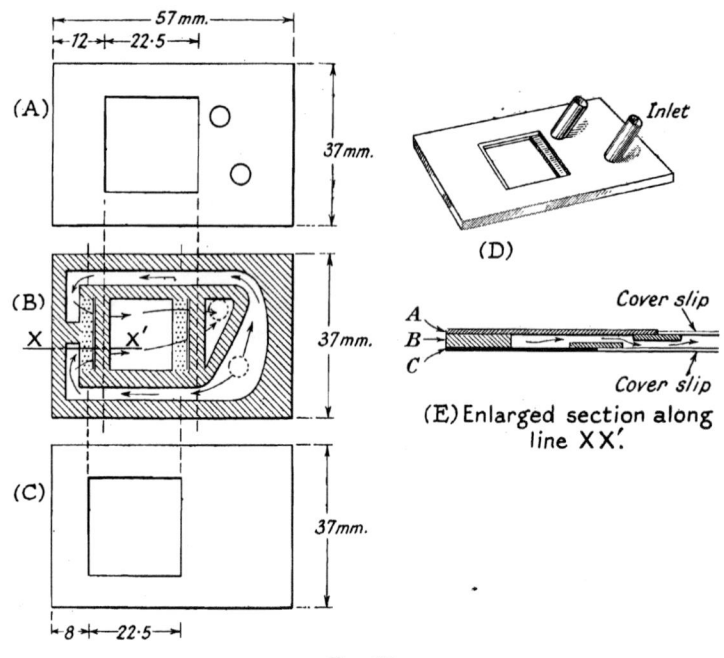

FIG. 291.

same amount is removed from the areas immediately to the right of these areas and the narrow slits (see Fig. 291, E). The three plates are thinly covered with solder and sweated together, inlet and outlet tubes are soldered on as shown, and the square openings above and below closed with No. 2 cover glasses cemented into position by means of a waterproof, heat-resisting cement such as Schaar & Co.'s Cementyte D, or "Durofix".

Another difficulty to overcome in all cells for this method is that the thin upper window breaks under cleaning, which is difficult to do effectively, and the immersion liquids tend to attack the cement around the edge of the window. To avoid these difficulties and to be able to use the thinnest possible glass commensurate with the necessary strength, one of us (A. S.) designed a cell which makes

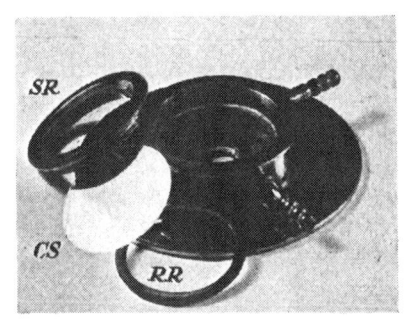

FIG. 292.—Warm Cell with Removable Upper Window.

the changing of the upper window a matter of a few seconds. Little need be said about the general construction, as the illustrations (Figs. 292, 293) are self-explanatory. The upper window consists of a circular No. 2 cover slip with a diameter of 32 mm. This slip rests upon a solid rubber ring with the same outer diameter as the cover slip. The rubber ring is of circular cross-section, which has

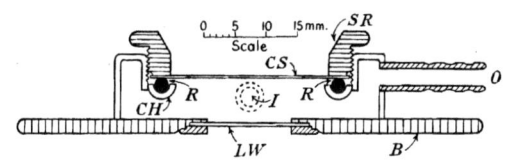

FIG. 293.—Section through Warm Cell (Fig. 292).

a diameter of 2·5 mm. This ring rests in a U-shaped channel, and a perfectly water-tight join results when the cover slip is gently clamped down by means of a screw ring above it.

For each determination a new slip may be inserted in a few moments by stopping the flow of water through the cell by means of the screw clip A (Fig. 289) which controls the supply, and then unscrewing the clamping ring (SR, Fig. 293). If a bubble trap is used the upper end of it should be closed, and no water will overflow from the cell. An air bubble finds its way into the cell during this operation, but it may be removed very simply by giving the cell a small tilt after restarting the flow of water. Vigfusson[*] has also described a thin cell, and one could be made like his, incorporating the window changing device of the one described above.

Ideal immersion liquids for index variation work should be pure, stable chemical compounds, not too volatile, with a high thermal variation in index. The following twenty liquids have been recommended by Emmons (*op. cit.*) for single variation work. A number of them are rather volatile and the mount needs replenishing often. Other liquids can be selected and used if their refractive indices are measured by means of a refractometer in series with the cell, or if the value for dn/dt has already been determined.

Single Variation Liquids.	n_D at 25° C.	dn/dt
1. Methylene iodide	1·737	0·00068
2. Methylene iodide + iodobenzene	1·715	0·00065
3. α-Iodonaphthalene	1·699	0·00047
4. α-Iodonaphthalene + α-bromonaphthalene	1·675	0·00047
5. *o*-Bromiodobenzene	1·661	0·00048
6. Phenylisothiocyanate	1·647	0·00056
7. *s*-Tetrabromethane	1·634	0·00053
8. Iodobenzene	1·616	0·00057
9. Bromoform	1·594	0·00060
10. Aniline	1·583	0·00052
11. *o*-Toluidine	1·569	0·00051
12. Nitrobenzene	1·549	0·00048
13. Ethylene bromide	1·535	0·00056
14. Propylene bromide	1·516	0·00054
15. Pentachlorethane	1·501	0·00048
16. Methyl furoate	1·485	0·00045
17. Methyl thiocyanate	1·466	0·00054
18. Isoamylsulphide	1·451	0·00045
19. Ethyldichloroacetate	1·434	0·00047
20. Ethylmonochloroacetate	1·419	0·00047

To make a determination proceed as follows: A rough preliminary determination of the refractive index is made in order to

[*] Vigfusson, *Amer. Min.*, 1940, 25, 763.

decide which of the index variation liquids to use. Estimate at this stage (1) whether the double refraction of the substance is low, medium, or high, and (2) if the last, it is as well to determine approximately the limits of the highest and lowest indices. If the birefringence is less than about 0·02 the chances are that one mount in a well chosen liquid will suffice to determine all the indices, whereas if the double refraction lies between this value and 0·04, at least two or possibly three liquids might be necessary, depending upon how the refractive indices of the substance " sit " on the index variation curves of the liquids. Fig. 294 illustrates these points.

After selecting the liquid to be used, a mount is made in the usual way on the upper window of the warm stage and a drop of the liquid placed on the refractometer at the same time. The water is made to circulate first at a low temperature, and as observations are made upon the crystals the temperature is raised in small steps by giving a slight turn to the screw control clips first gradually decreasing the flow of water. Soon certain of the crystals will begin to match in one extinction position while in the other extinction position they will show a lower refractive index, which can be matched by raising the temperature still further. On a match being obtained, the refractive index of the liquid is read off on the refractometer using the same light source as for the microscope. It is necessary to make sure when a match has been effected that the thermometers on either side of the warm stage agree at least to within one degree. The temperature of the stage is taken as the average of the two readings. It is not difficult, after the apparatus has had time to stabilise, at a given setting for the flow of water and gas to get the thermometers in good agreement.

It is useful to look for isotropic sections of uniaxial substances, or β-sections of biaxial substances, and if a cell is being used which makes it impossible to view interference figures, the statistical method (p. 318) of determining indices is necessary, or a Johannsen bulb (p. 276) can be used to find principal sections. The advantages of using a thin cell will be obvious.

It is possible to make use of this variation method without a refractometer in series with the cell. Indeed, for liquids with an index above 1·7, it may be necessary to do so, because a refractometer reading above this value may not be available. If pure liquids are used and the dn/dt values are known, the refractive indices may be plotted against temperature as in Fig. 294 and the refractive indices read off for the temperature at which a match

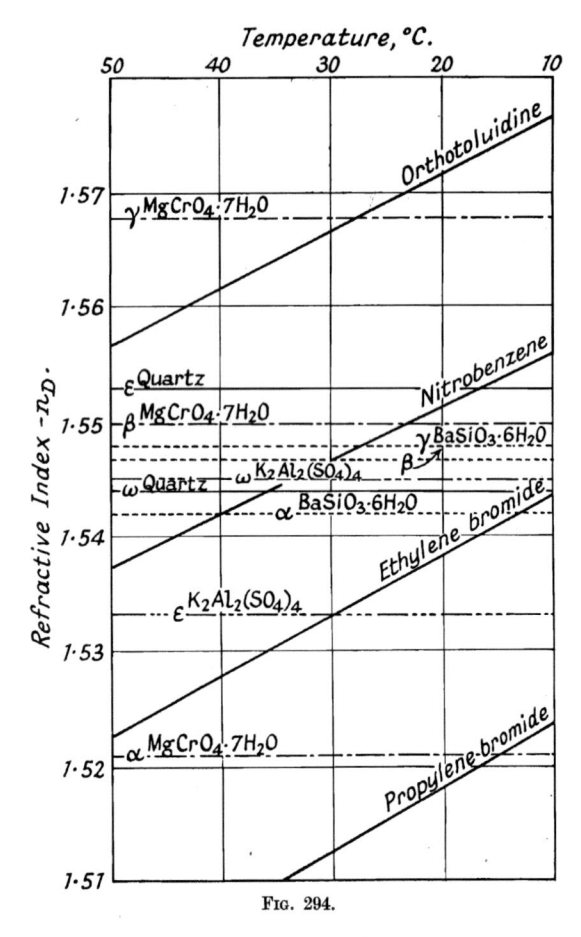

FIG. 294.

is obtained. It is, of course, necessary to use monochromatic light for temperature variation work (see p. 186).

An accuracy of at least ± 0.001 is possible with the single variation method although greater precision has been claimed.

The Double Variation Method. The general arrangement of

the apparatus is the same as for the single variation method just described, except that by the addition of a monochromator the wave-length of the incident illumination may be continuously varied throughout the visible spectrum, and the refractive index read off at the same wave-length when a match is made. Reference to Fig. 288 will remind the reader that by combining wave-length and temperature variation, each liquid can be made to cover a wider range of refractive indices and hence fewer liquids are required. If a monochromator is used an arc is necessary to give sufficiently bright illumination. However, if a monochromator is not available alternative expedients can be made to serve. A set of " monochromatic " filters arranged side by side in a slide so that they can be made to pass before a high-efficiency lamp will do, or isolation of the various lines of a mercury, or mercury-cadmium lamp by means of filters as described on p. 186 will give a series of separate but pure beams. In using a source of light giving such a discontinuous series of colours, the match must be made at each wave-length by altering the temperature in the appropriate direction. As refractometers of the Abbé type are constructed to give the correct value in sodium light, a correction must be made from tables supplied by the makers for other wave-lengths.

In making determinations by the double variation method the same preliminary examination is necessary in order to select a suitable immersion fluid, and the sample is mounted in the cell as before. It will be seen from Fig. 288 that if the temperature of the liquid (o-toluidine in this case) is 40° C., borate flint glass can be matched at a wave-length in the blue, of about 4895 Å, as at W. Now if the temperature is reduced to 30°, a match will be obtained at a longer wave-length, namely at 5500 Å (X), and if this process is repeated a match is made at 20° C. for a wave-length of about 6100 Å (Y). From these three points plotted on the dispersion diagram the dispersion curve of the solid can be drawn in, the refractive index of the solid for any wave-length found, and the dispersion $n_F - n_C$ estimated. The case just given as an example was a simple one in that only one refractive index at each wave-length and temperature had to be determined, but for crystalline substances two determinations must be made on each crystal section examined, as described in Chapter VIII. The two dispersion curves for the ω and ε rays for quartz are shown in Fig. 288. Grains of flint glass in an appropriate liquid, and quartz in o-nitro-toluidine make a very good pair of substances for a first try out of the method.

A list of double variation immersion liquids as recommended by Emmons (*op. cit.*) is given below.

Double Variation Liquids.	n_C at 50° C.	n_D at 50° C.	n_F at 50° C.	n_D at 25° C.	dn/dt.
1. Methylene iodide	1·711	1·721	1·747	1·737	0·00068
2. α-Iodonaphthalene	1·678	1·687	1·714	1·699	0·00047
3. α-Iodonaphthalene + α-bromonaphthalene	1·654	1·663	1·687	1·675	0·00047
3a. o-Bromoiodobenzene	1·640	1·647	1·665	1·660	0·00052
4. Phenylisothiocyanate	1·624	1·633	1·659	1·647	0·00056
4a. s-Tetrabromethane	1·611	1·619	1·635	1·634	0·00057
5. Iodobenzene	1·596	1·602	1·620	1·616	0·00057
6. Bromoform	1·574	1·579	1·592	1·594	0·00060
7. o-Toluidine	1·550	1·557	1·572	1·569	0·00051
8. o-Nitrotoluene	1·525	1·531	1·548	1·544	0·00049
8a. Ethylene bromide	1·519	1·522	1·535	1·536	0·00054
9. Propylene bromide	1·499	1·502	1·511	1·516	0·00054
9a. Pentachlorethane	1·483	1·489	1·499	1·502	0·00048
10. Methyl furoate	1·469	1·473	1·485	1·485	0·00045
11. Methyl thiocyanate	1·449	1·452	1·459	1·466	0·00054
12. Trimethylene chloride	1·431	1·434	1·439	1·446	0·00049
13. Ethyl monochloracetate	1·405	1·407	1·412	1·419	0·00047

Butler [*] gives data for a series of liquids of intermediate refractive index for variation work, using mixtures of α-monochloronaphthalene and a number of kerosene fractions.

General Discussion of the Methods. The advantages of variation methods are, first, that they need fewer liquids to cover a given range of refractive indices, and hence time is saved once the preliminary examination is completed, because fewer mounts need to be made. Secondly, they are more accurate, the single variation method giving results to at least ± 0·001. Winchell [†] claims a possible greater accuracy than ± 0·0004 for the double variation method. It is doubtful whether this degree of accuracy is possible for all workers, because the capacity of the human eye

[*] *Amer. Min.*, 1933, 18, 394.
[†] A. N. Winchell, *Microscopic Characters of Artificial Inorganic Substances* (Wiley, 1931), 150.

for detecting subtle changes in relief near the matching point varies greatly from person to person.

It is instructive, with reference to the applicability and convenience of these methods, to notice the distribution of the value of the birefringence in inorganic substances. A study of the artificial compounds (just over 900) described by Winchell (*op. cit.*) shows that about 27% are isotropic, about 40% have a birefringence up to a value of 0·02, and that the vast majority have birefringences below 0·04. As regards naturally occurring substances, Larsen's tables * (which list about 1100 to 1200 species) show that the majority of these have a birefringence between about 0·01 and 0·03, with a peak at about 0·018. From these two collections of data it is estimated approximately that over 60% of artificial inorganic substances and about 36% of minerals could be completely determined by the single variation method of Emmons, using only one liquid having the appropriate range of refractive index for each substance. The double variation method should be capable of dealing satisfactorily with the vast majority of artificial substances and with nearly 70% of minerals, in one carefully selected liquid for each substance.

So far as organic substances are concerned the methods do not seem to be so readily applicable. At the higher temperatures, organic substances are more readily soluble and may be unstable. Great care must also be used when studying hydrates by these methods because they may change to a lower hydrate at the higher temperatures.

The greater speed of determination which results from the use of these methods, after the apparatus is set up and working, is offset to some extent by the necessity of making a preliminary examination, as well as by the time taken up in stabilising the temperature near the matching point after a rise or fall. The saving of time theoretically possible, may therefore become illusory.

Taking everything into consideration, the experienced worker will be inclined to reserve these methods for special problems, of which the following are examples to which they are peculiarly suited.

The examination of aggregates composed of many different substances, such as dust or soil samples, is greatly facilitated,

* E. S. Larsen, " Microscopic Determination of the Non-opaque Minerals ", *U.S. Geol. Surv. Bull.* No. 848 (1934).

because the material can be studied in a series of mounts covering a continuous range of refractive indices, each separate constituent being identified in turn, and since there are no gaps in the sequence of indices, no substance need be missed.

The old method of studying the dispersion of refractive indices of crystals or glasses (the difference in refractive index for the F and C lines of the solar spectrum, expressed by $n_F - n_C$) was by using polished prisms and finding the refractive indices for different wave-lengths by the method of minimum deviation. By the double variation method the dispersion can be measured on crushed grains of either glass or crystals with a great saving in time, and with little loss of accuracy for all practical purposes. In the identification and comparison of glasses in forensic investigations, for example, the measurement of dispersion might prove to be a differentiating factor for glasses of which the refractive indices were very near.

A property which appears to be useful in identifying members of a complicated anisotropic solid solution series, is that of the birefringence-dispersion ratio. The birefringence of the substance for the D line (expressed by B_D) is divided by the difference between the birefringence for the F line and that for the C line (expressed by $B_F - B_C$) thus:

$$\frac{B_D}{B_F - B_C}$$

Winchell and Meek * have shown that this optical property is useful in determining by optical means the various ternary mixed crystals formed by members of the calcite group of minerals namely, $CaCO_3$, $MgCO_3$, $FeCO_3$, $MnCO_3$, $ZnCO_3$, and $CoCO_3$. In such studies the double variation method is the only one feasible, because crystals of such a size and perfection from which prisms can be cut are rare, and because by this method all that is needed in such pioneer studies is sufficient material for a chemical analysis and for making one or two mounts.

The Universal Stage

General Introduction. The universal stage is an instrument designed to orientate a crystal or crystal plate in any position by rotation about a number of axes which are mutually perpendicular.

* A. N. Winchell and W. B. Meek, *Amer. Min.*, 1947, **32**, 336.

THE UNIVERSAL STAGE 401

In the final form as designed by its inventor Federov, the stage has four axes of rotation, and as improved by Emmons,* five axes. The microscope stage adds one more movement around a vertical axis to both types. The arrangement of the axes of rotation in both the Federov and Emmons stages is shown in Fig. 295 (a) and (b) respectively. In the four-axis type of stage, beginning at the centre, is the inner stage axis, A_1, which is used first of all to bring the crystal to extinction. This rotates parallel to the microscope stage and around the optical axis of the microscope, when the other axes are at zero positions. Next is a horizontal N–S axis, A_2, by means of which an optical symmetry plane is brought vertical and parallel to the N–S cross-wire. The third axis, A_3,

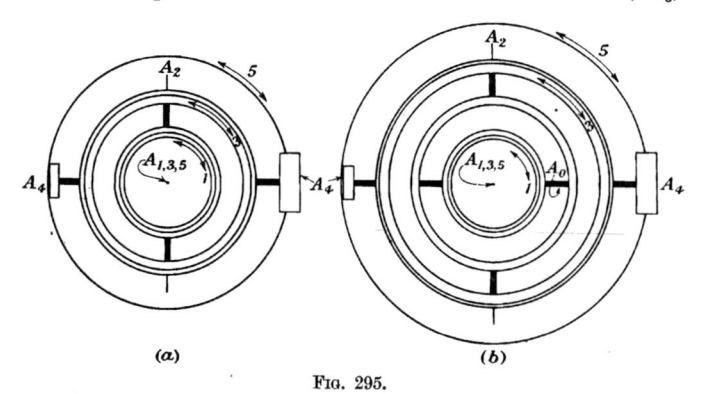

(a)   (b)

FIG. 295.

is also parallel to the microscope axis, and is used, after setting a principal vibration direction parallel to the microscope axis, to rotate each symmetry plane in turn into the N–S position, in order to search them for the emergence of the optic axes, and if these are found, to measure the optic axial angle on the horizontal axis, A_4, which carries a drum graduated in degrees and reading by a vernier to 6 minutes of arc. When measuring the optic axial angle, the crystal is turned into the 45° position on the microscope stage, the axis of which is designated A_5. The Emmons stage differs from that of Federov, by having an extra E–W horizontal axis, A_0 (Fig. 295 (b)), upon which the inner stage can be turned. This enables one symmetry plane to be arranged vertically, leaving the

* R. C. Emmons, *Amer. Min.*, 1929, **14**, 441.

N-S, A_2, axis free to be used in bringing the second optical symmetry plane vertical.

Nomenclature of Universal Stage Axes. Up to the present there has been no widespread agreement as to how the various axes should be designated, and hence some confusion is caused.* Berek † calls the inner stage A_1, and each axis in turn outwards is numbered consecutively as in Fig. 295 (a). This seems to us to be a logical scheme, because the first axis to be used in practice is that of the inner stage, and the E-W horizontal and microscope stage axes last, in the measurement of the optic axial angle and fast and slow directions. This arrangement is easy to remember as it makes all vertical axes odd, and all horizontal axes even. The introduction by Emmons of the additional E-W axis spoilt this simple and satisfactory scheme, for if the axes are labelled beginning at either the innermost or the outermost axis, one of the horizontal axes becomes odd and one vertical axis even. It would seem highly desirable to keep Berek's nomenclature for identical axes of both types of stage, and label the additional inner E-W axis of Emmons, A_0. The same axes of each type of stage would then bear identical symbols and confusion would be banished. This scheme of numbering the axes is adopted here.

General Description of the Stage. The general appearance of an Emmons type of stage is shown in Fig. 296 (a) and (b). In (a) the outer stage circle is nearly horizontal, and the object glass is tilted downwards towards the front on A_0, and downwards to the left on A_2. Two of the graduated arcs, of which there are four (hinged on the outer stage circle), are raised. These are used to measure in degrees the tilt around the A_0 and A_2 axes, which can be fixed after rotation, by means of the screw clamps, which can easily be seen in the figure. The clamp by which the outer E-W axis, A_4, is fixed, is seen in Fig. 296 (b) in front of the drum, and the knurled head of the screw by means of which the outer stage circle, A_3, is fixed, is to be seen to the left front of the horizontal circle (Fig. 296 (a)). Both this circle and the drum of A_4 are graduated in degrees, and by verniers read to 6 minutes of arc.

A circular glass plate holds the object glass, upon which the specimen is mounted, and two glass segments are fixed one above

* See for example D. J. Doeglas, *Amer. Min.*, 1940, 25, 286, for a comparison of different schemes.
† M. Berek, *Neues Jahrbuch. Min. Beil. Bd.*, 1923, 48.

FIG. 296 (a).

FIG. 296 (b).

the specimen and the other below the glass plate. These segments are of such a size that ideally they will form a sphere when separated by the thickness of the glass plate, object glass, and cover glass

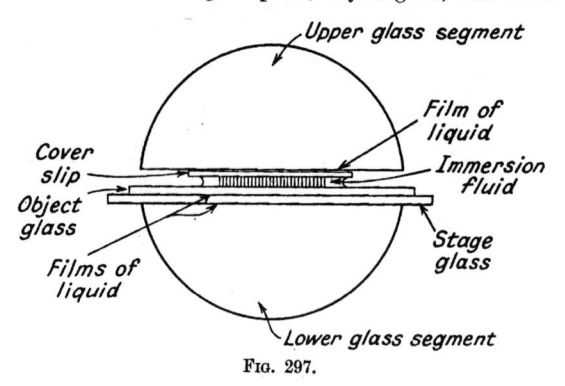

FIG. 297.

(Fig. 297). The upper segment is screwed down upon the object glass and the lower one is held by two projecting lugs. Contacts between the various members of this composite sphere should be

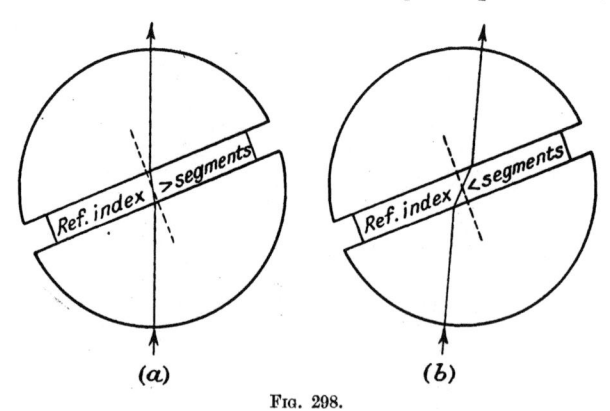

FIG. 298.

made with a liquid having as nearly as possible the same refractive index as the glass, so that rays travelling along the axis of the microscope will pass straight through the preparation without total reflection at any of the various junctions. The passage of light through the object is shown diagrammatically in Fig. 298 (a)

THE UNIVERSAL STAGE

and (*b*), from which it will be appreciated that if the refractive index of the crystal is much different from that of the segments, the angles read off on the arcs or drum will not represent true angles within the substance, and as these are what are required, a correc-

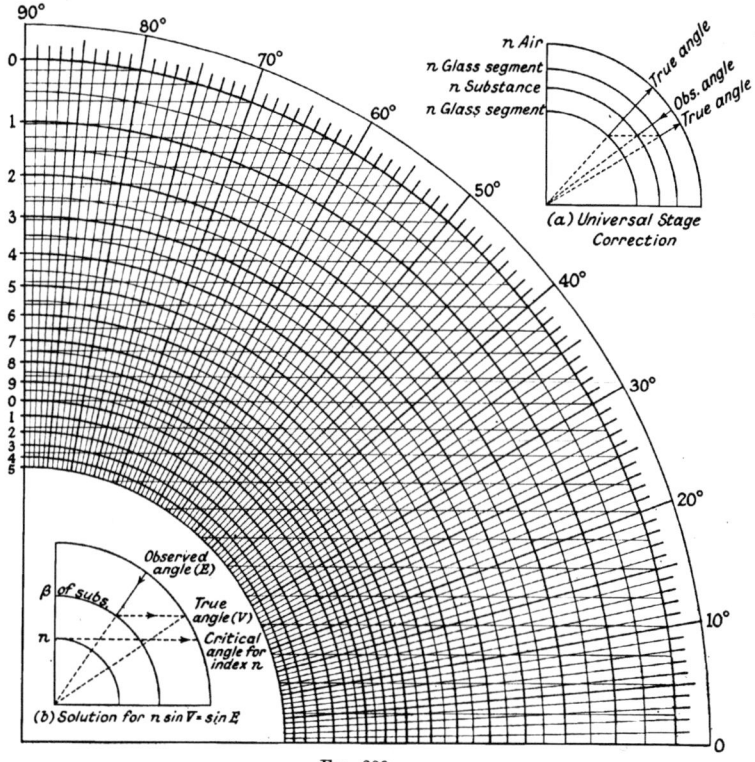

FIG. 299.

tion has to be applied to the readings, in the same way as the angle E is changed to V (see p. 288).

The formula for this is:

$$\frac{\beta \text{ (or mean index of substance)}}{n \text{ (of segment)}} = \frac{\text{sine of observed angle}}{\text{sine of true angle in substance}}$$

A graphical solution for this is accurate enough and is given in Fig. 299. At least two pairs of segments of different refractive indices are supplied with each stage and the most appropriate pair must be used. Typical values are 1·516 and 1·649. A second pair of segments is seen in Fig. 296 (a).

In spite of using segments of different refractive indices and carefully selecting the liquids for all contacts, total reflection may take place at large angles of tilt which vary according to the constants of the microscope, and the refractive indices of segments, crystal, and immersion liquids. It goes without saying that all liquids used for contacts should be non-corrosive. Beware of potassium mercuric iodide which is tempting to use for organic compounds soluble in the usual organic immersion liquids. It is very easy to ruin a stage by allowing a few drops of a corrosive fluid to run into the bearings of the inner stage (A_1) or the bearings of the arcs, and whatever liquids are used, the instrument must be thoroughly cleaned after use and never allowed to stand without attention after a series of determinations.

Adjustment of Stage for Use. Most modern research microscopes have long enough coarse adjustment racks to accommodate a universal stage, and in addition a large central aperture (to allow free rotation on the A_4 axis) to the microscope stage, normally filled by a central ring. For accurate work the condenser should have two diaphragms (one below the polariser), in order that a narrow pencil of light can be used. Further, it is best to use a carefully centred objective of low magnification and small numerical aperture with a good working distance above the stage so that on tilting the arcs will not foul the objective. It will be realised that owing to the size of the upper hemisphere, high powers cannot be used[*] and because, in addition, the preparation is high above the condensing system, interference figures cannot be seen, and the orientation of crystals must be ascertained by means of extinction phenomena.

To adjust the universal stage on the microscope, proceed as follows. Place the microscope vertical, raise the objective and remove the central ring from the stage. Place the instrument upon the microscope stage with the drum to the right, and hold it loosely in position by means of the two set-screws, which must not be tightened up until the final centring is done. Now place the central glass plate in position, and see that the scale readings for the A_1, A_0, A_2, and A_4 axes are all at zero. A_0 and A_2 are adjusted

[*] Higher powers can be used if the upper hemisphere is made very small.

by means of the arcs which should be lowered after the set-screws have been tightened. Turn the outer stage (A_3) until 270° coincides with 0 on the vernier scale, thus bringing A_2 normal to A_4. All the scales are now in their proper zero positions.

To centre the stage, place a few crystals on an object glass (which must be cut short to about $1\frac{1}{2}''$ long), and by rotating the outer stage, A_3, to find its centre of rotation, centre the universal stage by small lateral movements and tighten the set-screws. It is next necessary to see that the horizontal axes are properly adjusted to the eyepiece cross-wires. This can be done as follows. Fix an upper segment on the universal stage and, focussing on a piece of dust on the surface of the segment, rotate on A_4 after loosening its set-screw, and notice whether the particle moves parallel to the N–S cross-wire. If it does not, the *microscope stage* should be turned a little until it does so, and then clamped. The microscope stage vernier reading is then the true zero position for A_5.

Because of the narrow pencil of light normally used, a strong source of light should be provided. A high efficiency 40-watt, 6-amp. point-source lamp with a focussing lens is more satisfactory than an arc. When using monochromatic light or rather dense filters, the diaphragms above and below the condenser may need to be opened for satisfactory illumination.

Mounting Material for Examination. Place upon the glass plate of the universal stage a drop of liquid of about the same refractive index as the plate, and place the object glass upon it so that its length is normal to the line joining the set-screws of the upper hemisphere. Make a mount in the usual way upon the object glass, centre a crystal, and place upon the cover slip a drop of the same immersion liquid as that used in the mount. This is to ensure that one immersion liquid will not become contaminated with another. Now carefully screw the upper hemisphere in place, watching the mount the while, to see that the crystal selected is not broken by too much pressure, or not displaced too far from the centre of the field. Place a drop of the liquid, which was placed between the *object glass and the glass plate*, upon the upper surface of the lower hemisphere, and carefully fit it below the stage, fixing it by giving a slight turn to engage in the clips. If the preparation for examination is a thin slide mounted in Canada balsam with a cover slip already in place, it is possible to use the same liquid between each surface throughout, since no contamination of the

408   SPECIAL METHODS

mounting medium is possible. The hemispheres nearest in refractive index to that of the substance under examination should be used.

In mounting specimens care must be taken not only to centre a crystal properly but also to see that it lies exactly in the plane of the horizontal axes of the stage. If the grain is above or below these axes it will not remain centred on rotation about them, but will plunge away from the centre. The central glass plate of most modern stages can be adjusted by turning a threaded ring in which it is mounted, thus raising or lowering it sufficiently to compensate for different thicknesses of object glasses. If no such screw adjustment is possible, it is necessary to experiment with object glasses of different thicknesses. If an object glass of the correct thickness cannot be obtained, a thin one may be made thick enough by cementing a cover glass beneath it with Canada balsam.

Optical Examination. The essence of the method of examining crystals with the universal stage is the recognition and orientation of optical symmetry planes. Uniaxial crystals can be recognised by the presence of a unique optic axis parallel to which is an infinite number of optical symmetry planes and, normal to it, only one. Biaxial crystals on the other hand, have only three planes of optical symmetry, but have two optic axes. These facts enable the two classes of crystals to be readily differentiated by their extinction phenomena.

Uniaxial Crystals. In general the optic axis of the crystal will lie in a random position inclined at some angle to the microscope axis and not horizontal, as in Fig. 300 (a) (unless well-formed prismatic crystals are being examined, when the universal stage method will not be necessary). Proceed as follows. On A_1 turn the crystal to extinction. Now rotate on the E–W axis, A_4. If the crystal remains dark, an optical symmetry plane containing the optic axis lies in the N–S vertical plane. Return A_4 to its zero position and turn the microscope stage, A_5, through 45°. The crystal will not now be extinguished. Finally rotate again on A_4 until the crystal once more extinguishes (assuming that this is possible—see next paragraph). Rotation around A_5 will now reveal that the optic axis is vertical, because the section is isotropic. Each of these steps is shown by means of stereographic projection in Fig. 300 (a) to (d). If a crystal does not remain extinguished when rotated on A_4 after being brought to the extinction position, it means that the optic axis must lie in the vertical plane containing A_4 as shown in Fig. 300 (e). If this is so, the crystal must be turned through 90° about A_1, and then, if possible, the optic axis brought vertical as already described.

The optic axis may, however, have been so nearly horizontal that it is not possible to bring it vertical (the maximum angle of tilt being normally about 65°). It may then be placed parallel to A_4 as follows. After the crystal has been turned to extinction on A_1, it is rotated on A_2 until rotation on A_4 does not relieve the extinction. (If the rotation on A_2 relieves the extinction, it means that the optic

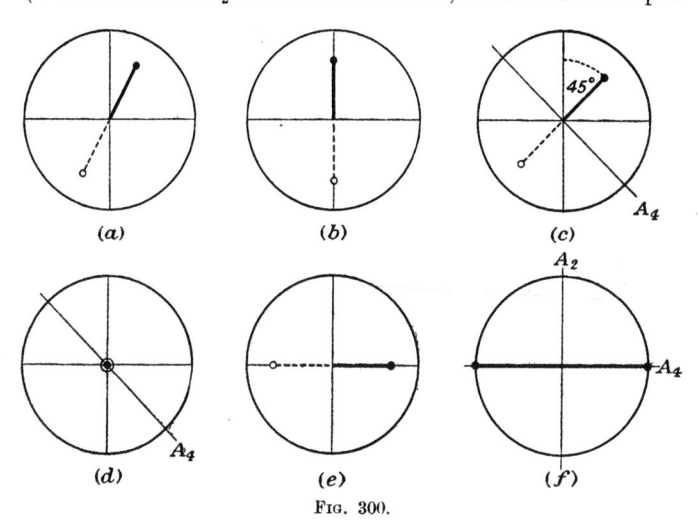

FIG. 300.

axis is in the vertical plane containing A_2, and in that case the crystal should be turned through 90° on A_1.) In this position the optic symmetry plane normal to the optic axis is vertical, and the optic axis lies along A_4 as in Fig. 300 (f). In the first position described, with the optic axis vertical, the refractive index ω alone can be compared with the immersion liquid, and in the last position the refractive indices ε and ω, and hence the optical sign, can be determined.

Biaxial Crystals. The three optical symmetry planes, especially if the material is in thin slice or crushed grains, will most often lie in a random and unsymmetrical manner with respect to the axes of the microscope and stage. This case will now be considered. When one or two optical symmetry planes already lie vertically the fact will very quickly be recognised as the orderly sequence of the examination proceeds. Fig. 301 (a) shows in stereographic projection a possible case in which the three symmetry planes of

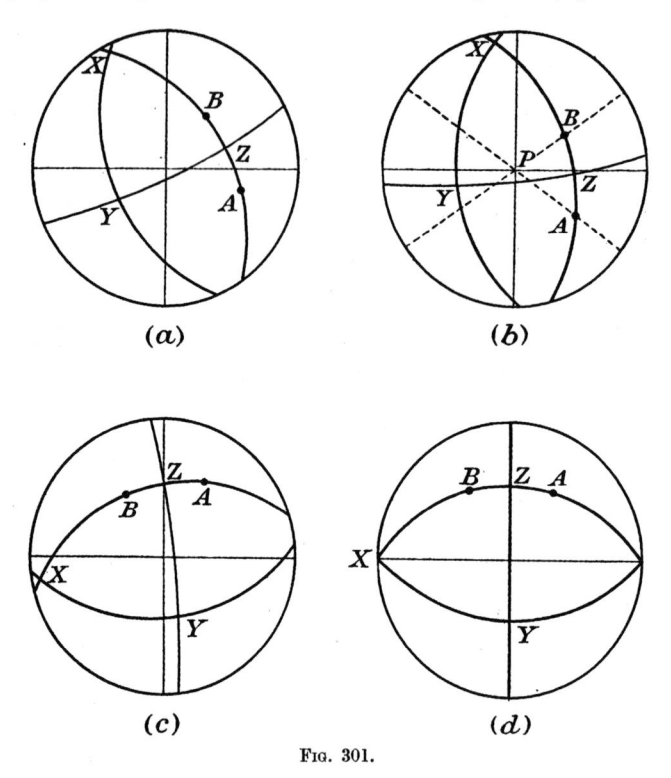

FIG. 301.

the biaxial ellipsoid are lying in a random manner. The symmetry planes are shown as great circles 90° apart and their intersections mark the points of emergence of the ellipsoid axes X, Y, and Z. A and B are the optic axes.

The first step in the determination of a biaxial substance is to locate an optic symmetry plane by making it vertical. If a Federov (4 axis) type of stage is being used proceed as follows. As in the uniaxial case turn the section to extinction (Fig. 301 (b)—crosswires bisect angles between planes through optic axes and P, the normal to the section) by rotating it on the inner stage axis, A_1, after making sure that all axes are at zero and A_3 and A_5 are clamped. Test the extinction directions by a rotation about an axis normal to each, and by means of A_1 turn that direction which shows the smaller change in birefringence parallel to the N–S axis (Fig. 301 (c)). Tilt a small amount on the N–S axis, A_2, and restore the extinction by a small rotation on the inner stage, A_1. Rotation on A_4 should show that the section is now more nearly vertical, i.e. it is more nearly extinguished than before, if the tilt on A_2 was in the right direction. Repeat these operations until rotation on A_4 does not relieve the extinction. One symmetry plane of the ellipsoid is now vertical and parallel to the N–S axis (Fig. 301 (d)). Take the readings on the inner stage and on the N–S axis, and return all axes to zero once again.

With the Federov stage it is necessary to plot the position of this plane on a stereographic projection and then start again to find a second symmetry plane, using the same method. Suppose the final readings on the inner stage A_1 and A_2 were 20° in a clockwise direction and 30° on the right-hand arc respectively, obviously if the A_2 is returned to zero the vertical plane will lie 30° (uncorrected) to the right of the N–S central line as shown on the stereographic projection in Fig. 302 (a); and when the inner stage is brought back to zero in a counter clockwise direction, the optical symmetry plane will move in sympathy as in Fig. 302 (b). After locating a second symmetry plane and plotting it on the stereographic projection, the co-ordinates of the point of intersection (either X or Y or Z) of the two planes can be found from the projection. These two co-ordinates are (i) a rotation on the inner stage A_1 and (ii), a tilt on A_2, bringing the two symmetry planes vertical. If, for example, X (Fig. 302 (c)) is the position of the point which locates the intersection of the two symmetry planes when all the scales have been returned to zero, X may be placed in the axis of the microscope by a rotation about A_1 of angle ϕ (Fig. 302 (d)) and another around A_2 of ρ (Fig. 302 (e)), the angles being read from the projection as shown. If the section is now turned to extinction by means of the microscope stage, the position of Fig. 302 (f) is attained.

The labour of determining each symmetry plane separately, and

412   SPECIAL METHODS

then finding the angular co-ordinates of the line of intersection, either by calculation or by graphical means, is greatly reduced by the use of the Emmons type of stage. With this, the first symmetry plane is orientated vertically by the use of the inner stage axis, A_1, and the inner E–W axis, A_0. This leaves the A_2 and A_4 horizontal axes still at zero, and so a second plane may, by a rotation about A_2, be placed vertical without disturbing the extinction of the first plane. The labour of calculation or of making a projection is

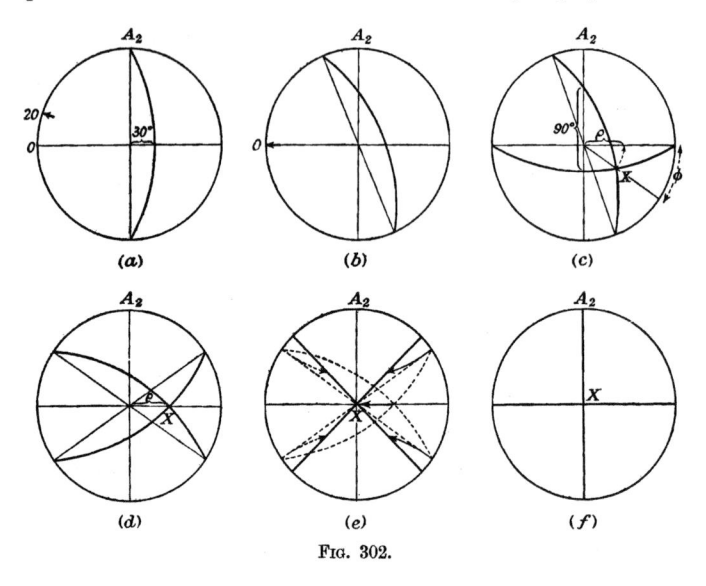

FIG. 302.

thus avoided, with a great saving of time. Using an Emmons stage, the method is as follows.

On the inner stage, A_1, turn the section to extinction. Test each direction of extinction by rotating on the horizontal axes normal to each extinction direction (*i.e.* on A_2 and A_4) and place that direction which departs least from extinction parallel to the E–W cross-wire, by turning once more if necessary on A_1. This brings the most steeply dipping symmetry plane into a position in which it is striking roughly E–W and dipping either north or south. The plane Y–Z in Fig. 301 (*b*) represents such a position. It can be seen from the figure that a tilt on the E–W axis, A_0, coupled with a slight adjustment on A_1 to restore the extinction, will bring this plane into a vertical E–W position. It is easy to decide which way to turn the section on A_0, because each adjustment, if in the right direction, will bring the section more nearly to a position of minimum change on making a test of rotation about A_2, and it will ultimately be found that a rotation on A_2 will not disturb the extinction, One symmetry plane is now vertical and lies E–W. The other two planes must strike N–S and dip east and west respectively. Rotation on the N–S, A_2 axis, must bring one or other of these two planes vertical. To accomplish this with least trouble, turn the section off the extinction position by a small rotation, say 10°, around A_4, and restore the extinction by a rotation on A_2. Bring back A_4 to its zero position and two optical symmetry planes will now be vertical. This may be confirmed by rotating separately on A_4 and A_2, bringing each axis back to its original position before rotating on the other. Take the readings on A_0, A_1, and A_2. X, Y, or Z now lies in the microscope axis, and it remains to find out which of these principal directions it is.

This is done by searching for the emergence of the optic axes in one or other of these two vertical planes, and by testing the section with a gypsum plate or quartz wedge to determine the directions of the fast or slow rays. To do this, turn the section into the 45° position by means of the microscope stage, A_5. Now rotate on A_4, and see whether a position of extinction is reached. It must be remembered that the so-called "extinction" along the optic axis of a biaxial substance is not necessarily the blackness of isotropism, but merely a uniform tint on rotation. To locate an optic axis, the microscope stage may be turned back and forth a little on either side of the 45° position as rotation is made on A_4. If such an extinction position is encountered, the section presented is a β section, and an optic axis is vertical. Take the reading on the drum. The difference between this reading and the position in which both symmetry planes were vertical is the apparent angle V (acute or obtuse), which may be corrected to the true angle by means of the formula on p. 405 or by using Fig. 299. If both optic axes can be found a measure of 2V can be made. If the first plane searched reveals no optic axes, the second plane should be treated similarly after the section has been turned through 90° on the outer stage axis, A_3. Should the optic axial plane be vertical and a measure obtained of either the acute or the obtuse angle, the

414   SPECIAL METHODS

optic orientation with reference to the microscope axis is known after the directions of the fast and slow rays have been found, because β lies normal to the plane containing X and Z, and the other refractive index in the plane of the section must be either α or γ.

If neither of the planes reveals the emergence of an optic axis,

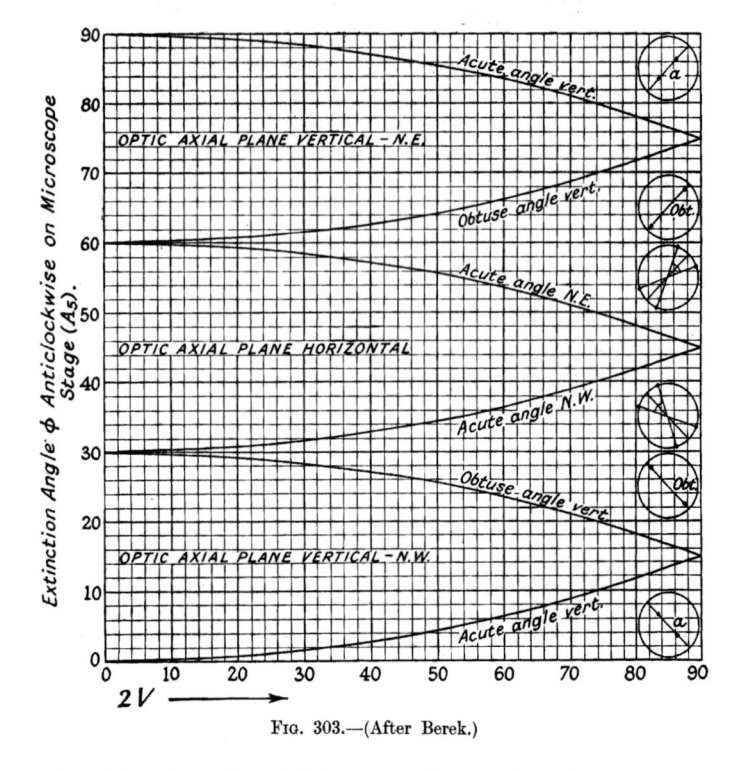

FIG. 303.—(After Berek.)

then either the optic axial plane must lie normal to the microscope axis, or the optic axes are beyond the angular limits to which search could be made on A_4, and the optic orientation and the size of the optic axial angle must be found by means of a method devised by Berek.* Briefly the method is as follows. After

* M. Berek, *Neue Wege zur Universalmethode: Neues Jahrb. f. Min. Geol. und Pal.*, Beilage Band, 1923.

placing two symmetry planes vertical, the crystal is turned 45° on the outer stage axis, A_3. On A_4 the section is turned by an amount which is equivalent to a true angle of 54·7° within the crystal. (The amount by which A_4 must be turned is determined from Fig. 299 taking into account the refractive index β of the substance and that of the hemisphere.) The direction of rotation on A_4 should be that which will result in the smaller angle between A_1 and A_5. The third step is to restore the extinction by a rotation of the microscope stage in the opposite direction to that in which the crystal was turned into the 45° position. Usually the turn into the 45° position (resulting in the so-called *standard position* for this method) is made in a clockwise direction on A_3 and in an anticlockwise direction on A_5. The angle, ϕ, through which the microscope stage is turned is used in Fig. 303 to find the size of the optic axial angle and the optical sign, and the orientation of the optic angle with respect to the axis of the microscope. The diagram gives approximate values for 2V. Dodge* has discussed the determination of 2V by this method, and gives a series of graphs which enable more accurate determinations to be made.

Determination of Refractive Indices with the Universal Stage. It will be appreciated, that while it is a great advantage to be able to turn a crystal into almost any desired position and so obtain principal vibration directions and their corresponding indices, the labour of making a number of successive determinations in a series of different liquids on the universal stage would be tiresome. The advantages of the instrument are therefore best appreciated when it is used in conjunction with the double variation method, for it is then often possible to determine all the optical constants in one mount. For this purpose Emmons uses a hollow lower hemisphere heated with water ; in a design made by E. Leitz the upper hemisphere is heated electrically. The measurement of refractive indices on the universal stage tends to be more accurate than with the usual arrangement, because the narrow pencil of light from the polariser in conjunction with a low-power objective of small N.A. makes the Becke line more prominent.

It sometimes happens that the measurement of one, or two, of the refractive indices cannot be easily made, because they lie beyond the limits of the immersion liquids available, or because an extreme platy habit makes the determination of the refractive

* *Amer. Min.*, 1934, **19**, 62.

index for the vibration direction normal to the plate very difficult. The universal stage can be used with advantage in solving both these problems. The first case, in which refractive indices lie beyond the range of those of the immersion liquids, can be solved by the method of extrapolation already described in Chapter VIII, but by using the universal stage for tilting the crystal, instead of the apparatus described by Wood and Ayliffe. The practical details are as follows. Suppose it has been possible to measure only the lowest refractive index α. The crystal is mounted on the universal stage in a liquid of known refractive index greater than α, and two optic symmetry planes are made vertical in the usual way, preferably so that the optic axial plane is vertical. Turn this plane parallel to the vibration plane of the polariser which should be orientated N–S, and so that Z is vertical. Now rotate on A_4 until a match is achieved and take the reading. The corrected value for the angle of rotation is the angle θ of Fig. 264, and γ can be estimated as shown in Fig. 265 using the folder at the back of this book. β may then be similarly estimated on either of the other two planes of the indicatrix since both α and γ are known, or may be calculated from α, γ, and 2V if a determination of the latter has been made.

The second problem is rather the converse of the one just dealt with. It is the physical habit in this case that is the difficulty and not the high value of the refractive index. In the case of the micas,* where the vibration direction X is very nearly normal to the basal plane (001), two of the refractive indices β and γ can be measured directly by ordinary immersion methods. A grain is then mounted on the universal stage in a liquid of known intermediate refractive index and a match obtained as before by rotating the optic axial plane in the plane of the polariser and extrapolating after correcting the reading on A_4 to give θ. A check can be made by doing a rotation with the $\alpha\beta$ plane N–S, if a liquid is used intermediate in value between β and α. If the substance is platy or tabular and a principal vibration direction does not emerge normal to the plate (as for example in a monoclinic substance with an excessive development of (001)), the two optic symmetry planes must be orientated correctly first, one refractive index in a horizontal direction determined, and then another mount made in a liquid of

* This method is described for the micas by R. B. Ferguson and M. A. Peacock, *Amer. Min.*, 1943, 28, 563.

appropriate refractive index, and the same optical symmetry planes made vertical before rotating to make a match.

It will be appreciated that for such problems the double variation method is ideal, because once the crystal is orientated in an appropriate liquid, a number of values of θ can be obtained after changing the temperature, or both temperature and wave-length. It might also be pointed out at this stage that it is not necessary to know even one of the principal refractive indices, for if the liquid in which the crystal is immersed has an intermediate refractive index between the two principal ones, a series of points (corresponding to given values of θ and μ, Fig. 265) on a straight line may be plotted on the graph, θ being measured from the standard position, where two optic symmetry planes are vertical. Both the lower and higher refractive indices may thus be measured, one of which might be inaccessible because of habit and the other because of value.

Orientation of the Ellipsoid with respect to Morphological Characters. So far, in this chapter, the biaxial indicatrix has been treated without reference to its orientation within the crystal with the main object of determining the refractive indices and optic axial angle of the crystal. However, the orientation of the ellipsoid with respect to crystallographic directions as shown by cleavage, composition planes, and crystal edges or faces, is of importance in some studies and for record purposes. For example, in certain isomorphous series, such as the plagioclase family of minerals, the end members of which are albite, $NaAlSi_3O_8$, and anorthite, $CaAl_2Si_2O_8$, the ellipsoid changes its orientation (and shape) within the crystal, in sympathy with the change in chemical composition. Fig. 304 illustrates this point, the optic axial plane for albite and anorthite, and one intermediate member labradorite (Ab50,An50), being shown shaded. Cleavage is indicated by broken lines.

For such an isomorphous series, the spatial relationship of the ellipsoid to the cleavage and twinning plane (in the case just quoted usually (010)) reveals the position of the crystal in the series and so enables an estimate of the chemical composition to be made.

The orientation of the ellipsoid with respect to the morphological characters is revealed when these are plotted in a stereographic projection together with the vibration directions X, Y, and Z. It is usual to use a thin section of the material, but cleavage fragments or idiomorphic crystals can be used instead.

418 SPECIAL METHODS

To orientate a cleavage plane, first turn the trace of the plane parallel to a cross-wire, say the E–W one, on A_1. Now rotate on A_4 until the trace of the plane is sharp and no lateral movement can be detected on altering the focus of the microscope, this indicating that the plane is vertical. The position of this plane in the projection can be plotted from the readings on the axes which have been used, in the same way as was an optical symmetry plane. Composition planes are treated similarly, being brought (between crossed Nicols) parallel to the same cross-wire as was used for the cleavage, and, the analyser having been removed, the

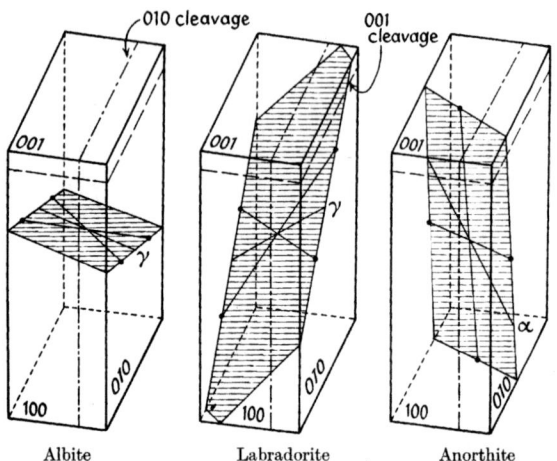

FIG. 304.—Some Members of the Isomorphous Plagioclase Series showing Optic Orientation and Cleavage.

planes are turned to the sharpest position. After measuring and plotting the morphological directions, return the axes to zero, and plot the positions of the principal planes of the ellipsoid on the same projection.

The measuring of the positions of crystal faces is somewhat more uncertain. It may generally be assumed that a pinacoid or other face on which the crystal rests is parallel to the plate of the inner stage. A test can be made to see whether all the edges of such a face come into focus simultaneously. Sloping faces can be dealt with as were cleavage planes, by bringing them parallel to the E–W cross-wire and turning them vertical, if this is possible.

It might be as well at this point to give some practical details as to methods of plotting morphological and optical characters.

When all the horizontal axes are at zero, A_1 and A_3 are vertical (A_5 is always so), and their positions would be given at the centre of a projection. As the work proceeds, it is as well to keep track of the migration of the inner axis, A_1, by plotting each successive rotation as it is made. Rotations around horizontal axes move the inner vertical axis in a direction normal to them, (*i.e.* along a *small circle* normal to the axis of rotation), whereas rotations on, say A_3, do not move A_1 if it is centred, but do rotate all the rest of the characters in the projection. Suppose all the axes are at zero. A rotation of 30° on A_4 to the south would bring A_1 to P

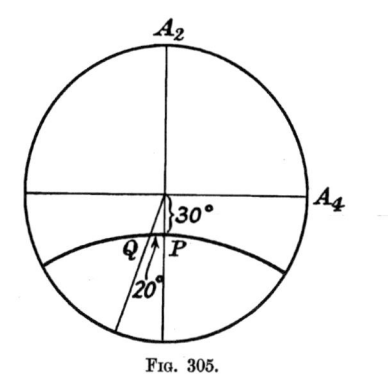

Fig. 305.

in Fig. 305 and a subsequent rotation around A_2 in a westward direction of 20° would move P to Q. The angle which the inner axis now makes with the microscope axis is given by the angle between Q and the centre of the projection measured along a radius.

A useful expedient is to use a piece of transparent paper pinned above the centre of a Federov or Wulff net. A circle is drawn on this paper, of the same diameter, and coincident with that of the primitive circle of the net, and an arrow is marked on the transparent paper at the zero point on the margin of the net. Each move of the universal stage around a vertical axis can be followed by rotating the transparent paper around the pin through a corresponding angle. The great circles representing crystallographic planes may be drawn in by rotating the tracing above the net, until a great circle comes into an appropriate position. Movements along small circles, normal to a horizontal axis, are done in the same way, by turning the tracing paper until the axis of rotation lies above the E–W or N–S axis of the Federov net and moving the point the proper amount along a small circle at right angles to the axis. The projection gradually builds up on the transparent paper which

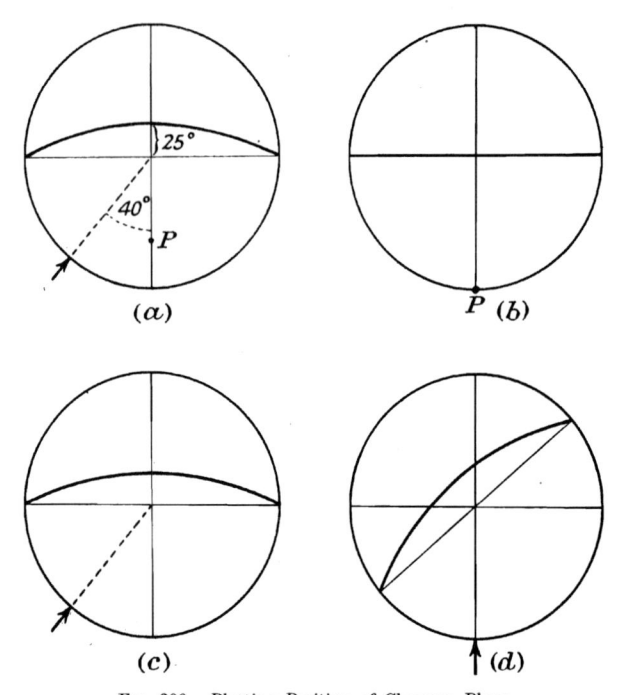

Fig. 306.—Plotting Position of Cleavage Plane

The corrected readings on A_1 and A_4 are respectively 40° and 25°.
(a) The tracing paper turned 40°, bringing cleavage trace E–W.
(b) Rotation on A_4 brings trace of cleavage vertical. P is the pole of the cleavage plane.
(c) A_4 returned to zero.
(d) A_1 and tracing returned to zero, showing true orientation of cleavage plane.

is brought back to the zero point between the determination of each plane and when all the data have been drawn in. Fig. 306 shows the steps in drawing a great circle representing a cleavage plane.

By such methods of plotting either morphological or optical characters, not only can the relations of optical and morphological properties to each other be studied in single crystals, but it is

THE UNIVERSAL STAGE 421

possible to study statistically in polycrystalline bodies *directional* properties which may have been induced by forces such as crystallisation while convection currents were operative, or as a result of external stresses. For example, by plotting the poles of the optic axes of quartz crystals in sandstones and other rocks, or of mica cleavage in schists, light has been thrown on the direction and intensity of earth movements in past geological time. The stresses have induced the minerals to assume a marked preferred orientation. In these studies it is necessary to make careful note of the orientation of the thin slice used with respect to the parent mass of which it formed a part. The reader who is interested will find the methods of orientational studies dealt with in English in a book by Knopf and Ingerson,* and by Fairbairn,† and in this country the method is exemplified in two papers by Phillips ‡ on the structural features of the Moine Schists.

In the above treatment of the universal stage and its uses, no attempt has been made to be comprehensive. The aim has been merely to sketch the methods and some of the uses to which they are applied. For further information the reader is referred to the references already given in this chapter and the list for further reading below.

RECOMMENDED FOR FURTHER READING

" Universalmethode und Feldspatstudien ", E. S. Federov, Z. *Krist.*, 1897, **27**, 337.
Universal Drehtischmethoden, M. Rheinard (Wepf et Cie Basel, 1931).
Die Federow-Methode, W. Nikitin (Berlin, 1936).
Elements of Optical Mineralogy, Part I, A. N. Winchell (Wiley, 1931).
Artificial Inorganic Solid Substances or Artificial Minerals, A. N. Winchell (Wiley, 1931). Chapter VIII by R. C. Emmons.
" The Universal Stage ", R. C. Emmons, *Geol. Soc. Amer. Mem.* 8, 1943.
Mikroscopische Mineralbestimmung mit Hilfe der Universaldrehtischmethoden, M. Berek (Berlin, 1924).

* E. B. Knopf and E. Ingerson, " Structural Petrology ", *Geol. Soc. Amer. Mem.* 6, 1938.
† H. W. Fairbairn, *Introduction to Petrofabric Analysis*, Dept. of Geology, Queen's Univ., Kingston, Canada.
‡ F. C. Phillips, *Q. J. Geol. Soc.*, 1937, **93**, 581, and *Geol. Mag.*, 1945, **82**, 205.

CHAPTER XII

EXAMPLES OF THE USE OF THE POLARISING MICROSCOPE

This chapter presents a number of specific examples of the use of the polarising microscope in pure and applied research and in routine chemical practice. Though very far from being exhaustive, they will serve to demonstrate the wide range of problems to which optical crystallographic methods have been applied, and may suggest to the reader how these methods could be most profitably employed in his own work.

Systematic Identification by means of Optical Properties.

The optical properties of several series of closely related organic compounds have been determined by various authors with the object of obtaining data by which the compounds can be differentiated and identified under the microscope. The value of such studies was originally pointed out by F. E. Wright in the course of an important paper on the use of the polarising microscope in analysis,* and in illustration he recorded the optical properties of the pentosazones and hexosazones, and of a number of the alkaloids.

As subjects for such studies, the tendency has naturally been to select families of compounds which it is difficult or tedious to differentiate by other means, e.g. because the melting-points are indefinite, or because the compounds are very liable to be encountered in an impure state or mixed together. In most cases the authors have used the most distinctive optical and morphological characters of the compounds to build up a table of systematic tests called a determinative table. These tests usually consist in mounting the specimen to be identified in a series of liquids, until one of the indices commonly shown by the crystals is matched, e.g. if the habit is acicular, the index for light vibrating along or across the

* J. Amer. Chem. Soc., 1916, 38, 1647.

length. Confirmation of the identity is then obtained from some other distinctive property, e.g. another commonly presented index, or characteristic abnormal polarisation colours. An example of such a determinative table is given on p. 424. This is taken *verbatim* from a paper by Hann and Keenan,* which describes the optical properties of the hydrochloride and mono- and di-sulphonates of benzyl-ψ-thiourea, crystallised from acidified aqueous solutions (mostly 0·2N HCl). The main purpose of the work was to afford a means of distinguishing the parent sulphonic acids, of which the benzyl-ψ-thiourea salts are readily prepared derivatives.

It is clear that tests of this kind depend very largely on the assumption that the crystals develop a particular habit, and it is therefore very important that every determinative table be accompanied by details of the method or methods used by the author to crystallise his compounds. In particular, the solvent used should be stated, for other solvents may yield not only different habits but crystals containing solvent of crystallisation. Some authors have not given sufficient consideration to these possibilities and their tables should therefore be used with caution, confirmation of the identity of the substance being sought wherever possible from properties which are independent of its habit, e.g. size of the optic axial angle, or a principal refractive index.

Two papers by Wherry † and Wherry and Yanovsky ‡ respectively on the optical identification of alkaloids emphasise the points just made. The first paper is devoted to a description of experimental procedure. The second paper describes the optical properties of cinchonine, cinchonidine, quinine, and quinidine, crystallised from alcohol, and from benzene. Separate determinative tables are given for crystals obtained from these two sources. Solvent of crystallisation is taken up by quinidine from alcohol, and by cinchonidine and quinine from benzene. Two experiments are also described to show the practical application of these results. In one, a mixture of equal weights of the four alkaloids was dissolved in alcohol, and the solution crystallised in five fractions, to obtain the last of which it was taken almost to dryness. The fractions were dried and examined optically. Each fraction contained one of the alkaloids in a preponderating amount with

* J. Phys. Chem., 1927, 31, 1082.
† U.S. Dept. Agr. Bull. 679 (1918).
‡ J. Amer. Chem. Soc., 1918, 40, 1063.

BENZYL-ψ-THIOUREA SALTS of NAPHTHALENE MONO- AND DI-
SULPHONIC ACIDS.

DETERMINATIVE TABLE (after Hann and Keenan).

Index.	Description of Crystals and Confirmatory Data.	Compound.
1·570	Material consists of rods and needles, examined in ordinary light. This index value is shown crosswise on rods and needles. Confirm by mounting in liquid 1·670, which matches γ shown lengthwise on rods and needles	2 : 6
1·645	When examined in ordinary light, the material consists of irregular fragments and thin, six-sided plates. Confirm by mounting in liquid 1·570, which matches α	2 : 7
1·670	Material consists of rods and irregular fragments when examined in ordinary light. This value is commonly found lengthwise on rods. Confirm by immersing in liquid 1·565, which matches α and is shown crosswise on rods	1 : 6
1·672	The material consists of rods when examined in ordinary light. This value is commonly found and occurs lengthwise on rods. Confirm by immersing in liquid 1·680, which matches γ and is shown crosswise on rods	α
1·682	Material in ordinary light consists of rods, many of which have oblique terminations. This index value occurs lengthwise on rods. Confirm by immersing in liquid 1·577, which matches α shown frequently crosswise on rods	β
1·689	Material in ordinary light consists of fibrous, rod-like crystals. This index matches γ. Confirm by immersing substance in liquid 1·631, which matches α and is shown crosswise	Hydrochloride
1·690	Material consists of irregular fragments when examined in ordinary light. This index value is shown on fragments extinguishing sharply in parallel polarised light (crossed Nicols). Confirm by immersing in liquid 1·670, which matches β, and is frequently shown	1 : 5

smaller quantities of one or two of the others. No difficulty was experienced in identifying the substances by means of the appropriate determinative table. The same experiment was repeated with benzene as the solvent, with similar results. No difficulty is therefore likely to arise in identifying the constituents of mixtures of these alkaloids, if they are partially separated by fractional crystallisation.

In the second experiment, a few pills of a proprietary remedy, presumed to contain the alkaloids as sulphates, were rubbed with potassium carbonate to liberate the alkaloids, suspended in water, and stirred with chloroform to dissolve them. The chloroform solution was allowed to evaporate, but the residue, which contained a considerable amount of nondescript gummy matter, was noncrystalline. The alkaloids were therefore dissolved out with dilute sulphuric acid, precipitated with dilute ammonium hydroxide, and extracted with benzene. The benzene layer was separated, concentrated by slow evaporation at the ordinary temperature, and a drop of the solution allowed to crystallise on a microscope slide. Under the microscope three entirely distinct crystalline substances could be recognised with the greatest readiness. Their optical properties indicated that they were quinine, cinchonidine, and cinchonine respectively. This was confirmed by allowing several drops of the solution to evaporate slowly to dryness on separate slides, and immersing the crystals in the necessary refractive liquids to apply the appropriate determinative table.

In certain series of related compounds the differences in optical properties between some members has proved to be too small to enable them to be distinguished very readily from one another, if the observations are confined to white light or to a single wavelength. This was the case with some adjacent and isomeric members in a series of 3 : 5 dinitrobenzoates of the aliphatic alcohols studied by W. M. D. Bryant.* Bryant and J. Mitchell jun. have since shown that very considerable differences in the nature and extent of the dispersion of the optic axes with wave-length may exist in a series of related compounds. Thus in a study of the picrates of 32 amines † (increased actually to 39 crystal species owing to the occurrence of polymorphism) thirteen cases of crossed axial plane dispersion, fifteen of crossed dispersion, three of inclined dispersion, two of horizontal dispersion, and at least three of axial dispersion

* *J. Amer. Chem. Soc.*, 1932, **54**, 3758. † *Ibid.*, 1943, **65**, 128.

with change of sign were found, with marked quantitative differences in the dispersion as well. It is evident therefore, as the authors have concluded, that measurements, or even qualitative observations, of dispersion may add very considerably to the precision with which substances may be differentiated optically. Unfortunately, Bryant and Mitchell do not appear to have studied the dispersion of the alcohol derivatives mentioned above.

A list of references to studies of the kind discussed in this section, including for completeness those already given, will be found in the Appendix which follows this chapter. Mention may also be made here of A. N. Winchell's *Optical Properties of Organic Compounds* (see Appendix) which includes a graphical determinative table covering all the compounds described in the work (about a thousand). Winchell has also constructed a similar determinative table for the inorganic compounds described in his *Microscopic Characters of Artificial Inorganic Solid Substances or Artificial Minerals* (again see Appendix).

Characterisation and Identification of Single Compounds by Optical Properties. Under this heading we shall consider some of the very large number of instances in which the optical properties of a compound have been used to confirm its identity, to examine it for impurities, or to ascertain whether it is a new species or not. Such problems frequently involve distinguishing between isomers, polymorphs, or different combinations of the same ions, so that chemical analysis is of little use. Even where orthodox methods are applicable, however, the optical method may be able to give the required answer more expeditiously.

Three simple examples in inorganic chemistry taken from the experience of one of the present authors (N. H. H.) may first be quoted.* A supply of pure potassium mercuric iodide ($K[HgI_3]$) was required. A specimen submitted by a firm was examined under the microscope, when it was at once seen that, in addition to the yellow birefringent crystals of the complex salt, colourless isotropic crystals were present. On mounting a little of the specimen in a liquid of index 1·66 the isotropic crystals become almost invisible, but showed a slightly greater index than the liquid. They were thus confirmed as being potassium iodide, which had been suspected, for this salt is cubic and has a refractive index of 1·67. This impurity could not have been detected by qualitative

* *J. Soc. Chem. Ind.*, 1933, **52**, 367.

chemical analysis, since in solution the complex salt behaves as a mixture of potassium, mercuric, and iodide ions. Even quantitative chemical analysis would have failed to distinguish between $K[HgI_3]$ contaminated with KI, and $K[HgI_3]$ contaminated with $K_2[HgI_4]$ which was also a possibility. A specimen of "A.R." diammonium hydrogen phosphate received from a firm was very different in appearance from that already in stock, which was known to be satisfactory. The presence of ammonium and phosphate ions in the former specimen was shown by chemical analysis, but that did not afford complete proof of its identity ; it might have been, for example, the dihydrogen salt. Under the microscope, however, crystals of both specimens frequently presented single optic axial sections which yielded the same value of 1·51 for β. The identity of the two specimens was therefore confirmed.

A specimen of ammonium dihydrogen phosphate was examined under the microscope for impurity. Most of the crystal fragments observed had one index equal to 1·525, which is the value for ω for this salt. A few, however, showed indices which were *both* less than 1·52, and moreover gave biaxial interference figures. Their optic angle and β value showed them to be the diammonium salt. Here again, the impurity could not have been detected by qualitative chemical analysis.

In a summarising paper on the use of the polarising microscope as a time and labour saver in organic chemistry, Benedict * describes how, in studying methods for the preparation of amino acids, he was greatly helped by Keenan's previous determinations of their optical properties (see Appendix), particularly in view of the fact that amino acids have such indefinite melting-points. The present authors make no apology for quoting Benedict's paper at some length, for the examples he records are so very much to the point. He writes :

" A drop of the reaction mixture can be smeared on a microscope slide and examined between crossed Nicols to see whether or not any birefringent material is present. Usually a few other simple optical properties, such as extinction angle and sign of elongation, can be ascertained. These data alone are frequently sufficient and can be determined on the crude mixture. One can tell whether or not it is worth while to go through a laborious isolation process.

* *J. Ind. Eng. Chem. (Anal.),* 1930, **2**, 91.

This is particularly valuable in testing short-cut methods. In preparing alanine * it was possible to show that alanine was present in the reaction mixture without any attempt at purification. The alanine was isolated by dissolving the hydrochloride in alcohol and precipitating with aniline.

"Two obvious questions arose.

"(1) Was all of the alanine extracted from the sodium and ammonium chlorides by the alcohol ? Examination of the residue between crossed Nicols showed no birefringence, indicating that removal had been complete. (Sodium and ammonium chlorides are isotropic.)

"(2) How complete was the precipitation of the alanine by the aniline ? In one run, when aldehyde ammonia was the starting material, some of the alcoholic mother liquors were concentrated and gave a crystalline material. Without purification a little of this under the microscope showed needle-like crystals, closely resembling alanine, but with the opposite sign of elongation, tested merely by introducing a selenite plate between the crossed Nicols. It could not be alanine. Later it was found to be iminodipropionic acid, . . . the optical properties of which are not on record. This example shows that it is not always necessary that the optical properties be known in advance.

"Work has also been done on the synthesis of serine . . . and the ability to determine whether any serine is present in the crude reaction mixture has cut the work in half because no time is wasted in working up reactions which give no yield of serine. In preparing some intermediates for this synthesis, it was thought that ethoxyacetic acid was a by-product in amounts too small to be readily purified for identification. A little copper ethoxyacetate was prepared and some of its optical properties were determined. When copper carbonate was added to the solution suspected of containing the ethoxyacetic acid and a drop or so concentrated on a slide, crystals corresponding in properties to those previously prepared were observed. Later, in large-scale runs, it was found possible to work up this portion and obtain a yield of the acid.

"In an attempt to prepare a substituted aspartic acid, a compound identified as ethylphthalimido maleate was obtained. A great saving in time resulted from finding that the hydrolysis

* By hydrolysing a mixture of acetaldehyde, aqueous ammonium chloride, and sodium cyanide with concentrated hydrochloric acid.

products of this material were phthalic anhydride and ammonium chloride. The phthalic anhydride was extracted with ether. The remaining material was not at all birefringent. This meant that it belonged to the cubic system, and as organic compounds are very rarely found in this system, the material was inorganic. The featherlike appearance of the crystals indicated that it was ammonium chloride."

Some of the other applications of the optical method recorded by Benedict are as follows:

"The students in a biochemistry class prepared cystine which crystallises out in hexagonal plates. One student turned in some boric acid which also crystallises in what appear to be hexagonal plates. But cystine gives a uniaxial interference figure with crossed Nicols and convergent light, while boric acid gives a biaxial figure with a small optic axial angle. The point to be stressed is that the material turned in was shown not to be cystine by a simple microscopic test, which also gave evidence as to its identity, an additional advantage of the method.

"A colleague brought to this laboratory a small amount of some pasty material which was thought to contain benzaldehyde bisulphite. A minute's work was required to prepare some benzaldehyde bisulphite and see with the polarising microscope that the two materials were identical.

"To distinguish between . . . butyn and procaine, sodium chloride was added to butyn to yield the crystalline chloride, which is insoluble. The form is long platelets with pointed ends, having an extinction of about 20° and negative elongation. Procaine, on the other hand, is unaffected by sodium chloride solution but yields long fine needles, with parallel extinction and negative elongation, when treated with bromine water.

" . . . an unlabelled bottle was suspected of containing sodium hydrogen maleate. Some of this substance between crossed Nicols showed long crystals with pointed ends, parallel extinction, and a very large optic axial angle, with the axial plane at right angles to the elongation. The material in the unlabelled bottle had the same properties. . . . The time consumed in testing these two materials was less than five minutes. Yet the evidence was as conclusive as that obtained by spending several hours on an analysis."

The polarising microscope has been used by G. M. Bennett and his co-workers in a number of organic researches, for example as an aid in differentiating between the stereoisomeric dithian dioxides,* picrates of dithian monoxide methyl sulphonium salt,† dithian monoxide sulphilimines,† di-*p*-tolylethane dioxides,‡ 1 : 3-dimethylthiol benzene dioxides,‡ trimethylene trisulphide dioxides § (all Bell and Bennett), and ethylene dithiolacetic acid dioxides ‖ (Bennett and Statham). In some of these cases final proof of the difference between the isomerides was obtained from goniometrical measurements of the crystals.

A particularly interesting example relates to the isomeric trithioacetaldehydes. The existence of two isomerides, α- and β-, had been definitely established, and no more were required by stereoisomeric theory. An alleged third isomeride described by Mann and Pope proved on examination under the polarising microscope, and by other methods, to be a eutectic mixture of the α- and β- forms (Bell, Bennett, and Mann).¶

Harrison, a colleague of Bennett, discovered an oxidising enzyme which converts glucose into gluconic acid.** In the early stages of the work the identification of the gluconic acid by ordinary methods was impossible owing to lack of material. The phenylhydrazide was therefore prepared and compared with that of authentic gluconic acid. When crystallised under the same conditions the two compounds were found by Bennett to form similar square-ended plates with straight extinction. This practically established their identity, which was later proved by microanalysis and polarimetric examination when more material became available.

An alleged triethylene trisulphide described by Ray, was shown by microscopic examination to be identical with dithian (Bennett and Berry).††

The identity of a small amount of dichlorodiethyl sulphide formed in a reaction was proved by converting it into diphenoxydiethyl sulphide, and comparing it with an authentic specimen of this compound. Melting-point, mixed melting-point, and microscopic determinations showed that they were the same (Bell, Bennett, and Hock).‡‡

The polarising microscope was much used by Venable in studying the action between hydrogen peroxide and uric acid under

* *J. Chem. Soc.*, 1927, 1798.　　† *Ibid.*, 1928, 86.
‡ *Ibid.*, 1928, 3189.　　§ *Ibid.*, 1929, 15.
‖ *Ibid.*, 1931, 1684.　　¶ *Ibid.*, 1929, 1462.
** *Biochem. J.*, 1932, 26, 1928　　†† *J. Chem. Soc.*, 1925, 910.
‡‡ *Ibid.*, 1927, 1803.

IDENTIFICATION OF SINGLE COMPOUNDS

different conditions.* By its means the course of the reactions could be followed without the necessity of isolating the individual products, except in those cases where quantitative determinations were needed. But here also microscopic examination was found to furnish the best criteria of the purity and homogeneity of the product, for in the purine group sharp melting-points are exceptional. In a continuation of the work by Moore and Thomas * the polarising microscope was again used.

One of the present authors (A. S.) contributed some optical

	HMX.				RDX.[†]
	α-form.	β-form (stable form).	γ-form.	δ-form.	
System and habits	Orthorhombic; blades and rods	Monoclinic; equant, tabular, or prismatic	Monoclinic rhombs with {001} and {110}	Hexagonal prism and pyramid, acicular	Orthorhombic
α_D	1·562	1·589	1·537		1·580
β_D	1·563	1·595	1·585	$\left(\begin{matrix}\epsilon_D\ 1·566\\ \omega_D\ 1·607\end{matrix}\right)$	1·595
γ_D	1·735	1·773	1·663		1·600
Sign	Positive	Positive	Positive	Negative	Negative
2 E	—	37°	—	—	87½°
2 V	16°	22°	75°	—	51½°
Optic orientation	Not determined	$\beta \wedge c = 44\frac{1}{2}°$ in acute $\angle\beta$; optic plane normal to (010)	$\alpha \wedge c = 42°$ in acute $\angle\beta$; $\gamma = b$; optic plane normal to (010)	—	Not determined
Remarks		Crossed axial plane dispersion			$\rho > v$ (marked)

crystallographic investigations as part of a research programme on explosives during the late war. In the manufacture of cyclotrimethylene trinitramine (RDX), a by-product, cyclotetramethylene tetranitramine (HMX), was formed, and certain unstable polymorphs of this latter compound occurred. Four polymorphs were found (of which two, the α- and β-, had been previously reported) and these were studied and characterised by optical methods. During later research, these methods were used with considerable success as an aid to the control of the process. The optical proper-

* J. Amer. Chem. Soc., 1918, **40**, 1099, 1120.
† Bachmann, W. E., and Sheehan, J. C., ibid., 1949, **71**, 1842, and cf. Terpstra, P., Zeit. Krist., 1926, **64**, 150.

ties of the polymorphs, and, for comparison, those of RDX are given on p. 431.

The differences are very marked, and it will be appreciated that they enabled the crystal species in different batches of material to be identified with ease and certainty. Some typical samples of

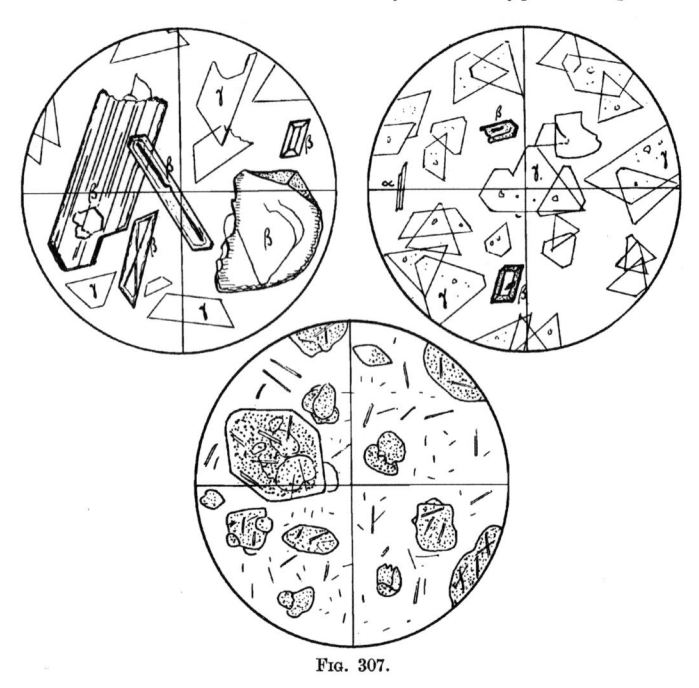

FIG. 307.

HMX as seen under the microscope are shown in Fig. 307 (top two drawings). One important fact which emerged as a result of the microscopic study was that crystals of α-HMX usually appeared first and subsequently became incorporated as inclusions in the crystals of RDX (see Fig. 307 (bottom)). It was therefore extremely difficult to get rid of them. It is worth noting in passing that this is the sort of information which can only be obtained by *optical* crystallographic methods. X-ray spectroscopy can reveal the different crystal species present in a mixture, but it cannot give any information about the state of aggregation of the crystals with respect to one another.

In the course of the same series of studies on explosives, two specimens, supposedly a specific compound (Fig. 308 (a)) related to RDX, were submitted for confirmation of their identity. The first did not correspond in crystallographic properties to (a) though it had nearly the same refractive indices. It did, however, show other optical properties which identified it as " BSX " (an allied compound ; see Fig. 308 (b)), namely, the central emergence of an optic axis on the basal plane, an optic axial angle of 62° and a positive sign. The edge angles of the basal plate were also identical with those of BSX.

The second specimen (Fig. 308 (c)) was also very interesting in

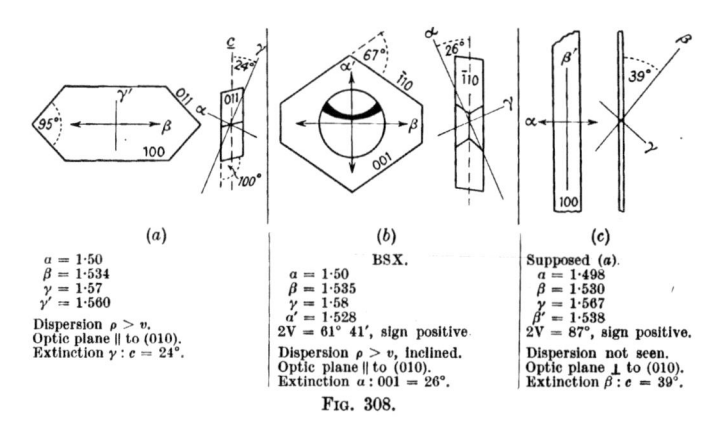

(a)
$a = 1\cdot50$
$\beta = 1\cdot534$
$\gamma = 1\cdot57$
$\gamma' = 1\cdot560$
Dispersion $\rho > v$.
Optic plane ∥ to (010).
Extinction $\gamma : c = 24°$.

(b) BSX.
$a = 1\cdot50$
$\beta = 1\cdot535$
$\gamma = 1\cdot58$
$a' = 1\cdot528$
$2V = 61° \ 41'$, sign positive.
Dispersion $\rho > v$, inclined.
Optic plane ∥ to (010).
Extinction $a : 001 = 26°$.

(c) Supposed (a).
$a = 1\cdot498$
$\beta = 1\cdot530$
$\gamma = 1\cdot567$
$\beta' = 1\cdot538$
$2V = 87°$, sign positive.
Dispersion not seen.
Optic plane ⊥ to (010).
Extinction $\beta : c = 39°$.

FIG. 308.

that, while the values of the refractive indices were close enough to those of (a) to suggest identity, the optic orientation was entirely different. Thus the optic axial plane was normal to (010) instead of parallel to it as in (a) ; the maximum extinction angle and the optical sign were also different. The optical properties, apart from the refractive indices, were also quite different from those of BSX (*cf.* Fig. 308 (c) with Figs. 308 (a) and (b)). This problem was one in which it was necessary to use the universal stage in order to fix the relationship of the optical ellipsoid to the plane of the crystal plate ((100) in (c)).

The Investigation of Heterogeneous Equilibria. Some of the best examples of the value of optical crystallographic determinations are to be found in researches on heterogeneous equilibria. In particular, the investigations of high-temperature systems of refractory oxides, which have been carried out over a number of years at the Geophysical Laboratory and the Bureau of Standards in America, and at the Building Research Station in Britain, would not have been possible without the use of the polarising microscope. Such systems cannot be studied by the method of thermal analysis used in metallurgy, since crystallisation is too sluggish for any arrest points to be observed on cooling curves. Heating curves have a limited usefulness, e.g. for checking inversion points of compounds exhibiting polymorphism, but the primary method of study is as follows. Mixtures of known composition are maintained at definite temperatures until equilibrium is established, and are then rapidly quenched. In each case the resulting solid is crushed and examined under the polarising microscope. Any liquid phase in the original equilibrium mixture quenches to a glass, which may be recognised by its isotropic character. The solid phases are differentiated, and, in the case of previously known compounds, identified by means of their optical properties. The composition of a new solid phase is given by that of any equilibrium mixture which is shown by the microscope to consist of that phase alone. By these methods the phase composition of any mixture at any temperature may be found, and the thermal diagram constructed.

A knowledge of the equilibria in systems of this type is of fundamental importance in petrological research and in the study of cement, slags, and refractories. The first and classical example was the system $CaO-Al_2O_3-SiO_2$, studied by Rankin and Wright at the Geophysical Laboratory.* The results of this investigation have shed much light on the constitution and behaviour of Portland cement, as well as other technological materials largely composed of these oxides, e.g. silica bricks used for furnace linings.† Other systems studied at the Geophysical Laboratory have included $MgO-Al_2O_3-SiO_2$ (Rankin and Merwin),‡ $CaO-MgO-SiO_2$ (Ferguson and Merwin),§ $Na_2SiO_3-Fe_2O_3-SiO_2$ (Bowen, Schairer, and Willems),‖ $CaO-Al_2O_2-MgO$ (Rankin and Merwin),¶ $K_2O-CaO-SiO_2$

* Amer. J. Sci., 1915, (4), 39, 1.
† In this connection see *Special Reports on the Mineral Resources of Great Britain*, Vol. XVI (1920), by Thomas, Hallimond, and Radley, pp. 47 f.
‡ Amer. J. Sci., 1918, (4), 45, 301.   § Ibid., 1919, (4), 48, 81.
‖ Ibid., 1930, (5), 20, 405.   ¶ J. Amer. Chem. Soc., 1916, 38, 568.

(Morey, Kracek, and Bowen).* At the Bureau of Standards the systems $CaO-Al_2O_3-Fe_2O_3$ (Hansen, Brownmiller, and Bogue)† and $CaO-Na_2O-Al_2O_3$ (Brownmiller and Bogue) ‡ have been investigated.

At the Building Research Station the quaternary system $CaO-2CaO,SiO_2-5CaO,3Al_2O_3-4CaO,Al_2O_3,Fe_2O_3$ was studied by Lea and Parker.§ This together with the ternary systems $CaO-Al_2O_3-SiO_2$ and $CaO-Al_2O_3-Fe_2O_3$ (see above) are the most important in connection with the study of Portland cement.

A compilation of phase-rule diagrams for these refractory systems has been made by Hall and Insley.||

Optical methods have also proved invaluable in the study of aqueous systems as a means of characterising, and confirming the identity of, solid phases. In conjunction with X-ray analysis they have been used for this purpose in investigations on the highly complex systems $CaO-Al_2O_3-H_2O$ and $CaO-SiO_2-H_2O$, which are both of importance in connection with the problem of the hydration of cements, whilst the former is of interest in other fields, such as water purification and soil equilibria. On account of the low solubility of the solid phases, equilibrium is not readily achieved, and investigations along orthodox lines (solubility and "moist residue" determinations) are therefore very difficult and some of the results not free from doubt. However, by a combination of chemical and crystallographic evidence, a number of compounds have been identified, some of which occur in nature. The work of the numerous investigators on these systems has been reviewed by Bessey,¶ and a more recent contribution to the problem has been published by Wells, Clarke, and McMurdie.**

In a paper on the system $Fe_2O_3-SO_3-H_2O$ by Posnjak and Merwin †† the authors state: " The establishment of the composition of the solid phase required the use of both chemical and microscopical methods and varied somewhat to meet the different conditions.' After the compounds encountered in this system were established

* Trans. Soc. Glass Tech., 1930, 14, 149.
† J. Amer. Chem. Soc., 1928, 50, 396.
‡ Amer. J. Sci., 1932, (5), 23, 501.
§ Phil. Trans. Roy. Soc., 1934, 234, No. 231, 1.
|| J. Amer. Ceram. Soc., 1933, 16, (10), 459; 1938, 21, (4), 113.
¶ Symposium on Chemistry of Cements, Stockholm, 1938, 178.
** J. Res., Nat. Bur. Standards, 1943, 30, 367.
†† J. Amer. Chem. Soc., 1922, 44, 1965.

the optical identification was relied upon, as this eliminated a great deal of analytical work." Optical identification was also used by Posnjak and Tunell in investigating the system CuO–SO_3–H_2O,* together with identification by means of X-rays in the case of those solid phases which were too fine-grained for microscopic methods.

A study of the system K_2CO_3–Na_2CO_3–H_2O at 100 and 150° was undertaken by Ervin, Giorgi, and McCarthy † in connection with an investigation of a process for the extraction of potassium carbonate from wyomingite, a leucite-bearing rock. The solid phases were (i) Na_2CO_3, (ii) Na_2CO_3,H_2O, (iii) K_2CO_3,Na_2CO_3, and (iv) $K_2CO_3,3/2H_2O$, and they were identified by their optical properties, and in some cases by X-ray analysis. The optical properties of (i) and (ii) were already on record in the literature but those of (iii) and (iv) had to be specially determined. The moist residue method was also used to some extent, but proved to be considerably less accurate than the crystallographic methods, owing presumably to difficulties associated with the high temperatures at which the system was studied.

In an investigation of the ternary systems $CaCl_2$–$MgCl_2$–H_2O, $CaCl_2$–KCl–H_2O, and $MgCl_2$–KCl–H_2O at 75° by Lightfoot and Prutton ‡ the identity of all moist solid phases was confirmed by examination under the polarising microscope, using an electrically heated hot stage to prevent further crystallisation by cooling from the equilibrium temperature.

The value of the polarising microscope in work of this kind cannot be overemphasised. The instrument does not, of course, avoid the necessity of first establishing the composition of some, at least, of the solid phases by analysis, but it may save a great deal of time by reducing the volume of analysis in the whole system, and, in any event, will provide evidence in confirmation of the analytical results. This is especially necessary in those by no means infrequent cases, where, in applying the moist residue method, the point of intersection of the tie lines is somewhat uncertain, as in the work on the system K_2CO_3–Na_2CO_3–H_2O cited above. To take a general case, suppose that a ternary system of two salts MX, NX, and water is to be investigated. MX and NX, or their hydrates, will exist as solid phases in the system and their optical properties can be determined beforehand, if they are not already on record. So too can the

* Amer. J. Sci., 1929, (5), **18**, 1.
† J. Amer. Chem. Soc., 1944, **66**, 384.   ‡ Ibid., 1947, **69**, 2098.

INVESTIGATION OF HETEROGENEOUS EQUILIBRIA 437

optical properties of any known double salt, which may be expected to occur in the system. These properties will usually enable the salts to be recognised readily under the microscope, and it will therefore be unnecessary to analyse all of the moist solid phases. Solid phases not previously known must first be established in the usual way, but in subsequent work on the system, e.g. at other temperatures, their optical properties may be used to identify them. Microscopic examination is also necessary to confirm cases of suspected polymorphism.

A study of the phase equilibria of acid soaps by McBain and Field * illustrates a somewhat different kind of application of the polarising microscope in phase rule work. In order to determine the critical points on heating curves in the system lauric acid–potassium laurate, use was made of the fact that acid potassium laurate is strongly, and normal potassium laurate weakly, birefringent. Between crossed Nicols the transition between these salts could therefore be sharply detected; so too could the commencement of melting by the appearance of the isotropic liquid phase.

Slags and Refractories. In the previous section, reference has been made to the importance of certain high-temperature phase rule investigations in connection with the study of slags and refractories. In the present section, some examples will be given of the application of optical crystallographic methods to the study of slags and refractories as such.

In metallurgy a knowledge of the chemical reactions which are taking place in the furnace is of vital importance. A study of the solid phases present in the slag at different stages in the process throws much light on these reactions. It is here that the polarising microscope (as in the phase rule studies already mentioned) can play an important part in supplementing the information given by bulk chemical analyses. For a short bibliography of papers on this subject, and as an example of what can be done, reference may be made to a paper by S. O. Agrell,[†] who lists the optical properties of 26 known phases which occur in basic slags, and shows how microscopic examination of thin sections and crushed material enables the phases in slags of various compositions to be identified,

* *J. Phys. Chem.*, 1933, **37**, 675.
† " Mineralogical Observations on Some Basic Open-Hearth Slags ", *J. Iron and Steel Inst.*, 1945, No. 11, 19P.

and the interactions (*e.g.* solid solubility) and order of crystallisation of the phases to be studied.

The optical method plays an important part in the study of refractories, both as regards the selection of raw materials and in the study of the behaviour of the finished product in use. For the manufacture of silica bricks, for example, the best material should not only be almost pure silica, with a very low content of alkalis (less than 2%), but should in addition consist of small angular grains of quartz cemented with secondary silica, *i.e.* silica deposited from solution. This information is gathered at once by sectioning the rock and examining under the polarising microscope. The presence of the common alkali-bearing minerals, the felspars, which are undesirable, is readily detected by their optical properties and their relative amount can be assessed by means of the Rosiwal method of measuring random linear intercepts on an eyepiece scale.*

Silica bricks are used for many different purposes and a variety of properties is therefore required. These depend upon the nature of the raw material, the crushing, the bonding (usually effected by adding about 2% of lime), and the firing. Bricks may vary greatly in such properties as strength, porosity, density, and coefficient of expansion. The user can check the suitability of the brick for his particular purpose by examination of a section under the microscope. Thus the density of the brick, which is closely related to its resistance to corrosion in a furnace, depends upon the ratio of the amount of unchanged silica to that of the higher temperature modifications, and this ratio can be estimated between crossed Nicols. Again, the structure of the brick can be determined, *e.g.* whether or not the ground mass consists of a strong felted scaffolding of tridymite crystals, in which case only a small expansion in use is to be expected, and the refractory character is greatly increased. The changes undergone by the brick in the furnace can also be studied by microscopic examination. In a soft-burned brick before use the silica is present mainly as quartz, distinguishable from the other

* The Rosiwal method for estimating the relative amounts of substances by volume in a thin section, is to make a series of traverses across the section, preferably using a mechanical stage, tabulating *random* intercepts for each substance. If the sums of random intercepts for, say, three substances are x, y, and z, respectively, the percentage volume of each is $100x/(x + y + z)$, etc. The relative weight of each is found by multiplying the volume of the substance by its density. See E. S. Larsen and F. S. Miller, *Amer. Min.*, 1935, **20**, 260.

modifications of silica by its marked double refraction. During the use the silica in the inner and hottest of the brick becomes converted to cristobalite, which is isotropic, and thus can be immediately recognised. Adjoining this zone is a somewhat cooler one, where the silica has been converted to tridymite, which has a low double refraction and characteristic habit. The widths of these zones show the temperature gradient to which the brick has been subjected, and this is an important factor in the life of the furnace lining.

For a comprehensive treatment of the application of optical methods to the study of refractories, see, for example, *Memoir of the Geological Survey, Special Reports on the Mineral Resources of Great Britain*, Vol. XVI, by Thomas, Hallimond, and Radley (1920). An example of the application of these methods to the study of raw materials for steel moulding sands is afforded by the work of Davies and Rees on sandstones of the North of England.* The necessary properties for this purpose, such as grain size and shape, lack of strain, and mineral composition, could only be determined under the microscope.

Textile Fibres. Cell Walls of Plants. Most natural and artificial textile fibres, and the walls of plant cells other than those used in textiles, are anisotropic to a greater or less degree, and they have been much studied by means of the polarising microscope. In order that the objects and results of such studies shall be properly appreciated, however, it will be necessary to include in this section some account of the main characteristics of these materials, since owing to the complexity and variability of their structure, they may show in certain respects significant differences in optical behaviour from that of true crystals, with which we have so far been mainly concerned.

The anisotropy of fibres was originally attributed to internal strain,† but it is now agreed that it is due to the fact, revealed by X-ray studies,‡ that fibres are wholly or mainly composed of long chain molecules which tend to be arranged more or less parallel to the fibre axis. In cotton, flax, hemp, and plant fibres in general,

* *J. Iron and Steel Inst.*, 1943, No. 11, 11P.
† See Harrison, *British Research Association for the Woollen and Worsted Industries*, Publication No. 2, 1918 ; also *Proc. Roy. Soc.*, 1918, **94**, 460.
‡ See for example, *Fundamentals of Fibre Structure*, W. T. Astbury (Oxford U.P., 1933).

the main structural constituent is cellulose,* the molecules of which consist of some hundreds of β-glucose residues linked together as shown below.

$$\cdots\text{—O—CH}\underset{\underset{\text{OH OH}}{|\ \ |}}{\overset{\overset{\text{CH}_2\text{OH}}{|}}{\overset{\text{O—CH}}{\diagup}}\diagdown}\text{CH—O—CH}\underset{\underset{\text{CH}_2\text{OH}}{|}}{\overset{\overset{\text{OH OH}}{|\ \ |}}{\overset{\text{CH—CH}}{\diagup}}\diagdown}\text{CH—O—CH}\underset{\underset{\text{OH OH}}{|\ \ |}}{\overset{\overset{\text{CH}_2\text{OH}}{|}}{\overset{\text{O—CH}}{\diagup}}\diagdown}\text{CH}-$$

Cellulose, or one of its derivatives, also constitutes those artificial fibres which are prepared from cotton, e.g. viscose rayon (regenerated cellulose) and acetate rayon (cellulose acetate). In protein fibres, such as wool and silk, and regenerated protein fibres, such as those made from casein, the molecules are long polypeptide chains—

$$\cdots\diagdown_{\text{CO}}\diagup\overset{\overset{\text{R}'}{|}}{\text{CH}}\diagdown_{\text{NH}}\diagup\overset{}{\text{CO}}\diagdown\underset{\underset{\text{R}''}{|}}{\text{CH}}\diagup\overset{}{\text{NH}}\diagdown_{\text{CO}}\diagup\overset{\overset{\text{R}'''}{|}}{\text{CH}}\diagdown\cdots$$

where R′, R″, etc. stand for about two dozen different kinds of side groups, which may be distributed along the main chain in many different ways. Fibres made of the important synthetic polymer nylon consist of long chain molecules formed by the repeated condensation of dicarboxylic acids and diamines, adipic acid and hexamethylene diamine being commonly used and resulting in a chain of the following type:

$$\cdots\diagup\underset{\underbrace{\qquad\qquad\qquad\qquad\qquad}_{\text{Adipic acid residue}}}{\overset{\text{CO}}{\diagdown}\text{CH}_2\diagup\overset{\text{CH}_2}{\diagdown}\text{CH}_2\diagup\overset{\text{CH}_2}{\diagdown}\text{CO}\diagup}$$

$$\underset{\underbrace{\qquad\qquad\qquad\qquad\qquad\qquad}_{\text{Hexamethylene diamine residue}}}{\diagup\overset{\text{NH}}{\diagdown}\text{CH}_2\diagup\overset{\text{CH}_2}{\diagdown}\text{CH}_2\diagup\overset{\text{CH}_2}{\diagdown}\text{CH}_2\diagup\overset{\text{CH}_2}{\diagdown}\text{NH}\diagup}\cdots$$

---

* Incrusting substances (lignin, pectin, etc.) occur in varying amounts in many plant fibres and other types of plant cells. Unless present in considerable proportion, however, they appear to have little effect on the optical behaviour. A high degree of lignification reduces the double refraction.

The preferred orientation of the molecules in the direction of the fibre axis, which exists in most fibres, approximates to an arrangement of parallel rods and so results in positive birefringence (see Chap. IV, p. 148). The lateral arrangement of the long molecules is, when averaged over the whole cross-section of the fibre, a random one, though a more regular arrangement exists over small elements of volume in some cases. The fibre therefore behaves as a positive uniaxial medium, with a maximum refractive index for light vibrating parallel to the length and a minimum one for all directions of vibration normal to the length. In fibres of highly acetylated or highly nitrated cellulose, however (cellulose triacetate and cellulose trinitrate), the birefringence is negative, presumably because the acid groups projecting sideways from the main chain destroy its rod-like character. In wool fibres the birefringence though positive is small, owing, it is believed, to the polypeptide chains being kinked in some way. (On stretching the fibres, a different X-ray picture is obtained, which is interpreted as being due to the chains becoming fully extended.) A similar explanation may be given for the very small birefringence which is shown by casein fibres. (See values for the birefringence of cellulose acetate, wool, and casein fibre given in the table on p. 442.)

The orientation of the molecules in artificial fibres arises from the way in which they are spun in the manufacturing process. In general a viscous solution of the polymer is extruded under pressure through a fine orifice. As the solution emerges, the solvent is either removed by evaporation or destroyed by chemical action according to its nature, leaving the polymer in the form of a thin filament which is reeled off as fast as it is formed. Thus in the viscose process, the cellulose is dissolved as the xanthate by treatment with caustic soda and carbon disulphide, and this solution after passing through the orifice, enters a bath of sulphuric acid which splits off the xanthate radical and regenerates the cellulose. In the cellulose acetate process, the solvent is acetone and this merely evaporates from the filament. In either case, the flow of the solution through the orifice and the slight tension to which the fibre is subjected as it is drawn away combine to orientate the long molecules parallel or nearly parallel to the fibre axis. The spinning of silk by the silkworm is a process essentially similar to that just described, and the molecules are particularly well orientated in this fibre.

The orientating influence of the above factors is shown by the fact that if a viscose solution is extruded very slowly through a coarse jet (1 mm. in diameter) without applying tension, the resulting fibre is isotropic or very nearly so.

Approximate values for the refractive indices (n_D) of some typical fibres are given below.

	Index for Light Vibrating Along the Length. (n_γ)	Index for Light Vibrating Across the Length. (n_α)	Birefringence and Sign.
Cellulosic fibres:			
Ramie, flax . .	1·59–1·60	1·53	0·06–0·07 (+)
Cotton * . . .	1·58	1·53	0·05 (+)
Mercerised cotton	1·56	1·52	0·04 (+)
Regenerated . .	1·55–1·57	1·52–1·53	0·02–0·05 (+)
Cellulose acetate .	1·47–1·48	1·475	0·005 (−) (when fully acetylated) to 0·005 (+)
Silk	1·59	1·54	0·05 (+)
Nylon	1·58	1·52	0·06 (+)
Wool	1·55	1·54	0·01 (+)
Casein fibre . .	1·54	1·54	Very small (+)

* In strict botanical terminology, cotton is a seed hair and not a fibre. From the practical point of view, however, it would be academic to insist on this distinction, and we shall not do so here.

Two general points regarding the refractive indices of fibres may be made. The first is that fibres of quite unrelated types may have refractive indices which do not differ very much (*e.g.* flax and silk, see table above). The second is that the refractive indices of fibres of one and the same type may show significant variations, affecting the third or even the second decimal place, depending on the degree of orientation of their molecules and their water, or other liquid content (see p. 443). A few stand out sharply, *e.g.* cellulose acetate with its low indices and low double refraction, but in general it must be said that the positive identification of fibre type by refractive indices *alone* is not possible. However, fibres differ greatly in their morphological characteristics (compare, for example, the flattened and twisted hollow cotton fibre with the scaly cylindrical wool fibre), and many may be differentiated by their reaction to dyes which often produce characteristic pleochroic effects due to orientated adsorption of the dye molecules ; so that where identification of an unknown fibre under the microscope is called for, as in certain forensic problems, there is usually no difficulty.*

Fibres possess the property of absorbing water and other liquids of low molecular weight in varying degrees. This suggests that the fibre material is not continuous, but that the long molecules are grouped into bundles (*micelles* or *crystallites*) with spaces between them into which the liquid can penetrate. This theory was originally put forward by Nägeli, who pictured the micelles as discrete building blocks in the structure. It has received general support from X-ray studies of fibres, and an estimate of the size of the micelles can be made from the width of the spots on the X-ray diffraction photograph. In native cellulose fibres they appear to be about 60–70 Å wide and at least 600 Å long, and figures of this order are suggested by other evidence derived from studies of the internal surface of the fibres (dye and gas absorption experiments). The modern view, however, is in favour of a structure consisting of bundles of parallel molecules which merge *continuously* into intervening amorphous regions in which the chains are in disarray, rather than one consisting of discrete blocks. That is to say, individual molecules can, at different parts of their length, occupy both crystalline and amorphous parts of the structure.

The absorption of water by a fibre lowers both its refractive indices. This is illustrated by the following values obtained by

Grams H_2O per gram of Cellulose.	$(n_\gamma)_D$	$(n_x)_D$
0·023	1·574	1·534
0·038	1·573	1·535
0·055	1·570	1·531
0·092	1·565	1·529
0·135	1·555	1·519
0·198	1·541	1·506

* In connection with the general microscopy of fibres, the reader is referred to *Modern Textile Microscopy*, J. M. Preston (Emmott, London, 1933), and *Handbook of Chemical Microscopy*, Vol. I, E. M. Chamot and C. W. Mason (Wiley, 1938). A useful collection of photomicrographs of natural and synthetic fibres with explanatory notes, *Textile Fibres under the Microscope*, has been published by Imperial Chemical Industries Ltd. (1939.)

P. H. Hermans * on specimens of a viscose rayon containing different amounts of water.

It is evident therefore that precise values for the refractive indices of fibres can only be obtained if (*i*) their moisture content is definite, (*ii*) there is no interchange of moisture between the fibres and the liquids used as immersion media, and (*iii*) these liquids do not themselves penetrate the fibres. We shall return to these points later.

A fibre having a micellar structure as pictured above, and containing absorbed liquid, should owe its double refraction in part to a cause other than the optical anisotropy of the molecules. It was shown by Wiener † that a body consisting of parallel rod-shaped particles, composed of isotropic material having a refractive index n_1, embedded in an isotropic medium having a different refractive index n_2, will be anisotropic provided that the rods are narrower than the wave-length of light. According to Wiener, the double refraction of the body, $n_\gamma - n_a$ (where n_γ is the index for light vibrating parallel to the rods, and n_a that for directions normal to them), is given by the expression :

$$n_\gamma - n_a = \frac{\delta_1 \delta_2 (n_1^2 - n_2^2)^2}{(\delta_1 + 1)n_2^2 + \delta_2 n_1^2}$$

where δ_1 and δ_2 are the relative volumes of the rods and the embedding medium respectively, so that $\delta_1 + \delta_2 = 1$. Thus for a given value of δ_2, the double refraction should vary with the refractive index of the embedding medium.

It is difficult to test the Wiener equation experimentally, mainly owing to uncertainties in the determination of δ_2, and to the fact that the birefringence due to the structural anisotropy of the molecules is very much larger than the effect being sought. The equation assumes that the total volume remains constant, *i.e.* that imbibition of liquid is not accompanied by swelling of the fibre, whereas this does, and theoretically must, take place. Also the properties of the liquid in the fibre are somewhat different from those of the free liquid. It has a higher density, a lower vapour pressure, and a different refractive index. Moreover there is the possibility that it may itself acquire an anisotropic character as a result of orientated

---

* *Monographs on the Progress of Research in Holland during the War. Contributions to the Physics of Cellulose* (Elsevier, 1946).
† *Abh. Sachs. Ges. (Akad.) Wiss.*, 1912, **32**, 507.

adsorption on the surfaces of the micelles. According to P. H. Hermans,* the initial adsorption of water by a cellulose fibre is a process of molecular dispersion of water molecules in the amorphous regions, so that the optical behaviour follows the law of molecular mixtures rather than the Wiener theory. At high percentages of water, however, layers of water several molecules thick will be formed, and there is evidence of a small Wiener component in the total birefringence, though smaller than that calculated. Frey † has carried out experiments on grass haulms from which the organic matter has been removed, leaving only a silica skeleton. When this skeleton was caused to imbibe liquids of different refractive indices, the birefringence was found to vary in rough agreement with Wiener's equation, and became zero when a liquid having the same refractive index as the silica was used, as the equation requires. To sum up, such evidence of this kind as exists is not out of harmony with the idea of a structure consisting of micelles of sub-microscopic dimensions.

We shall now consider the values given for the refractive indices of cellulosic fibres in the table on p. 442. Comparing first the indices of ramie and flax with those of cotton, it will be noted that whereas the cross index, n_a, is the same in all three cases, the length index, n_y, for cotton is considerably less than the corresponding figure for the other two. The reason lies in the fact that in ramie and flax, the micelles are well orientated parallel to the fibre axis, whilst in cotton they are arranged in a spiral manner and make an angle of about 30° with the fibre axis.‡ This is explained more fully in Fig. 309 (a). The cotton fibre is hollow, and in any region of the wall the optical indicatrix XOZ is inclined so that the major axis OZ makes an angle (of about 30°) with the fibre axis, since the direction of OZ is determined by the direction in which the long axes of the micelles lie. In making a refractive index determination by the immersion method, it is the *edge* of the fibre which is observed and which disappears when a match with the liquid has been

---

* *loc. cit.* † *Jahrb. wiss. Bot.*, 1926, **65**, 200.

‡ This statement regarding the cotton fibre strictly refers only to the so-called *secondary wall*, which, however, comprises by far the greater bulk of the material in the mature fibre. The very thin outer *primary wall* is believed to contain cellulose micelles, possibly widely spaced and criss-crossed, together with incrusting substances. In any event it seems to be isotropic or nearly so, and to have no appreciable effect on the optical behaviour of the fibre as a whole, though the reason for this is not fully understood.

obtained. The refractive index for light vibrating parallel to the length of the fibre at its edge is that corresponding to the major axis of the section of the indicatrix which is parallel to this direction, namely the axis OZ' of the section XOZ' in the figure. The minimum index n_a * is not affected by the inclination of the indicatrix, as will be clear from the figure. Thus this index is the same in cotton as in ramie and flax. The value of the length index, n'_γ, for cotton can be calculated from the spiral angle as determined by X-ray analysis, and the length index, n_γ, of ramie or flax by

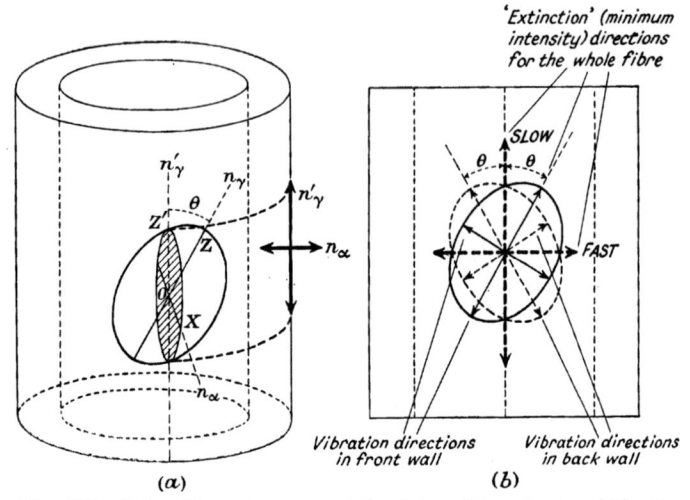

Fig. 309.—Optical Characters presented by Cotton Fibres (a) in a Refractive Index Determination, (b) between Crossed Nicols (diagrammatic).

means of the following equation based on the geometry of the ellipsoid, and the result is in satisfactory agreement with observation.

$$(OZ')^2 \cos^2 \theta/(OZ)^2 + (OZ')^2 \sin^2 \theta/(OX)^2 = 1$$

or

$$n'^2_\gamma \cos^2 \theta/n_\gamma^2 + n'^2_\gamma \sin^2 \theta/n_a^2 = 1$$

(The symbols have the same significance as in Fig. 309 (a).) Con-

* Although a fibre as a whole behaves as a uniaxial medium, the symbols α and γ are commonly used in preference to ω and ε, since within a micelle, *i.e.* within a single crystalline region of the substance, the structure is, in general, that of a biaxial crystal. The cellulose crystal is monoclinic for instance.

versely the spiral angle can be calculated from the observed values of n' and n_y.

When a cotton fibre is examined between crossed Nicols, it does not extinguish completely in any position, but the light transmitted by the analyser has a minimum intensity when the fibre is parallel to a cross-wire. This results from the spiral arrangement of the micelles. The light passes through both the front and back walls of the fibre, and the vibration directions for these two walls are at an angle, which, for a perfectly cylindrical fibre, reaches a maximum value of 2θ (between the " slow " directions) along the centre line (see Fig. 309 (b)). We have here an example of the general case of two superposed crystal plates at different orientations between crossed Nicols, which is treated in most textbooks of advanced optics.* When the two plates are of the same thickness and structure, and so produce the same relative retardation (see Chap. VII, p. 244) as may be assumed in the present case, it may be shown that the intensity, I, of the light transmitted by the analyser is given by:

$$I = a^2[\sin 4\theta \sin\{4(\alpha - \theta)\} \sin^2 \frac{\delta}{2}$$
$$+ \sin 2\alpha \cos^2 2\theta \sin\{2(\alpha - 2\theta)\} \sin^2\delta]$$

where a is the amplitude of the incident light, 2θ the acute angle between the slow vibration directions for the two plates, α the angle between the vibration direction of the incident light and the slow direction of one of the plates, and δ is the *phase difference* for each plate, *i.e.* the difference in phase between the fast and slow rays emerging from the plate. Phase difference is expressed in angular measure, and is obtained at once from the relative retardation or path difference (Chap. VII, p. 244) by multiplying this by $360/\lambda$ or $2\pi/\lambda$, according as the answer is required in degrees or radians, λ being the wave-length of light used. The equation gives a minimum value for I at $\alpha = \theta$, $(90 + \theta)$, etc., *i.e.* when the cross-wires of the microscope bisect the angles between the vibration directions for the two plates, the slow direction for the compound plate (or major extinction position as this is sometimes called) being the bisectrix of the angle 2θ. A further consequence of the equation is that if $2\theta = 90°$, the intensity is zero for all values of α, *i.e.* the compound plate behaves as though it were isotropic, a conclusion

* See for example, *The Theory of Light*, T. Preston (Macmillan, 1928).

which should be obvious on general grounds, since the plates are in the position where the slow direction of one is parallel to the fast direction of the other and *vice versa*. Compound plate phenomena, of which the cotton fibre affords a simple example, are commonly encountered in the study of plant tissues, and further cases are cited later.

In the above treatment the cotton fibre has been represented as though it were a perfectly cylindrical tube. Actual cotton fibres, unless swollen with liquid, are always very much flattened, with numerous twists and convolutions along their length, and show an irregular cross-section. This does not, however, materially affect the conclusions reached above, if observations are confined to untwisted portions of the fibres.

We may next compare the refractive indices of the natural cellulosic fibres, such as ramie, flax, and cotton, on the one hand with those for mercerised cotton (*i.e.* cotton treated with caustic soda solution followed by washing) and regenerated cellulosic fibres on the other. As the figures in the table on p. 442 show, both n_γ and n_a are lower for the second group than for the first. The reason is that the processes of mercerisation and regeneration produce another form of cellulose (cellulose hydrate) with a different crystal structure from that of native cellulose. The long chains are still parallel to one another, but the side-to-side packing is different.

The practical applications of the optical study of textile fibres range from the simple use of polarised light to bring out macro-structural features, such as the nodes and dislocations in flax and hemp,* to the differentiation of allied types of fibres,† and to detailed investigations of the fine structure or of the relationship between the optical and other physical properties, such as strength. Only a few of the more interesting examples can be noticed here.

J. M. Preston ‡ has described a test for mercerisation, based on the fact, mentioned above, that this process lowers both refractive indices of the fibre. Tests for the maturity of cotton fibres may be made, depending on the fact that the immature hairs have a very low birefringence, and show the characteristic blue and yellow colours with the first order selenite plate between crossed Nicols,

* Ambronn, *Kolloid Z.*, 1925, **36**, 119.
† A. Herzog, *Die Unterscheidung der Flachs und Hanffaser* (Berlin, 1926); *Die mikroskopische Untersuchung der Seide und der Kunstseide* (Berlin, 1924); *Technologie der Textilfasern* (1930).
‡ *Trans. Far. Soc.*, 1933, **29**, 65 ; *Modern Textile Microscopy, op. cit.*

whereas the mature fibres have a much stronger birefringence.* The progressive nitration of cellulose fibres results in a change from positive to negative birefringence at about 12% of nitrogen (fully nitrated cellulose, i.e. cellulose trinitrate, contains about 15%) and thus a simple microscopic test is afforded of the approximate degree of nitration. Undesirable irregularities in the nitration of the fibres can also be readily detected under the microscope.†

Herzog (in the first work cited in the footnote on p. 448) has described a means of distinguishing between hemp and flax fibres. This cannot be done by means of their refractive indices for these are too close, and they also show the same polarisation colour between crossed Nicols. On insertion of the first order selenite plate (in the usual 45° position), however, they behave differently as follows. When the length direction of the fibre is parallel to the vibration direction of the analyser, hemp shows a first order orange, and flax a colour between first order red and second order violet. On turning the stage through 90°, these colours are reversed. No difference in colour is observable when the fibres are either parallel or normal to the slow vibration direction of the selenite plate. The test evidently depends on a comparatively small structural difference between the two materials.

R. Meredith ‡ has made a careful study of the relationship between the refractive indices of 36 samples of raw cotton, drawn from a wide range of sources, and their tensile properties. He finds that about 80% of the variation in tensile strength and initial Young's modulus can be attributed to variation in the molecular orientation (spiral angle of the micelles) as measured by the double refraction (see equation on p. 446). The refractive indices had to be determined with the greatest possible accuracy, since the differences were small; the maximum and minimum values of n'_γ obtained were 1·582 and 1·574 (D), corresponding to spiral angles of 27° and 35° respectively. The measurements were made in an air-conditioned room with the temperature controlled to $\pm 1°$ and the relative humidity to $\pm 2\%$. Both fibres and immersion media were given time to come to equilibrium with the humidity of the

* See, e.g., Schwartz and Shapiro: *Rayon Text. Monthly*, 1938, **19**, 371, 421, 480, 570.
† Olsen, *Colloid Symp. Mon.*, VI, 1928.
‡ *Shirley Institute Memoirs*, 1944–5, **19**, 365.

room so that there should be no exchange of water between them during a determination. The liquids used were mixtures of liquid paraffin and α-monobromonaphthalene, which had previously been found suitable for cotton fibres, *i.e.* they did not penetrate the fibres. A further control of the temperature was effected by carrying out all determinations on a constant temperature stage, consisting of a small glass cell through which water at constant temperature was circulated. Determinations were made on a

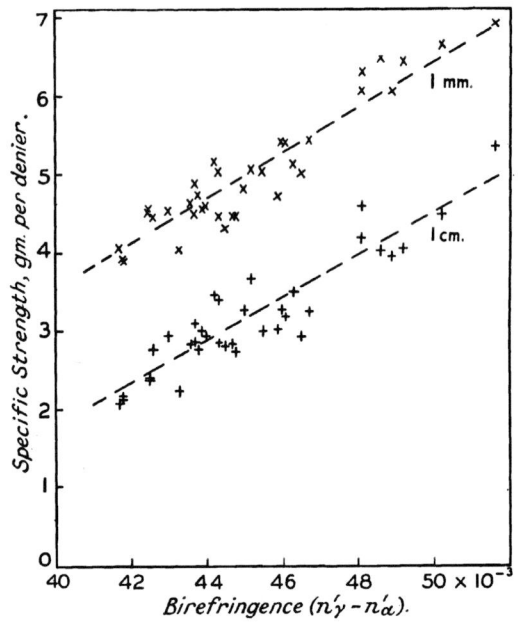

FIG. 310.—Relation between Birefringence and Specific Strength of Cotton Fibres (1 mm. and 1 cm. lengths).

number of fibres for every sample examined. (This is necessary in all refractive index work on fibres, even when the accuracy required is much less than in the case now described, for individual fibres of a sample are always likely to show significant variations.) Meredith's specific strength/birefringence graphs are shown in Fig. 310.

It is appropriate to mention here other immersion media which have been found suitable for fibres. On grounds of permanence, P. H. Hermans (*op. cit.*) recommends mixtures of butyl stearate and tricresyl phosphate (1·4446 to 1·5586), and tricresyl phosphate and diphenylamine with 1% hydroquinone as an antioxidant (1·5586 to 1·602) for natural and artificial cellulosic fibres. He states that technical butyl stearate contains some other volatile ester to lower the freezing-point, but that this can be removed by distillation *in vacuo*. R. D. Preston (see later) has used mixtures of cedar oil and α-monobromonaphthalene, and xylol and α-monobromonaphthalene for cellulosic materials, but the latter mixture has the disadvantage that it loses xylol rather rapidly unless the mount is sealed against evaporation.

In general it may be said that any immersion medium, which is satisfactory on other grounds, may be used in fibre work provided that (*i*) it is without solvent action on the fibres, and (*ii*) it does not cause them to swell. Swelling indicates imbibition of the liquid into the inter-micellary spaces, and makes the measurement of the refractive index of the fibre in its original state impossible.

The walls of plant cells other than those used for textile purposes have been much investigated from the botanical standpoint by optical and other methods, with the object of elucidating their structure and gaining an insight into the mechanism of their growth. As in the plant textile fibres, the main structural constituent is cellulose. Such walls often consist of a number of layers, not necessarily of equal thickness, in which the orientation of the micelles is different in adjacent layers, so constituting a compound plate of a more complex type than that illustrated by the cotton fibre (Fig. 309 (*b*)). From an abundant literature on the subject, the major contributions to which include papers by A. Frey-Wyssling, I. W. Bailey, and G. van Iterson Jr., we may mention some of the researches of R. D. Preston, which are characterised by a particularly sound appreciation of the physical principles involved. Preston has used both X-ray and optical methods, which are largely complementary for the solution of problems of this kind. In the following accounts, the botanical implications of the researches are not dealt with.

As the first example we will take Preston's study of the cross walls of xylem vessels of *Helianthus annuus* (sunflower).* These walls consist of bars or rods having an approximately circular cross-

* *Proc. Leeds Phil. Soc.*, 1935, **3**, (II), 102.

section and composed of lignified cellulose. The major extinction position (m.e.p.) of the bars is parallel to their length, but this observation by itself is no indication that the cellulose chains necessarily run in the same direction (compare the m.e.p. of cotton fibres, for example, Fig. 309 (b)). However, after the lignin had been removed by treatment with dilute sodium hydroxide solution, the bars showed refractive indices which were very nearly the same as those of ramie fibre, in which it is known from X-ray analysis that the cellulose chains are almost parallel to the length. Thus it could be inferred that the orientation in the bars was similar.

The birefringence of the bars, as obtained from the direct determination of the refractive indices, was confirmed by measurements of their phase difference, by means of a Sénarmont compensator (see note at end of chapter), and their diameter, d, by means of a micrometer scale, whence from the relation $\dfrac{\delta\lambda}{2\pi d} = (n_\gamma - n_a)$ the birefringence could be calculated.

Preston and W. T. Astbury have investigated the wall structure of the green alga *Valonia ventricosa*.* The wall of this bubble-like cell consists of numerous thin layers, and under the microscope it shows two sets of striations crossing each other at an angle of rather less than 90°, one being usually more easily visible than the other. By means of an X-ray microspectrometer (*i.e.* one designed to give the X-ray spectrum of a very small area of the material) applied to sections of the wall, which were marked so that they could be subsequently identified, it was established that the striations were parallel to two sets of cellulose chain directions, and that the more easily visible set of striations always corresponded to the set of cellulose chains giving the more intense diffraction spots, *i.e.* to the set present in greater thickness. The m.e.p. varied for different parts of the wall but always lay in the acute angle between the two sets of striations, and nearer to the more prominent set. This is just what is to be expected from a compound plate consisting of many layers showing two different orientations, the total thickness of the plates having one orientation being different from that of the plates having the other orientation. The laminated nature of the wall could be clearly made out at the edges of torn fragments, and occasionally on such edges very thin lamellæ

* *Proc. Roy. Soc.*, B, 1937, 122, 76.

could be seen, which only showed one set of striations parallel to the m.e.p. ; these were probably single layers. Also on these torn edges, it was possible to observe the change in m.e.p. with the number of layers through which the light passed.

The importance of these observations in connection with the general study of plant cell walls by means of the microscope was that (i) they supported the view, which was already generally accepted, that whenever striations are visible on walls of the higher plants (phloem and xylem fibres, tracheids, and cotton hairs) they correspond closely to the structure of the wall, and (ii) they demonstrated that the m.e.p. of a wall can only be taken as an indication of the cellulose chain direction, or mean direction, if this is the same throughout.

Preston and Astbury also studied the cell wall of the green alga *Cladophora*,* and found it to have a very similar structure to that of *Valonia*.

Preston has been able to show by optical means alone that the cellulose chains in the walls of parenchyma in oat coleoptiles † are arranged in a spiral, the angle of which in relation to the longitudinal axis of the cell varies from about 75° for cells which have a small length to breadth ratio, to about 30° for the most elongated ones. These angles were obtained from the m.e.p.'s on single walls, which in this case could be taken as parallel to the cellulose chain directions, since transverse sections of the walls appeared uniform in polarised light, indicating uniformity of structure. The single walls were obtained by mounting whole cells on a slide by means of albumin, and then scraping the upper wall away with a fine needle. In agreement with the above results, whole cells with a small length to breadth ratio showed a transverse m.e.p. (*i.e.* a negative sign of elongation), whilst the longest ones had the m.e.p. along the length (positive sign of elongation). At an intermediate degree of elongation the whole cell appeared to be isotropic. The transverse m.e.p.'s were often not quite at 90° to the length axis, and this indicated that the back and front walls in these cases were not quite identical, possibly in thickness, m.e.p., or structural arrangement of the cellulose. Fig. 311 illustrates the above results diagrammatically.

A number of Preston's researches have been concerned with the orientation of the cellulose micelles in conifer tracheids and wood

* *Ibid.*, 1940, **129**, 54. † *Ibid.*, 1938, **125**, 372.

fibres.* In transverse section between crossed Nicols, the walls of these cells typically show three layers, the two outer ones appearing bright and the middle one dark. The X-ray evidence pointed to only one mean direction for the cellulose micelles in the wall as a whole, and Preston attributed the difference in optical behaviour of the three layers to differences in angular dispersion, though others, notably I. W. Bailey, considered that in the outer (bright) layers the micelles were arranged transversely while in the

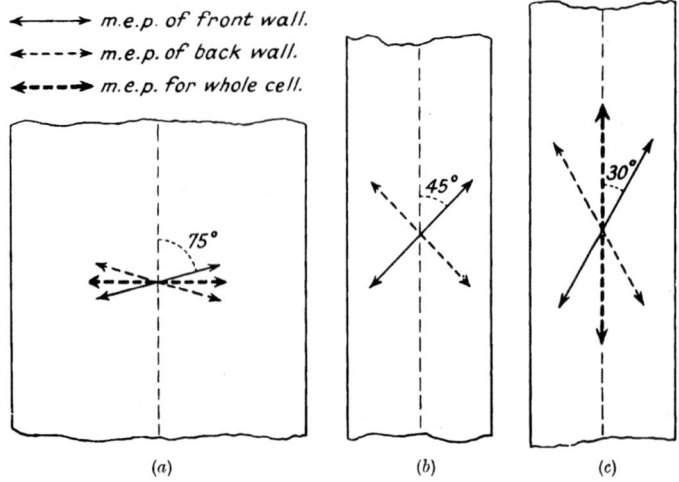

⟶ m.e.p. of front wall.
⟵ --- ▶ m.e.p. of back wall.
◀--- ▶ m.e.p. for whole cell.

(a)   (b)   (c)

FIG. 311. — Major Extinction Positions of Parenchyma in Oat Coleoptiles (diagrammatic).

(a) Short wide cell. Sign of elongation negative. (b) Intermediate case. Isotropic. (c) Long narrow cell. Sign of elongation positive.

middle (dark) layer they were arranged longitudinally. For two species, one of *Picea* and another *Nothofagus cunninghami*, the question seems to have been decided recently by experiments by A. B. Wardrop and Preston † in which the birefringence of sections cut at different angles to the longitudinal axis have been determined by means of the Sénarmont compensator and measurements of the thickness. If the mean direction of the chains were the

* *Proc. Leeds Phil. Soc.*, 1939, **3**, (IX), 546. *Proc. Roy. Soc., B.*, 1946, **133**, 327. *Ibid.*, 1947, **134**, 202.
† *Nature*, 1947, **160**, 911.

same throughout all three layers they should show a maximum birefringence at the same angle of sectioning. This proved not to be the case. In the *Picea* species, the outer layer showed a maximum birefringence at about 50°, and the middle one at about 18°, to the long axis. In *Nothofagus* the corresponding angles were 60–90° and about 10°. The birefringence of the innermost layer was not determined, owing to difficulties of observation. Thus for these species at least, the orientation or mean orientation of the micelles is much more transverse in the outer than in the middle layer. There must, however, be other factors involved in the optical heterogeneity of the wall (of which differences in angular dispersion may be one), for the maximum birefringence of the outer layer is 0·02, whereas that of the middle layer is quite different, namely about 0·04.

Note on the Sénarmont compensator. The principle underlying this compensator is explained in various textbooks on advanced optics, though not usually under that name, and by Ambronn and Frey in *Das Polarisationsmikroscop* (Leipzig, 1926). The compensator consists of (i) an accurate $\lambda/4$ retardation plate, mounted above the specimen and with its slow direction of vibration *parallel* to the vibration direction of the polariser of the microscope (not at 45° as in the case of the usual $\lambda/4$ compensator), (ii) a rotatable analyser with a divided circle so that the amount of rotation can be measured. The analyser is first crossed with respect to the polariser, and the specimen is placed on the stage and turned into the 45° position. Monochromatic light must be used. The light from the polariser, OP (Fig. 312) is resolved into the two components OS (slow) and OF (fast) by the crystal, and these combine to give, in general, an elliptic vibration with semi-axes OX (parallel to the vibration direction of the polariser) and OY, as shown in the figure. If δ is the phase difference between the components OS and OF, then it can be shown that

$$\text{OY}/\text{OX} = \tan \delta/2$$

On entering the $\lambda/4$ plate, the elliptic vibration is resolved along the vibration directions of the plate, *i.e.* along the axes of the ellipse OX and OY. As may readily be shown, the phase difference between these components on entry is $\pi/2$ radians (equivalent to a retardation of $\lambda/4$). Passage through the plate increases the difference by another $\pi/2$ radians, so that on emergence the components differ in phase by π radians (equivalent to $\lambda/2$). They therefore combine to produce linearly polarised light OR. If now the analyser be turned into the position OA' perpendicular to OR, extinction will result. The angle θ (AOA') through which the analyser has been rotated is equal to the angle ORY, the tangent of which is OY/YR = OY/OX. Thus from the relation given

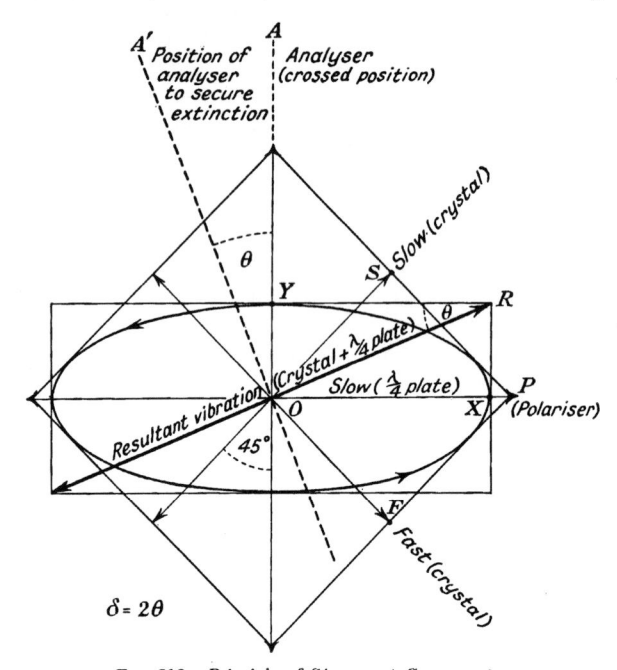

Fig. 312.—Principle of Sénarmont Compensator.

above, it follows that the required phase difference δ is equal to 2θ.

Instead of making a single determination of θ and multiplying by 2, a more accurate result may be obtained as follows. The position of the analyser to give darkness is determined as before, and the crystal then turned through 90°. The analyser will now have to be rotated to the other side of the crossed position to obtain extinction. The angular difference between these two extinction positions of the analyser is equal to δ.

It should be noted that by this method the same result is obtained for $\delta = x°$ as for $\{n(360) + x\}°$, where n is a whole number. This disadvantage is slight, however, for the order of the phase difference can usually be inferred from the polarisation colour in white light, and the value of n so obtained.

APPENDIX

TABLES AND COLLECTIONS OF OPTICAL CRYSTALLOGRAPHIC DATA

The following list is not exhaustive, but is only intended as a general guide.

Chemische Kristallographie, P. Groth, Vols. I–V (Engelmann, 1906–19). Both morphological and optical data are given in this work.

International Critical Tables, Vol. I, pp. 165, 279, 320; Vol. VII, pp. 12, 16.

Microscopic Characters of Artificial Inorganic Solid Substances or Artificial Minerals, A. N. Winchell, Part II (Wiley, 1931).

Microscopic Determination of the Non-opaque Minerals, E. S. Larsen, Bull. Geol. Survey U.S.A., No. 848 (1934), pp. 47 et seq.

The Optical Properties of Organic Compounds, A. N. Winchell (The University of Wisconsin Press, 1943).

Physikalisch-Chemische Tabellen, H. R. Landolt and R. Bornstein, Vol. II, pp. 922, 927, 932; Ergänz. I, pp. 486, 490, 499; Ergänz. IIb, pp. 719, 729, 746.

Tables Annuelles Internationelles de Constantes &c. (Paris).

Tables for the Microscopic Identification of Inorganic Salts, W. H. Fry, Bull. Dept. Agr. U.S.A., No. 1108 (1922).

In addition to the data scattered throughout the chemical journals, chemical substances are often described in the following:

American Journal of Science; American Mineralogist; Mineralogical Magazine; Zeitschrift für Kristallographie.

SOURCES OF CERTAIN MATERIALS MENTIONED IN THE TEXT

Index immersion media :
Baird & Tatlock (London) Ltd., Freshwater Road, Chadwell Heath, Essex.
British Drug Houses Ltd. (B.D.H. Laboratory Chemicals Group), Poole, Dorset.
R. P. Cargille, 118 Liberty St., New York 6, N.Y., U.S.A.
J. T. Rooney, P.O. Box 358, Buffalo, N.Y., U.S.A.

Mineral, rock, and orientated crystal sections :
Gregory, Bottley, & Co., 30 Old Church St., Chelsea, London, S.W.3.
The Geotechnical Laboratory, 33 Gower St., London, W.C.1.
W. Harold Tomlinson, 260 N. Rolling Rd., Springfield, Pa., U.S.A.
Ward's Natural Science Establishment Inc., P.O. Box 24, Rochester 9, N.Y., U.S.A.

Stereographic nets and crystal models :
Thomas Murby & Co., 40 Museum St., London, W.C.1.

OPTICAL CRYSTALLOGRAPHIC STUDIES ON GROUPS OF CLOSELY RELATED ORGANIC COMPOUNDS

(*Note*: The comprehensiveness of the studies varies considerably from case to case.)

Class of Compound.	Author(s).	Reference.
Alcohols:		
Heptitols	Wherry	*J. Biol. Chem.*, 1920, **42**, 377
3 : 5-Dinitrobenzoates of aliphatic,	Bryant	*J. Amer. Chem. Soc.*, 1932, **54**, 3758 ; 1933, **55**, 3201
Aldehydes:		
2 : 4-Dinitrophenyl-hydrazones of aliphatic,	Bryant	As above
Aliphatic acids:		
Bromanilides of the lower,	Bryant and Mitchell	*J. Amer. Chem. Soc.*, 1938, **60**, 1394, 2748
Salts of aconitic acid	Keenan, Ambler, and Turer	*J. Amer. Chem. Soc.*, 1945, **67**, 1
Naphthalene sulphonic acids:		
α- and β-Naphthylamine salts of,	Ambler and Wherry	*J. Ind. Eng. Chem.*, 1920, **12**, 1081
Benzyl-ψ-thiourea salts of,	Hann and Keenan	*J. Phys. Chem.*, 1927, **31**, 1082
Amino acids:		
Free acids	Keenan	*J. Biol. Chem.*, 1924, **62**, 163 ; 1929, **83**, 137
Picrolonates of,	Dunn, Inonye, and Kirk	*Mikrochemie*, 1939, **27**, 154

Amines:		
Picrates of, . . .	Bryant and Mitchell	*J. Amer. Chem. Soc.*, 1943, **65**, 128
Carbohydrates:		
Free compounds . . .	Wherry	*J. Amer. Chem. Soc.*, 1918, **40**, 1852; 1920, **42**, 125
		J. Wash. Acad. Sci., 1928, **18**, 302
Pentosazones and hexosazones .	Wright	*J. Amer. Chem. Soc.*, 1916, **38**, 1647
Phenylosazones of, . . .	Morris	*J. Amer. Chem. Soc.*, 1932, **54**, 2843
Alkaloids:		
Free compounds . . .	Wright	*J. Amer. Chem. Soc.*, 1916, **38**, 1647
	Wherry and Yanovsky	*J. Amer. Chem. Soc.*, 1918, **40**, 1063
Salts of strychnine . . .	Poe and Sellers	*J. Amer. Chem. Soc.*, 1932, **54**, 249
		J. Ind. Eng. Chem. (Anal.), 1932, **4**, (1), 69
Thiocyanates of, . . .	Keenan and Warren	*J. Amer. Pharm. Assoc.*, 1945, **34**, 300
Aromatic nitro- compounds:		
(*o*-, *m*-, and *p*-nitro-anilines, -phenols, -aldehydes, -acids, -toluenes, and -chloro- and -bromo- benzenes) . .	Davies and Hartshorne	*J. Chem. Soc.*, 1934, 1830
Miscellaneous:		
90 substances in U.S. Pharmacopœia	Keenan	*J. Assoc. Official Agr. Chem.*, 1944, **27**, 153
34 substances in N.F. VII . .	Keenan and Eisenberg	*J. Amer. Pharm. Assoc.*, 1946, **35**, 94

INDEX

Absorbing crystals, 124 et seq.
  indicatrix for, 125
  " rotation axes " in biaxial, 125
Absorption, Lambert's exponential law, 124
Absorption axes, 239
— coefficient, 124
— formula, 239
— of light, 124, 238
Acicular habit, 46
Acute bisectrix, 119
— — sections in convergent light, 284
Agrell, S. O., 437
Alanine, preparation of, 428
Albite,
  twinning of, 84
  twinning between crossed Nicols, 267
Alcohols, 3,5-dinitrobenzoates of, 425, 460
Aldehydes, 2,4-dinitrophenylhydrazones of, 460
Alkaloids, 422, 423, 461
Allotriomorphic crystals, 44
Allotropy, 16
Ambler, J. A., 460
Ambronn, H., 448, 455
Amines, picrates of, 425
Amino acids, 427, 460
Ammonium dihydrogen phosphate, 374, 427
— magnesium phosphate, 70
— nitrate, 12, 150, 364
  rotation of ions in, 12
— sulphate, 69
Amorphous state, 15
Analyser, 164
Angle between primary and secondary optic axes, 117
— of incidence, 94, 96
— of refraction, 95, 96
Angular aperture, 167
  effect of, on interference figure, 169, 270

Anhedral crystals, 44
Anisotropic substances, 89
Anisotropism, 14, 241
Anomalous interference colours, 249
Anorthic system, 67
Apatite, 73
" Araclor ", 213
Arago, F. J. D., 100
Aragonite,
  calculation of refractive indices of, 143
  stellate twin of, 266
Arrow-head twin, 82
Aspartic acid, preparation of, 428
Astbury, W. T., 439, 452, 453
Axes, crystal, 49
Axes of symmetry, 52
Axial ratios, 50
Ayliffe, S. H., 216, 219, 416

Bachmann, W. E., 431
Barium antimonyl tartrate and potassium nitrate, double salt of, 73
— chloride dihydrate, 377, 385
— nitrate, 76, 153, 377
Barker, T. V., 37, 44
Barlow, W., 43
Barrett, C. S., 37
Bartolinus, E., 89
Basal pinacoid, 63
— sections in convergent light, 278
— uniaxial interference figure, explanation of, 278
Bausch and Lomb microscope, " LD " model, 171
Becke lens, 164
" Becke line ", explanation of, 225
Becke method of refractive index determination, 225
Bell, E. V., 430
Benedict, H. C., 429
Bennett, G. M., 347, 429
Benitoite, 73
Benzaldehyde bisulphite, 429

INDEX

Benzene, estimation of refractive indices of, 151
Benzyl-ψ-thiourea, naphthalene sulphonates of, 423, 460
Berek, M., 159
Bernal, J. D., 323, 342
Berry, W. A., 430
Bertrand, E., 276
Bertrand lens, 164
Beryl, 73
Bessey, G. E., 435
Biaxial crystals, 109 et seq.
  determination of refractive indices of, 317
  optic axes of, 115
  orientation and dispersion of indicatrix in, 121
  wave surfaces of, 111
Biaxial crystals in convergent light, 284 et seq.
  acute bisectrix sections of, 284
  determination of sign of, 306
  obtuse bisectrix sections of, 288
  optic axial sections of, 289
  optic normal sections of, 289
  oblique sections of the indicatrix of, 290
Biaxial indicatrix, 117
Biaxial interference figures, 284 et seq.
  explanation of, 285
  study of, 370
  vibration directions in, 286
Biot and Fresnel, law of, 251, 253
Biotite,
  interference figure of basal sections of, 368
  pleochroism of, 363
  use for finding vibration direction of polariser, 363
Birefringence, 89
  determination of, 263
  of a section, 249
  sign of, 92, 119
Bladed habit, 46
Bogue, R. H., 435
Boracite, 85
Born, M., 143
Bose, E., 326
Bowen, N. L., 435
Brachy-axis, 63
Brachy-pinacoid, 63

Bragg, W. H., 1, 352
Bragg, W. L., 1, 25, 85, 143
Bravais, A., 38
Bravais mica plate, 259
Brookite, TiO_2, crossed-axial-plane dispersion of, 122, 299
Brownmiller, L. T., 435
Bryant, W. M. D., 232, 425, 460, 461
B.S.X., 433
Bunn, C. W., 25
Butler, R. D., 234
Butyn, 429

$CaCl_2$–KCl–H_2O system, 436
$CaCl_2$–$MgCl_2$–H_2O system, 436
Caesium magnesium selenate, crossed-axial-plane dispersion of, 301
—— sulphate, crossed-axial-plane dispersion of, 301
— selenate, 300
Calcite,
  anomalous interference colours of, 249
  calculation of refractive indices of, 143
  cleavage of, 77
  double refraction of, 89
  etch pits on prism of, 56
  habits of, 45
  Huygens' construction for, 95
  isomorphism of, with sodium nitrate, 24
  " twinkling ", 241
Calcium thiosulphate, 66
Canada balsam, mounting in, 210, 212
CaO–Al_2O_3–H_2O system, 435
CaO–Al_2O_3–Fe_2O_3 system, 435
CaO–Al_2O_3–MgO system, 434
CaO–Al_2O_3–SiO_2 system, 434
$2CaO,Al_2O_3,SiO_2$, 70
CaO–$2CaO,SiO_2$–$5CaO,3Al_2O_3$–$4CaO,Al_2O_3,Fe_2O_3$ system, 435
CaO–MgO–SiO_2 system, 434
CaO–Na_2O–Al_2O_3 system, 435
CaO–SiO_2–H_2O system, 435
Carbon dioxide, structure of solid, 138
Carnallite, optical study of, 367
Cassiterite, twinned crystal of, 266
Celestine, 69
Cellulose, nitration of, 449

Centre of symmetry, 54
Centring of objectives, 171
Chamot, E. M., 190, 218
Characteristic sections, optical study of, 355
Chart of birefringences, 247
Cholesteric mesophases, 325, 345 *et seq.*
  optical sign of, 345
  optical activity of, 345
  colours scattered by, 345
  effect of on circularly polarised light, 346
  " plans de Grandjean ", 347
Cinchonidine, 423
Cinchonine, 423
Cinnabar, optical activity of, 129
Circular sections of biaxial indicatrix, 118
Circularly polarised light, 108
Clarke, W. F., 435
Clausius-Clapeyron equation, 17
Cleavage, 76
Clerici solution, 212
Clino-axis, 63
Clino-pinacoid, 63
Cogswell, R. F., 235
Coincidence movements, 41
Colour filters, 187
Coloured solutions for use as light filters, 188
Comparison of two substances by optical examination, 361
Compensation, definition of, 260
Composition plane, 81
Compound cell, 40
— microscope, principle of, 160
Condenser, 160, 162
Conjugated systems, refractivity of, 137
Conoscope, 270
Contact twins, 80
Convergent light, dispersion effects in, 294
Converger, 162
Cooke microscopes,
  " Elementary " model, 180
  " Research " model, 180
Cooling cell, 185
Co-ordinate micrometer scale, 310
Co-ordination compounds, optical activity of, 127, 129

Copper ethoxyacetate, 428
— sulphate pentahydrate, optical study of, 372, 376
— pyrites, 71
Corundum, 72
Cotton, 445 *et seq.*
  extinction of, 447
  maturity of fibres, Schwartz and Shapiro's test for, 449
  spiral angle and refractive indices of, 446
  strength and birefringence of, 450
  test for mercerisation of, 448
Cox, E. G., 151
Cross wires, testing setting of, 364
Crossed or rotated dispersion, 297
Crowfoot, D., 323, 342
Crystal aggregates between crossed Nicols, 265
— angles, measurement of, 48
— classes, 41
— elements, 50
— form under the microscope, 220
Crystal structure, 1
  and optical properties, relationship between, 130 *et seq.*
  of benzene, 7
  of calcium fluoride, 5
  of diamond, 7
  of iodine, 7
  of magnesium, 3
  of mercuric iodide, 5
  of sodium chloride, 4
  of xenon, 2
Crystalline form and chemical constitution, 8
Cubic forms, 62, 75
— system, 1, 14, 74
$CuO-SO_3-H_2O$ system, 436
Cuprammonium sulphate solution as light filter, 185
Cuprite, 76
Cyclical twinning, 83
Cyclographic projection, 28
Cystine, 429
Czapski, S., 275

Dana, E. S., 314
Davies, E. S., 299, 461
Davies, W., 439
" Daylight " lamp and filters, 185
Deming, W. E., 151

INDEX 465

Denigès, G., 203
Determination of refractive indices of liquids, 233
  by comparison with cubic substances, 236
  by hollow prism, 234
  by Leitz-Jelley refractometer, 235
  by Wright stage refractometer, 237
Determinative tables, 422
Dextro- and lævo-rotation, 65, 127
Diammonium hydrogen phosphate, 427
Diamond, 7, 60
Diaphragms, 174
Dichlorodiethyl sulphide, 430
Dichroism, 238
1 : 3 Dimethylthiol benzene dioxides, 430
3,5-Dinitrobenzoates of alcohols, 425, 460
2,4-Dinitrophenylhydrazones of aldehydes, 460
Diopside, multiple twin of, 267
Dioptase, 73
Dippy, J. F. J., 378
Di-*p*-tolylethane dioxides, 430
Direction of single wave-velocity, 115
Directions image, 270
Dispersion, 120 *et seq.*, 294 *et seq.*
  crossed-axial-plane, 122, 299
  crossed or rotated, 297
  horizontal or parallel, 298
  inclined, 298
  in convergent light, 294
  measurement of, 301
  of brookite, 122, 299
  of indicatrix, 120
  of the bisectrices, 121, 297, 299
  of the optic axes, 294, 297, 299
Dithian, 430
— dioxides, 430
— monoxide methyl sulphonium salt, picrates of, 430
— — sulphilimines, 430
Dodecahedron, 75
Dodge, T. A., 415
Domes, 62
Double refraction, 89
Dynamic isomerism, 16

E, 2E, 288
Eder, R., 208

Edge (profile) angles, measurement of, 222
Electron polarisation, 131
Ellipsoid of rotation, 91
Elliptically polarised light, 108
Emmons, R. C., 238, 389, 401
Enantiomorphism, definition of, 64
Enantiomorphous crystals, 64
Epsomite, 69, 70
Equatorial symmetry, 55
Ervin, G., 436
Etch pits, 56
Ethylene dithiolacetic acid dioxides, 430
Ethylphthalimido maleate, 428
Euhedral crystals, 44
Evans, J. W., 175, 273, 274
Evans, R. C., 25
Experiments, 362 *et seq.*
External conical refraction, 114
Extinction, 242, 249 *et seq.*
  straight, symmetrical, oblique, 250
Extinction angles, 249 *et seq.*
  graphical determination of, 253
  in diopside, 252
  in monoclinic crystals, 252
  in orthorhombic crystals, 250
  in triclinic crystals, 252
  in uniaxial crystals, 250
  measurement of, 257
Extraordinary rays, 90
Eye lens of ocular, 160
— shade for microscope, 189
Eyepiece, 160, 172

Fairbairn, H. W., 421
Fajans, K., 140
v. Federov, E. S., 43, 177, 401, 421
Fe_2O_3–SO_3–H_2O system, 435
Ferguson, J. B., 434
Ferguson, R. B., 416
Ferrous sulphate, 68
Fibres, 439 *et seq.*
  anisotropy of, 439
  immersion media for, 450
  micelles, 443
  refractive indices of, 442
Field lens of ocular, 160
Field, M. C., 437
First median line, 119
Fluorspar (fluorite), CaF_2, 5, 76, 77
  twin of, 82

INDEX

Forms, 62
  closed and open, 62
Fracture, type of, 80
Frankenheim, M. L., 38
Frenkel defects, 10
Fresnel, A. J., 100, 117, 129, 251
Frey, A., 445, 455
Friedel, G., 322, 328, 336, 347, 352

Galena, PbS, 77
Giorgi, A. L., 436
Glan-Thompson prism, 166
Glass slides and cover slips, 192
  cleaning of, 193
  size of, 192
Glasses, 15
Glide planes, 57
Gluconic acid, 430
Goniometers, 48
Grandjean, F., 329, 340, 347
Gravity methods, separation by, 211
Greenockite, 73, 74
Gypsum, twin of, 81
Gypsum (or unit retardation) plate, 261

Haas, W., 208
Haase, M., 143
Habit, 45
  influence of environment of growing crystal on, 47
  terms descriptive of, 46
Hall, F. P., 435
Hallimond, A. F., 183
Hann, R. M., 423, 460
Hansen, W. C., 435
Harrison, D. C., 430
Harrison, W., 439
Hartshorne, N. H., 299, 378, 426, 461
Haüy, R. J., 38
Hemihedrism, definition of, 63
Hemimorphism, definition of, 64
Hemimorphite, 69, 83
Hemimorphous crystals, 64
Heptitols, 460
Hermann, K., 328
Hermans, P. H., 444, 445, 451
Herzog, A., 448
Heterogeneous equilibria, 433
Hexagonal forms, 74
— system, 73
Hexakisoctahedron, 75

Hexosazones, 422, 461
High refractive index media, 233
Hock, A. L., 430
Hodoscope, 270
Hendricks, St. B., 151
H.M.X., 431
Holmes, A., 209
Holoaxial symmetry, 55
Holohedrism, 41
  definition of, 63
Holosymmetry, 41
Hooke, R., 38
Horizontal dispersion, 298
Hot stage for liquid crystals, 349
" Hour glass " structure, 268
Hutchinson, A., 37
Huygens, C., 89, 91, 93, 99
Huygens' construction, 93
  for biaxial sections normal to optic axes, 115
  for isotropic substances, 93
  for uniaxial crystals, 95
Huygens ocular, 172
Hydrogen peroxide and uric acid, action between, 430
Hypersthene, 239

Iceland spar, 89, 164
Icositetrahedron, 75
Idiomorphic crystals, 44
Illuminating apparatus, 184
Illumination, 183
Immersion media, 229 *et seq.*
  discussion of, 229
  for fibres, 450
  for organic compounds, 232
  formula for mixing, 233
  list of, 230
  standardisation of, 233
Inclined dispersion, 298
Index variation immersion methods, 237, 387 *et seq.*
  apparatus for temperature control, 391
  double variation method, 396
  immersion liquids for, 394, 398
  single variation method, 389
  warm cells for, 391
Indicatrix, 103
  biaxial, 117
  orientation and dispersion of, 120
  uniaxial, 104

INDEX

Indices, Millerian, 50
Ingerson, E., 421
Insley, H., 435
Interaxial angles, 49
Intercepts, 50
Interference colours, 242 *et seq.*
 anomalous, 249
 between parallel Nicols, 247
 orders of, 245
Interference figures, 270 *et seq.*
 effect of thickness on, 281, 287
 general discussion of, 276
 in bubbles, 276
 methods of viewing, 271
 of biaxial crystals, 284
 of optically active crystals, 128, 281
 of small crystals, methods of isolating, 273
 of uniaxial crystals, 278
 summary of optical characters as given by, 277
Internal conical refraction, 116
Interpenetration twins, 80
Interstitial compounds, 14
Iodosuccinimide, 70
Iron pyrites, 56, 76
Isogyres or " brushes ", 277
Isomerism, 16
Isomorphism, 20
 mixed crystals and, 23
Isopropyl-ammonium choroplatinate, 85
Isotropic crystals, determination of refractive indices of, 224
Isotropism, 14, 241
 between crossed Nicols, 241
 of cubic crystals, 138
 of glasses, 16
 of liquids and gases, 135

Jelley, E. E., 159, 235
Jenkins, F. A., 130
Johannsen, A., 30, 159, 191, 209, 253, 264, 272, 276, 284, 311, 391
Johnson, B. K., 191
Jones, B., 347
Joos, G., 140

K_2CO_3–Na_2CO_3–H_2O system, 436
Keenan, G. L., 423, 460, 461
Klein, D., method of viewing interference figures, 272

Klein, G., 208
Klein solution, 231
Knopf, E. B., 421
K_2O–CaO–SiO_2 system, 434
Kolk, S. v.d. (see Schroeder van der Kolk), 225, 276
Kracek, F. C., 435
Krumbein, W. C., 209

Landolt, H., 188
Larsen, E. S., 233, 234, 238, 399
Laspeyres, H., 240
Lauric acid–potassium laurate system, 437
Law of Rational Indices, 49
Lea, F. M., 435
Lead chloride, 377
— nitrate, 153, 377
Lehmann, O., 322
Leitz microscope, " SY "-model, 171, 276
Lemniscate bands in interference figures, 284
Lenses, care of, 190
Leucite, 85, 266
 twinning in, 265
Lines of single ray velocity, 113
Long chain compounds, optical sign of, 149, 152

Macro-axis, 63
Macro-pinacoid, 63
Magnesium platinocyanide, 239
Magnesium sulphate heptahydrate, 70
 optical activity of, 127
Magnetite, 46
Mallard constant, 311
Mallard's formula, 311
Mallard's method (size of optic axial angle), 311
Malus, E. L., 99
Mann, F. G., 430
Martin, L. C., 191
Mason, C. W., 190, 218
McBain, J. W., 437
McCarthy, C. E., 436
McMurdie, H. F., 435
Mechanical stage, 175
Meek, W. B., 400
Melatopes, 284
Melting of crystals, 15

Meredith, R., 449
Mertie, J. B., 320
Merwin, H. E., 231, 388, 434
Mesomorphic properties in homologous series, gradation of, 347
— state, 322
Metastable equilibria, 18
$MgCl_2$–KCl–H_2O system, 436
MgO–Al_2O_3–SiO_2 system, 434
Mica plate ($\frac{1}{4}$ λ retardation), 260
Michel-Lévy,
chart of birefringences, 247
method of determining large optic axial angles, 313
Microcline, 84
Micrometer eyepiece, 173, 223
Microscope (see also polarising microscope)
changing orientation of crystals under, 176, 213, 400
crystal form as seen under, 220
magnification of, 160
Miers, H. A., 88, 176
Miller, F. S., 438
Millerian indices, 50
Mimetic twins, 83
and polymorphic forms, distinction between, 84
Mitchell, Jun., J., 425
Mixed crystals, 23
melting of, 15
structure defects in, 12
Mixtures of chlorides and nitrates of barium and lead, optical study of, 377
Mixtures, optical examination of, 361
Molecular refractivity,
for principal directions of anisotropic crystals, 139
of cubic crystals, 139
of liquids and gases, 135
Monochromatic light,
by means of electric discharge lamps, 186
by means of filters, 187
by means of monochromator, 186
Monoclinic forms, 63, 68
— system, 67
Moore, F. J., 431
Morey, G. W., 435
Morris, V. H., 461
Mosaic structure, 10

Mounting of crystals, 198 et seq.
permanent mounts, 212
without recrystallisation, 198
Multiple twinning, 83, 266
Muscovite mica, interference figure of, 287, 370

N_g, N_p, N_m, 110
Nägeli's micelle theory, 443
Naphthalene sulphonic acids,
α- and β-naphthylamine salts of, 460
benzyl-ψ-thiourea salts of, 423, 460
Na_2SiO_3–Fe_2O_3–SiO_2 system, 434
Negative birefringence, 92
— uniaxial crystals, wave surfaces of, 92
Nematic mesophases, 324, 341 et seq.
crystallisation " memory " of, 343
effect of electric and magnetic fields on, 327, 341
thread-like lines in, 343
Newton's colours, 245
Nickel sulphate, 71
Nicol prisms, 164
care of, 190
testing the setting of, 362
Niggli, P., 159
Nikitin, W., 421
Nitrates and carbonates, birefringence of, 148
m-Nitroaniline, optical study of, 382
o-Nitroaniline, optical study of, 378
anomalous interference colours of, 249
p-Nitroaniline, optical study of, 383
p-Nitrobenzaldehyde, crossed-axial-plane dispersion of, 299
Non-stoichiometric compounds, 14
Numerical aperture, 167

Object image, 160, 272
Objective, 167 et seq.
clutch, 170
diaphragm, 315
revolving nosepiece for, 170
Oblique illumination, double-diaphragm method of, 227
Obtuse bisectrix, 119
— — figure, 288
Octahedron, 46, 47, 75
Ocular, 160, 172

INDEX

Oil-immersion objectives, 169
Olsen, F., 449
Optic axes (biaxial crystals), 115
Optic axial angle, 119
  distinction between acute and obtuse, Dana's method, 314
  measurement of, 309
  real and apparent, 288
  relation of refractive indices to, 309
Optic axial plane, 119
Optic axial sections in convergent light,
  biaxial crystals, 289
  uniaxial crystals, 278
Optic axis (uniaxial crystals), 89
— binormals, 115
— biradials, 113
— normal sections, 281, 289
Optical activity, 126
  Fresnel's theory of, 129
Optical anomalies, 268
— data, presentation of, 357
Optical examination, 353 et seq.
  preliminary to chemical analysis, 360
  summary of, 219
Optical properties and crystal structure, relationship between, 130 et seq.
Optical sign of crystals, 92, 119
  determination of, in convergent light, 302
  determination of, in parallel light (uniaxial crystals), 263
Optical tube length, 160
Ordinary rays, 90
Orientation of indicatrix, 120
Ortho-axis, 63
Orthoclase, twin of, 82, 84
Ortho-pinacoid, 63
Orthorhombic cleavage, types of, 79
— crystals, oblique interference figures of, 290
— forms, 62
— system, 67, 69
Oxalates, calculation of refractive indices of, 151

Paraffins, higher, rotation of molecules in, 12
Parameters, 50
Parametral plane, 50

Parker, T. W., 435
Parry, J., 217
Penfield, S. L., 37
Pentosazones, 422, 461
Pericline twinning in plagioclase, 267
Pettijohn, F. J., 209
Permanent mounts of crystals, 212
Peacock, M. A., 416
Phase difference, 447
  determination of, by Sénarmont compensator, 455
Phase rule studies, use of polarising microscope in, 433
Phemister, T. C., 314
Phenylosazones, 461
Phillips, F. C., 59, 421
Pinacoid, definition of, 62
Plane of symmetry, 52
Planes of vibration, 100
Plant cells, walls of, 451 et seq.
Platy habit, 46
Pleochroism, 238
  and molecular structure, 241
Pleochroism formula, 239
Poe, C. F., 461
Point group, 43
— — and space group nomenclature, 56
— — and space group symmetry, relationship between, 59
Polar symmetry, 55
Polarisation and polarisability of atoms, 130
  by light waves, 131
Polarisation colours, 242
— of light, 90
Polariser, 162
  determination of vibration direction of, 363
Polarising microscope,
  chemical applications summarised, x
  general description of, 161
  systematic identification by means of, 422
  use and care of, 189
  use of in phase rule studies, 433
" Polaroid ", 163
Polymorphic forms and mimetic crystals, distinction between, 84

Polymorphic transformations,
 due to development of defect structures, 20
 in close-packed metal structures, 19
 in silica, 19
Polymorphism, 16 *et seq.*
 definition of, 16
 enantiotropism and monotropism, 17
 of mercuric iodide, 207
Polysynthetic crystals, 84
Poorly-formed crystals, optical examination of, 356
Pope, W., 430
Positive birefringence, 92, 119
— uniaxial crystals, wave surfaces of, 92
Posnjak, E., 435
Potassium chlorate, study of, 384
— chloride, optical study of, 366
— dichromate, study of, 385
— dihydrogen phosphate, 70, 376
— lithium sulphate, optical activity of, 127
— magnesium sulphate, 67, 355
— mercuric iodide, 231, 406, 426
— nitrate and barium antimonyl tartrate, double salt of, 73
— —, optical study of dimorphism of, 373
Potassium sulphate, 70
 interpenetration twin of, 82
 study of, 384
Potassium tetrathionate, 67
Preparation and mounting, apparatus for, 194
Preston, J. M., 443, 448
Preston, R. D., 451 *et seq.*
Preston, T., 447
Primary optic axes, 115
Principal refractive indices,
 of biaxial crystals, 110
 of uniaxial crystals, 108
Principal vibration directions of biaxial crystals, 110
Prism, definition of, 62
Prismatic habit, 46
Procaine, 429
Pyramid, definition of, 62
Pyritohedron, 75
Pyroelectricity, 64

Quarter wave mica plate, 260
Quartz, SiO_2, 73
 enantiomorphism of, 65
 interference figure of basal section of, 281
 optical activity of, 127
 rotatory dispersion of, 128
Quartz wedge, 262
Quinidine, 423
Quinine, 423

Radley, E. G., 439
Ramsden disc, 272
— ocular, 173
Rankin, G. A., 434
Rational Indices, Law of, 49
Ray, P. C., 430
Ray velocity, 93
Record purposes, optical examination for, 359
Recrystallisation, 200 *et seq.*
 by sublimation, 206
 from a melt, 208
 from organic solvents, 204
 from solution, 200
R.D.X., 431
Rees, W. J., 439
Refraction equivalents, 136
Refractive index, 95, 97
 determination of, in birefringent crystals, 316
 indirect determination of, 319
 of extraordinary ray, 97
 principles of determination, 224
Refractive indices of calcite and aragonite, calculation of, 143
Refractivity,
 atomic, 136
 molecular, 135
Refractometers,
 Abbé type, 233
 hollow prism type, 234
 Leitz-Jelley type, 235
 stage, 237
Refractories, 437
Refractory materials, preparation of thin sections of, 210
Relative retardation, 244
Relief or strength of border, 224, 241
Resolution, limit of, 167
Resolving power, 167
Revolving nosepiece, 170

INDEX

Rheinard, M., 421
Rhombic dodecahedron, 75
Rhombohedral system, 72
Rigby, G. R., 209
Rinne, F., 159
Rotating molecules or polyatomic ions, structures containing, 11
Rotation apparatus, 176, 213
Rotation-inversion axes, 52
Rotation of atom groups in crystals, effect of, on optical properties, 150
Rosenbusch, H., 159, 191, 253
Rosiwal analysis of thin sections, 438
Rubidium sulphate, 300
Rutile, TiO_2, 122, 299

Saylor, C. P., 227, 391
Scalenohedron, 72
Scapolite, 71
Schairer, J. F., 434
Scheelite, 70
Schönflies, A., 43
Schönite, 67, 355
Schottky defects, 10
Schroeder van der Kolk, 276
method of refractive index determination, 225
Screw axes, 57
Second median line, 119
Secondary optic axes, 113
Section cutting, 216
Selenite (or gypsum) plate, 261
Sellers, J. E., 461
Sénarmont compensator, 455
Serine, preparation of, 428
Shapes of crystals under the microscope, 220
Sheehan, J. C., 431
Sieves, 199
Sign of elongation, 263
Silver iodide, 11
— mercuric iodide, 11
" Sira " mountant, 212
Size and thickness of crystals, measurement of, 223
— of crystals for mounting, 198
Slags, 437
Slawson, C. B., 315
Slides and cover slips, 192
Smectic mesophases, 324, 327 *et seq.*
focal conic structure of, 331 *et seq.*

homeotropic structure of, 329
stepped drops in (" gouttes à gradins "), 329
Sodium chloride, effect of urea on habit of, 47
— hydrogen maleate, 429
— nitrate, study of, 384
Sodium periodate, 72
optical activity of, 129
Softening of crystals on heating, 16
Sohncke point systems, 43
Sonstadt (or Thoulet) solution, 231
Wherry's modification of, 231
Space groups, 43
nomenclature, 56
symmetry of, relation with point group symmetry, 59
Space lattice, 38
relation to crystal systems, 41
types of, 38
Spherical lenses for viewing interference figures, 276
— projection, 27
Spherulites, optical properties of, 265
" Spinel " twin, 82
Stage goniometer, Miers', 176
— refractometers, 237
Stand, microscope, 173
Statham, F. S., 430
Staurolite, twin of, 266
Stereographic nets, 35
Stereographic projection, 26 *et seq.*
constructions, 30
face normals in, 28
face poles in, 28
great and small circles in, 28
primitive circle in, 26
Strontium antimonyl tartrate, 73
Structure defects in real crystals, 9
Strychnine, salts of, 461
Strychnine sulphate, 70
Stuart, A., 378, 393, 431
Sucrose, optical activity of, 128
Sugars, optical activity of, 127
Supplementary twinning, 83
Swift microscopes,
" Dick " model, 171
" Lapidex " model, 177
" Survey " model, 179
Symmetry and internal structure, 56, 59

INDEX

Systematic examination of a single compound, 353
— identification by means of optical properties, 422

Tabular habit, 46
Tartaric acid, 65, 67
  optical activity of, 127
Taylor, E. W., 183
Terpstra, P. A., 431
Tesseral symmetry, 55
Tetartohedrism, definition of, 63
Tetragonal forms, 71
— system, 70
Tetrahedrite, 76
Tetrahedron, 75
Tetrahexahedron, 75
Textile fibres (see fibres),
Thin sections, preparation of, 210
Thomas, H. H., 439
Thomas, R. M., 431
Thoulet solution, 231
  Wherry's modification of, 231
Torbernite, dispersion of, 122
Tourmaline, 73
  pleochroism of, 239
Triaxial ellipsoid, 110
Triclinic forms, 62, 66
— system, 67
Triethylene trisulphide, 430
Trigonal forms, 72
— system, 72
Trimethylene trisulphide dioxides, 430
Trisoctahedron, 75
Trithioacetaldehydes, 430
Tsuboi, S., 388
Tunell, G., 436
Tutton, A. E. H., 8, 44, 85, 88, 215, 300, 301
" Twinkling ", 241
Twinned crystals, 80 et seq.
  between crossed Nicols, 265 et seq.
Twinning axis, 81
— plane, 81

Uniaxial crystals, 91 et seq.
  determination of refractive indices of, 317
  orientation and dispersion of indicatrix in, 121
  sign of, 108

Uniaxial crystals in convergent light, 278 et seq.
  basal sections of, 278
  determination of sign of, 303
  oblique sections of, 283
  optic normal sections of, 281
Uniaxial indicatrix, 103
Uniaxial interference figures, 278 et seq.
  explanation of, 278
  study of, 368
  vibration directions in, 280
Unit cell, 40
— retardation plate, 261
Universal stage, 400 et seq.
  adjustment of, for use, 406
  axes, nomenclature of, 402
  biaxial crystals, examination by, 409
  cleavage planes, plotting of, 419
  correction of readings on, 405
  Emmons type, 401
  Federov type, 401
  optic angle, determination by, 413
  uniaxial crystals, examination by, 408
Uric acid and hydrogen peroxide, action between, 430

V, 2V, 288
  determination of, 309 et seq.
Variation in relief, 241
Venable, C. S., 430
Vibration directions, 100
  distinction between, in crystal plates, 260
  in biaxial crystals, 111, 118
  in uniaxial crystals, 103, 107
Vigfusson, V. A., 394

Wandering atoms, lattices containing, 11
Wardrop, A. B., 454
Wasasterjerna, J. A., 140
Wave direction, 92
— front, 92
— normal, 92
Wave surfaces,
  of biaxial crystals, 111
  of isotropic media, 91
  of uniaxial crystals, 91
Wave velocity, 93

INDEX 473

Weatherhead, A. V., 209
Well-formed crystals, optical examination of, 354
Wells, A. F., 25, 48, 76
Wells, L. S., 435
Werner, O., 208
Wherry, E. T., 231, 423, 460, 461
White, H. E., 130
Wiener, O., 444
Willems, H. W. V., 434
Williams, A. F., 217
Winchell, A. N., 236, 386, 391, 398, 400, 421, 426, 459
Wollaston, W. H., 49
Wood, R. G., 216, 319, 416
Wooster, W. A., 149
Wright, F. E., 217, 232, 237, 258, 263, 310, 313, 320, 422, 461

Wulfenite, 70
Wülfing, E. A., 159, 191, 253

X-ray methods of structure analysis, 1
  use of optical crystallographic data in, xi, 150

Yanovsky, E., 423
Young, T., 100

Zinc blende, ZnS, 76
— sulphate heptahydrate, optical study of, 371
Zircon, 70
Zone, definition of, 33, 49
Zoning, 267